This book makes the connection between European agricultural history and the experience of developing countries, and shows how fascinating, informative and revealing the linkage can be.

Dr Paul Brassley, *Centre for Rural Policy Research, University of Exeter, UK*

Too few academics attempt to fill the gaps between disciplines. Jonathan Harwood sheds light on the potential for a second generation Green Revolution to improve the lives of small farmers in developing countries by appealing to the history of plant breeding in nineteenth and early twentieth century Europe.

Carrie A. Meyer, *Associate Professor of Economics, George Mason University, USA*

Europe's Green Revolution and Others Since

How poverty and hunger in the developing world can be alleviated is an issue which has preoccupied development experts and politicians since 1945. Success, however, has been limited. The well-known post-war 'Green Revolution' programmes managed to increase grain production substantially in Latin America and South Asia, but by 1970 critics were arguing that they had made little impact upon rural poverty because the agricultural technology introduced was better suited to large commercial farms than to small subsistence ones. From the 1970s onward it seemed as though this criticism would bear fruit as experts developed a number of new approaches to cultivation which were tailored to the needs of peasant farmers. But levels of rural poverty and malnutrition remain very high, prompting major development agencies in recent years to urge that more attention be given to the plight of small farmers.

From a European perspective this persistent failure to develop a technology suitable for smallholders is puzzling. From the late nineteenth century, for example, various European countries also underwent 'green revolutions' designed to boost agricultural productivity, and several states established plant-breeding stations whose express purpose was to serve the needs of peasant farmers. Some of these stations were very successful, making a substantial impact upon the regional economies they served.

Remarkably, however, this European episode is virtually unknown among development experts today. Indeed, since 1945 the development industry, despite struggling to make an impact upon poverty, has displayed an astonishing lack of interest in the success or failure of past development projects. After telling the story of Europe's peasant-friendly approach to agriculture and tracing the history of Green Revolutions since 1945, this book argues that state-funded agricultural research has the potential to make a large impact upon poverty in the developing world but is unlikely to do so as long as the development industry continues to ignore its past.

Jonathan Harwood is Emeritus Professor of the History of Science and Technology at the University of Manchester, UK.

Routledge explorations in economic history
Edited by Lars Magnusson
Uppsala University, Sweden

1 **Economic Ideas and Government Policy**
Contributions to contemporary economic history
Sir Alec Cairncross

2 **The Organization of Labour Markets**
Modernity, culture and governance in Germany, Sweden, Britain and Japan
Bo Stråth

3 **Currency Convertibility**
The gold standard and beyond
Edited by Jorge Braga de Macedo, Barry Eichengreen and Jaime Reis

4 **Britain's Place in the World**
A historical enquiry into import controls 1945–1960
Alan S. Milward and George Brennan

5 **France and the International Economy**
From Vichy to the Treaty of Rome
Frances M. B. Lynch

6 **Monetary Standards and Exchange Rates**
M. C. Marcuzzo, L. Officer and A. Rosselli

7 **Production Efficiency in Domesday England, 1086**
John McDonald

8 **Free Trade and its Reception 1815–1960**
Freedom and trade: Volume I
Edited by Andrew Marrison

9 **Conceiving Companies**
Joint-stock politics in Victorian England
Timothy L. Alborn

10 **The British Industrial Decline Reconsidered**
Edited by Jean-Pierre Dormois and Michael Dintenfass

11 **The Conservatives and Industrial Efficiency, 1951–1964**
Thirteen wasted years?
Nick Tiratsoo and Jim Tomlinson

12 **Pacific Centuries**
Pacific and Pacific Rim history since the sixteenth century
Edited by Dennis O. Flynn, Lionel Frost and A. J. H. Latham

13 **The Premodern Chinese Economy**
Structural equilibrium and capitalist sterility
Gang Deng

14 **The Role of Banks in Monitoring Firms**
The Case of the Crédit Mobilier
Elisabeth Paulet

15 **Management of the National Debt in the United Kingdom, 1900–1932**
Jeremy Wormell

16 **An Economic History of Sweden**
Lars Magnusson

17 **Freedom and Growth**
The rise of states and markets in Europe, 1300–1750
S. R. Epstein

18 **The Mediterranean Response to Globalization Before 1950**
Sevket Pamuk and Jeffrey G. Williamson

19 **Production and Consumption in English Households 1600–1750**
Mark Overton, Jane Whittle, Darron Dean and Andrew Hann

20 **Governance, The State, Regulation and Industrial Relations**
Ian Clark

21 **Early Modern Capitalism**
Economic and social change in Europe 1400–1800
Edited by Maarten Prak

22 **An Economic History of London, 1800–1914**
Michael Ball and David Sunderland

23 **The Origins of National Financial Systems**
Alexander Gerschenkron reconsidered
Edited by Douglas J. Forsyth and Daniel Verdier

24 **The Russian Revolutionary Economy, 1890–1940**
Ideas, debates and alternatives
Vincent Barnett

25 **Land Rights, Ethno Nationality and Sovereignty in History**
Edited by Stanley L. Engerman and Jacob Metzer

26 **An Economic History of Film**
Edited by John Sedgwick and Mike Pokorny

27 **The Foreign Exchange Market of London**
Development since 1900
John Atkin

28 **Rethinking Economic Change in India**
Labour and livelihood
Tirthankar Roy

29 **The Mechanics of Modernity in Europe and East Asia**
The institutional origins of social change and stagnation
Erik Ringmar

30 **International Economic Integration in Historical Perspective**
Dennis M. P. McCarthy

31 **Theories of International Trade**
Adam Klug
Edited by Warren Young and Michael Bordo

32 **Classical Trade Protectionism 1815–1914**
Edited by Jean Pierre Dormois and Pedro Lains

33 **Economy and Economics of Ancient Greece**
Takeshi Amemiya

34 **Social Capital, Trust and the Industrial Revolution: 1780–1880**
David Sunderland

35 **Pricing Theory, Financing of International Organisations and Monetary History**
Lawrence H. Officer

36 **Political Competition and Economic Regulation**
Edited by Peter Bernholz and Roland Vaubel

37 **Industrial Development in Postwar Japan**
Hirohisa Kohama

38 **Reflections on the Cliometrics Revolution**
Conversations with economic historians
Edited by John S. Lyons, Louis P. Cain, and Samuel H. Williamson

39 **Agriculture and Economic Development in Europe Since 1870**
Edited by Pedro Lains and Vicente Pinilla

40 **Quantitative Economic History**
The good of counting
Edited by Joshua Rosenbloom

41 **A History of Macroeconomic Policy in the United States**
John H. Wood

42 **An Economic History of the American Steel Industry**
Robert P. Rogers

43 **Ireland and the Industrial Revolution**
The impact of the Industrial Revolution on Irish industry and society, 1801–1922
Andy Bielenberg

44 **Intra-Asian Trade and Industrialization**
Essays in memory of Yasukichi Yasuba
Edited by A. J. H. Latham and Heita Kawakatsu

45 **Nation, State and the Industrial Revolution**
The visible hand
Lars Magnusson

46 **A Cultural History of Finance**
Irene Finel-Honigman

47 **Managing Crises and De-globalisation**
Nordic foreign trade and exchange 1919–1939
Edited by Sven-Olof Olsson

48 **The International Tin Cartel**
John Hillman

49 **The South Sea Bubble**
Helen J. Paul

50 **Ideas and Economic Crises in Britain from Attlee to Blair (1945–2005)**
Matthias Matthijs

51 **Bengal Industries and the British Industrial Revolution (1757–1857)**
Indrajit Ray

52 **The Evolving Structure of the East Asian Economic System since 1700**
Edited by A. J. H. Latham and Heita Kawakatsu

53 **German Immigration and Servitude in America, 1709–1920**
Farley Grubb

54 **The Rise of Planning in Industrial America, 1865–1914**
Richard Adelstein

55 **An Economic History of Modern Sweden**
Lennart Schön

56 **The Standard of Living and Revolutions in Russia, 1700–1917**
Boris Mironov; edited by Gregory L. Freeze

57 **Europe's Green Revolution and Others Since**
The rise and fall of peasant-friendly plant breeding
Jonathan Harwood

For H.I.H. (who was patient)

Europe's Green Revolution and Others Since

The rise and fall of peasant-friendly plant breeding

Jonathan Harwood

Routledge
Taylor & Francis Group
LONDON AND NEW YORK

First published 2012
by Routledge
2 Park Square, Milton Park, Abingdon, Oxon OX14 4RN

Simultaneously published in the USA and Canada
by Routledge
711 Third Avenue, New York, NY 10017

Routledge is an imprint of the Taylor & Francis Group, an informa business

British Library Cataloguing in Publication Data
A catalogue record for this book is available from the British Library

Library of Congress Cataloging in Publication Data
A catalog record has been requested for this book

ISBN: 978-0-415-59868-2 (hbk)
ISBN: 978-0-203-11804-7 (ebk)

Typeset in Times New Roman
by Wearset Ltd, Boldon, Tyne and Wear

Contents

Figures

Preface

The path leading to this book has been rather long and winding. During the 1980s my work was focused mainly upon the history of genetics, but through a collaborative project with the sociologist R. D. Whitley, and my colleague, John V. Pickstone, I became interested as well in the general question of the relations between science and technology. Subsequently the agricultural economist Colin Thirtle and I secured funding for a study of the science–technology relation in British plant breeding. The series of papers by Paolo Palladino which emerged from that project stimulated my interest in breeding yet further so that during the 1990s my interests gradually shifted toward agriculture, and I began to explore archival sources for a history of plant breeding in Germany. The book that I then envisaged was a kind of historical sociology of applied science which would analyse the ways in which German academic plant breeding had been shaped by different kinds of institutional context.

While a book along those lines did eventually appear (Harwood 2005), it didn't have much to say about plant breeding. And that was largely because my ideas about the history of plant breeding had by then been radically revised as a result of my involvement in a 2003 conference initiated by my colleague in international relations, Inderjeet Parmar. The conference was entitled 'Foundations of Globalisation', and one of its sessions was devoted to the role of US foundations in the Green Revolution. In order to line up speakers for the session, I read through the existing secondary literature and was struck by the parallels between agricultural transformation in Western Europe *c.*1900 and that promoted by Green Revolution (GR) programmes in the developing world after 1945. Equally significant, however, were the differences. For whereas some German plant-breeding stations established around 1900 were quite successful in assisting smallholder agriculture, a recurring criticism of GR programmes was that they failed to do so. At that point I began to look at the history of plant breeding with very different eyes, and the book I decided to write had to be a comparative one.

In the nine chapters that follow I am trying to do three things. One of them is to contribute to the emerging literature on the history of plant breeding. Over the last 10 to 20 years a series of works has begun to appear which does much more than simply recount the history of this body of practices (e.g. Bonneuil and Thomas 2009; Fitzgerald 1990; Kloppenburg 1988 and 2004; Maat 2001; Moser

2003; Palladino 2002; Paul and Kimmelman 1988; Wieland 2004). Instead these authors are connecting breeding in interesting ways to a range of more general issues. Arguably this literature is also important because it provides us with a perspective from which to better understand and assess current controversies surrounding the genetic modification of organisms and the patenting of life forms. Apart from just presenting new empirical material, however, what I have tried to do below is to move beyond the existing histories in two respects: to look more closely at the process by which breeders have come to serve particular constituencies and to analyse the changing relations between private- and public-sector breeding.

My second aim is to present an enlarged and revised history of the Green Revolution. 'Enlarged' because I locate the conventionally defined post-1945 Revolution on a longer trajectory which had already begun by the late nineteenth century with 'Europe's Green Revolution'. Furthermore, in contrast with most of the existing historiography, I devote more attention to later phases in the history of the Revolution, namely from the 1970s into the present. The advantage in stretching out the Revolution to cover this extended period is that the patterned nature of its history becomes more evident. Drawing upon the work of others, I argue that this long Revolution has gone through alternating phases in which its central concern has shifted between boosting production and securing social equity.

Third, I want to show how the history of Green Revolutions might usefully inform development policy. While the more general idea that history can inform policy has been around for decades, it is only quite recently that work arguing the policy relevance of the history of development has begun to appear (e.g. Chang 2002 and 2009; Bayly *et al.* 2011). The basic idea is that institutions have something to gain from reflecting upon their experience. That point seems obvious, but if the analysis that follows is correct, the development industry's lack of interest in its own past has been astonishing. Moreover, the need to 'look back' before taking decisions that will shape the future has gained urgency since the 1980s through the dominance in many countries of neo-liberal political thought. That markets should routinely be given preference to state action, however – not least in development policy – seems to me to be grounded far more in ideology than in the analysis of past action and its consequences. There is a great deal at stake here, and it is important that those of us who know something about the past should speak up.

Finally, in order to forestall a possible misunderstanding, I should make my position clear at the outset. On reading a draft chapter (in which I criticized the Green Revolution for failing to aid smallholders), one agricultural economist wrote back annoyed, citing the big impact of the Revolution's high-yielding plant varieties upon production. 'You people obviously prefer mass starvation', he complained, dismissing critics of the Revolution as 'Luddites'. Although he misunderstood my argument, one can perhaps understand why he did so. For there are, indeed, some Luddites about: critics of quasi-romantic inclination who attack the technology central to industrial agriculture for poisoning our food, contaminating the environment, and fatally undermining the family farm. For them the solution is seen to lie, not in improved technology, but in indigenous knowledge and 'traditional' cultivation practices. On the other hand there are many, like my critic, who champion the 'modern' agricultural science underpinning the Green Revolution as virtually the only practical way to stave off hunger and poverty in the developing world. Taking my cue from the practitioners of peasant-friendly technical innovation, however, it seems to me that both of these camps are mistaken because their vision is exclusive. They throw the baby out with the bathwater. The most promising solutions to rural hunger and poverty – whether one looks at Europe around 1900 or at late twentieth century peasant-friendly innovation for the global South – are instead those that are synthetic: they combine the methods of science and technology with the best that indigenous knowledge and practice have to offer. That is hardly a Luddite stance. It is instead an argument in favour of an *appropriate* technology, one that takes the diverse needs and circumstances of its users – above all, small farmers – seriously rather than insisting that one size fits all.

Acknowledgements

Needless to say, while researching the book I have benefited enormously from the expertise and generosity of many people. Given my subject matter, most of the work necessarily took place in Germany. Of the many archivists there who were helpful, I particularly thank Dr Ulrich Fellmeth (Director of the University Archive, Universität Hohenheim). During two extended visits to Munich I was able to stay at the Deutsches Museum, thanks to the support of Helmut Trischler, Director of the Museum's Forschungszentrum.

By far the most important location for my work over the last ten years or so, however, has been the Max-Planck Institute for the History of Science in Berlin. This remarkable institution has provided not only a physical base for my work, with access to the city's libraries and archives, but just as importantly an intellectual home. With its large number of resident scholars, steady stream of visitors from abroad, numerous seminars and workshops, and superb library, it offers an unparalleled resource. Several of the library staff were exceptionally helpful, in particular Ellen Garske, Matthias Schwerdt and Ruth Kessentini. That I enjoyed the privilege of working at the Institute at all, of course, was thanks to the support of Lorraine Daston, Veronika Lipphardt and especially Hans-Jörg Rheinberger in whose department I was a visiting scholar on many occasions.

I am also grateful to many individual colleagues for discussion and/or feedback on drafts. Among historians of plant breeding I thank Christophe Bonneuil, Harro Maat, Albrecht Meinel, Peter Moser, Nils Roll-Hansen and Thomas Wieland. Conversations with the historians of development, Robert Anderson and Daniel Speich, have helped me to situate my work on a larger canvas. And the comments on large sections of the manuscript by Paul Brassley and Heide Ingenohl Harwood are much appreciated. Given the book's transdisciplinary scope, however, it was also essential for me, as a historian of science and technology, to make contact with people in development studies who were sympathetic to the project and willing to take time for an outsider's questions. My most immediate conversation partners in Manchester were Andy Bowman, David Colman and Philip Woodhouse. Further afield I gained much from exchanges with Stephen Biggs, Dana Dalrymple, Ian Scoones and Jeff Tansey. And in view of the book's central theme, it was important for me to be in touch with people working on participatory plant breeding. I was fortunate, therefore, to be able to correspond with Salvatore Ceccarelli, David Cleveland and Shawn McGuire.

Several seminar series and workshops provided an opportunity to try out some of my ideas on knowledgeable colleagues. Particularly fruitful were James C. Scott's Agrarian Studies colloquium at Yale (October 2009); Harriet Ritvo and Deborah Fitzgerald's Seminar in Environmental and Agricultural History at Massachusetts Institute of Technology (MIT) (April 2010); a workshop entitled 'What can development policy learn from the history of development?', organized by Steve Sturdy and me at the Economic and Social Research Council (ESRC)'s Genomics Policy and Research Forum, University of Edinburgh (June 2010); and a winter school on the history of development held at Ascona (January 2011), organized by Daniel Speich and Sara Elmer of the Technical University in Zurich (ETH). Two of the chapters were published previously in different form. A version of Chapter 1 appeared as 'Research and extension in political context: rural unrest and the origins of the Prussian chambers of agriculture', in Nadine Vivier (ed.), *The State and Rural Societies: Policy and Education in Europe, 1750–2000* (Turnhout: Brepols, 2009), 135–157. A version of Chapter 5 appeared as 'The fate of peasant-friendly plant breeding in Nazi Germany', *Hist. Studies in the Natural Sciences*, vol. 40 (4), 570–604, 2010. My thanks to Brepols Publisher n.v. and to the University of California, respectively.

Finally, numerous visits to libraries and archives in Germany were made possible by several funding bodies. I thank the Royal Society for a number of small grants for the history of science, the Leverhulme Trust for a Study Abroad Fellowship in 2007–2008 (no. SAF/10844), and most recently the British Academy for a Small Grant in 2009–2011 (no. SG-51457).

Introduction

> ...the historical facts about the development experiences of the developed countries should be more widely publicized. This is not only a matter of 'getting history right', but also of allowing the developing countries to make informed choices about the policies and institutions that may be appropriate for them.
>
> (Chang 2002: 140)

How best to improve agricultural productivity in the developing world has been the subject of recurrent debate throughout the twentieth century. In the case of the foundation-sponsored 'Green Revolution' (GR) programmes after 1945, for example, critical voices began to mount from the late 1960s. Although cereal yields and production had increased substantially, rural poverty and malnutrition persisted on a large scale. The explanation usually given for this outcome was that the Revolution's plant varieties and cultivation techniques were more suitable for large commercial farms than for (the much larger number of) small peasant farms. Critics called accordingly for more appropriate forms of technology which were better suited to the needs of smallholders. In the course of the 1970s, 1980s and 1990s it seemed as though this criticism was beginning to bear fruit. A variety of more peasant-oriented approaches were developed, especially at some of the newly established international agricultural research centres (IARCs) which succeeded the original foundation-sponsored programmes.

Over the last decade, however, it has become clear that there is still a very long way to go. Numerous observers of the 'development industry' have concluded that most development projects fail to achieve their stated aims (e.g. Büschel and Speich 2009; Scott 1998; Lewis 2005; Rist 2008). Recent reports from development organizations point to continuing high levels of rural poverty and the urgency of addressing the needs of smallholders. The new peasant-friendly approaches since the 1970s seem to have made little headway. Some observers are saying that '... the war on hunger has not been won. It has not even been seriously engaged in numerous nations of the world' (Falcon and Naylor 2005: 1119). And few now expect that the U.N. Millennium Development Programme's goal – to halve the number of people in poverty by 2015 –

will be achieved in most countries. One is left with the uneasy feeling that little has been learned from the mistakes of the early GR programmes.

From a European perspective this failure to develop technology and institutions adapted to the needs of small farmers is puzzling. From the late nineteenth century, for example, various European countries also underwent 'green revolutions' in which a systematic attempt was made to devise appropriate technology and promote its adoption by peasant farmers. More particularly, around 1900 several Central European states established plant-breeding stations whose express purpose was to make high-yielding plant varieties as well as the basic techniques of plant breeding available to the small farmers who predominated in those regions. And the evidence suggests that some of these stations were very successful, producing varieties which were popular with smallholders and making a substantial impact upon the regional economies they served. The aims of this book, therefore, are threefold. I begin by telling the story of the 'European Green Revolution'. I then use the historical perspective thus acquired in order to reflect upon the strengths and weaknesses of the Green Revolutions since 1945. And finally, I examine recent calls for a 'second Green Revolution' – targeted at smallholders and drawing upon biotechnology – in order to establish to what extent the 'lessons of the past' have now finally been learned.

1 Agricultural crisis in the developing world

Poverty in the global South has been high on development agencies' agenda since the Second World War. And although progress has been made, the numbers living in poverty are still shockingly high. In 1993 one-half of the world's population was living on less than $2 per day ('poor' according to the World Bank's definition) and by 2008 about 20 per cent (1.2 billion people, 400 million of them in India alone) were living on less than $1 per day ('absolute poverty' by the Bank's definition) (Mazoyer and Roudart 2006: 483; McIntyre *et al.* 2009a: 21; Gupta 2009: 113).[1] Inevitably this means high levels of malnutrition. To be sure, nutrition has improved markedly since the mid 1960s, but approximately 900 million people are still chronically malnourished (FAO 2009a). The problem is actually getting worse in some 20 countries (FAO 2004: tab. A2), and in sub-Saharan Africa the *average* daily caloric intake is less than the minimum required for normal health (Pinstrup-Andersen and Schiöler 2001: 59).

The key fact about such poverty is that it is much worse in rural than in urban areas. It is estimated that about 75 per cent of the impoverished population in the developing world live in rural areas (FAO 2009b: 28; IFAD 2001: 20). In India, for example, about 30 per cent of the rural population live in poverty while in sub-Saharan Africa the proportion approaches 60 per cent (Ramaswami and Pray 2007: 157; Ndiritu 2000: 107). Given the high proportion of the labour force in the global South that is employed in agriculture – 54 per cent on average across the developing world but much higher in some regions (FAO 2004) – it comes as no surprise that poverty is especially severe among those living from agriculture.

A large majority of the world's absolute poor live on small farms (Birner and Resnick 2010: 1442; Wiggins *et al.* 2010: 1341), and the majority of the chronically malnourished are small farmers (Fukuda-Parr 2007c: 7; Pretty *et al.* 2006: 1114; Hazell *et al.* 2007: 1). If one wants to reduce poverty, therefore, agriculture is the place to start. Growth in agriculture, for example, is reckoned to be about twice as effective in helping the poor as growth in the non-agricultural economy (FAO 2009b: 3; Valdes and Foster 2010).

But improving agriculture is going to require not just an increase in production but changing the *way* in which food is produced. For the intensive cultivation methods ('industrial agriculture'), which were central to the Green Revolution and which continue to be promoted by development projects, have disadvantages which are becoming ever more serious. To begin with, they are heavily dependent upon fossil fuels which provide the raw material for fertilizers, herbicides and pesticides as well as the fuel for tractors and other machinery. This reliance upon fossil fuels makes them too expensive for many farmers now, and the cost will climb in future as sources become steadily more difficult to locate and extract.[2]

Ironically, industrial agriculture's dependence upon oil also makes it inefficient in an important respect. In our lifetimes we have become accustomed to hearing that high-input agriculture is very 'efficient' because both yields per acre and yields per unit of labour have increased enormously since the mid nineteenth century. While this is true, it overlooks the energy requirements of this process: namely, how much energy does it take to produce one calorie of food energy? The answer is sobering. In pre-industrial societies the expenditure of one calorie of energy in the production process (from sunlight, human labour and draft animals) has been estimated to yield from 10 to 40 calories of food energy while in modern high-tech agriculture this ratio of output to input is approximately one (Bayliss-Smith 1982). Of particular relevance to our concerns here is a study of Austrian agriculture which shows that nineteenth century technical improvements – improved crop rotations, stall-feeding, hand-operated machinery – actually *increased* the energetic efficiency of production (from a ratio of about five in 1830 to about 10 in 1910). It is only during the twentieth century – as agriculture became dependent upon fossil fuels and other commercial inputs, and both inputs and outputs required transport to distant markets – that the efficiency ratio declined. By the 1980s it had sunk below one (Krausmann 2006). To claim that only high-input methods are 'efficient' enough to secure food supplies for the future, as proponents of high-input agriculture routinely do,[3] is thus to take an exceedingly narrow view of efficiency. High productivity has been purchased at a hefty cost.

But oil is not the only non-renewable resource upon which intensive agriculture is dependent; there are now reports that supplies of mineral phosphorus are becoming scarce (Tomlinson 2011). Moreover a heavy reliance upon irrigation for food production in some areas of the global South since 1945 is rapidly becoming a liability. The World Bank estimates that between 15 and 35 per cent of water use in agriculture worldwide is unsustainable because more groundwater is being removed than is replaced by rain (World Bank 2007: 182). India's hydrologists estimate that 20 per cent of the country now suffers from this problem (though the situation is much worse in some areas) (Holt-Gimenez *et al.* 2006), and in China water experts are warning that aquifers in wheat-growing regions are rapidly becoming exhausted (Watts 2011). Furthermore, cost and dwindling supplies are not the only problems with agrochemicals; heavy use of chemical fertilizer over longer periods has reduced soil fertility while pesticides endanger human health and animal populations.[4]

Finally, the growth of intensive agriculture in the global South has led to serious social (and potentially political) problems with which governments there are now struggling. Already in the 1970s it was becoming apparent that the vision of agricultural development that dominated the 1950s and 1960s – according to which capital-intensive and highly mechanized production methods would release rural labour to supply urban industrialization – was much too optimistic. Few cities in the developing world have industrialized fast enough to be able to absorb the flow of labour displaced from rural areas, leading to the growth of huge slums occupied by unemployed, underemployed or homeless migrants (Frankel 1971; Johnston and Kilby 1975: xv; Araghi 2000; Davis 2006; Weis 2007; Otero 2008).

Faced with the hazards and unsustainability of high-input intensive agriculture, there have been numerous attempts to develop alternative cultivation methods (especially since the 1970s as we shall see in Chapter 7). One requirement of the alternative technologies, as experts have been arguing for some time, is that they should be labour-*intensive* in order to stem the flow of migration and take the pressure off the cities (e.g. Nair 1969; Wellhausen 1976; Flora and Flora 1989; Pinstrup-Andersen and Hazell 1985; Hazell *et al.* 2010). Furthermore, new cultivation methods need to be less reliant upon fossil fuels (Royal Society 2009), and food production should be decentralized in order to minimize transport costs. Finally, in view of the unfavourable growing conditions in which most farmers in the global South have to work, plant varieties and cultivation methods are needed which can cope with drought, poor soil, hilly terrain and periodic flooding (cf. Farrington 1988). In view of the limitations of high-input cultivation, all of this seems pretty self-evident.

2 The potential of peasant agriculture

One major feature of the new approaches, however, is perhaps not so obvious: namely, that the alternative methods must be appropriate to the needs of *small* farmers. Indeed, for those of us who have grown up in an industrial society, this may seem counter-intuitive. Given the benefits of scale, surely the most efficient way to produce more food is on large farms? According to some, anything else means 'holding back progress', 'backing losers', or 'romanticizing' the bygone world of traditional peasant life (e.g. Collier 2009). The superiority of large farms was indeed widely taken for granted in both North America and Europe during the 1950s and 1960s. But since the 1970s an increasing number of experts and major development agencies have concluded that a much more promising way to increase agricultural production, alleviate poverty, stimulate demand in the non-farm economy, reduce dependency upon fossil fuels, and reduce environmental damage in the developing world is to strengthen small farm agriculture (e.g. Johnston and Kilby 1975; Lipton 1978; Spencer 1986; Hayami 1998).

The argument in favour of small farms has several strands. To begin with, it is a relatively effective way to alleviate poverty. Merely increasing the amount of food produced will not be sufficient to alleviate poverty. As I will demonstrate in Chapter 7, hunger often exists despite huge food surpluses; the problem is instead to make sure that food reaches the needy.[5] Since, as we have just seen, poverty and malnutrition are concentrated on small farms, it makes sense to focus upon helping smallholders to grow more food. Small farms are quite good at creating jobs and boosting economic growth. Various studies have shown that raising income on small farms has a greater impact on rural poverty than doing so on large farms (Hazell *et al.* 2010; Lipton 2010). In Africa during the 1980s and 1990s, for example, the fastest rates of economic growth were recorded in countries whose farms were mainly small rather than in places like South Africa or Namibia where they tend to be large (Wiggins 2009). And small farms' stimulus upon job creation derives from the fact that they use more labour per hectare per year than large ones (Lipton 1978: 331).

Second, to focus upon small farms is not a matter of fiddling around with a minor part of the agricultural economy; most of the food consumed in the developing world is grown there. In Latin America, for example, small farms – on average two hectares (= five acres) in size – produce 50 per cent of the maize, 60 per cent of the potatoes, and nearly 80 per cent of the beans consumed domestically while in Asia the vast majority of rice is produced on farms of that size and smaller (Bello 2009: 15). Nor is small-farm production confined to subsistence; in various countries farms of only one or two hectares are still able to sell part of their harvest on the market (Nair 1969: 227; McCann 2005: 166–167), and as we shall see, the same was true in Germany *c.*1900. If one wants to make a major impact upon food production fairly quickly, therefore, small farms are the place to start.

Third, there is also a technical reason for devoting particular attention to small farms. Surprising though it might seem, they are almost always more productive per acre than are large ones. This has been shown in a variety of developing countries (Griffin 1974: 224; Lipton 1978: 324; Blyn 1983: 708–709; Bray 1986: 164; Wilkinson 1997: 43), but it also applies in Europe and North America (Schmitt 1991; Johnson and Ruttan 1994). Large farms (e.g. over 100 hectares) tend to be inefficient, partly because they often leave a good deal of land out of cultivation and sometimes do not make full use of the advanced technology that they own (Griffin and Khan 1978; Nair 1979; Lappe and Collins 1986; Kopsidis 2006).[6] Small farms' advantage is also attributed to the facts that they have little or no hiring and monitoring costs; the family members who work on them are more highly motivated than are hired labourers; they maintain more livestock per unit of land and grow more crops per year; and they economize in their use of scarce resources (Hayami and Ruttan 1985: 340–341; Harriss 1987: 228; Griffin 1999). Because of this advantage, the more equitable distribution of land via land reform has resulted in rapid increases in production (as well as productivity) in a wide range of countries (Staatz and Eicher 1998; Fan *et al.* 2002: 6; Wiggins *et al.* 2010), a fact acknowledged even by the World Bank (Huizer 1996; Lappe and Collins 1986: 72–74). Strengthening smallholders, therefore, is a more efficient way to boost agricultural growth.[7]

At this point, however, a sceptic might respond by saying 'This is all very well, but it assumes that small farmers can in fact be *persuaded* to adopt improved varieties and cultivation methods'. What underpins that scepticism is a long tradition, running throughout the nineteenth century and well into the twentieth, which has argued that peasant agriculture is 'traditional' (i.e. largely static), 'backward' or 'conservative'.[8] Over the last 50 years, however, it has become increasingly clear that this view is fundamentally mistaken.[9] The caution often observed among smallholders presented with new technology is due not to a hostility toward innovation but instead to risk aversion. A large farmer possessing plenty of disposable income can afford to take a chance on new technology that might fail; a smallholder cannot since it could mean starvation (Scott 1998: 20–25; Uekötter 2010: 59–63). Where risk is low, the technology is appropriate for small farm conditions, and the payoff from innovation is clear, peasant farmers have commonly taken an interest in new technology and have been quick to adopt (e.g. Schultz 1983; Stavis 1975: 76–77; Ruttan 1989; Bello 2009: 139–144).

There are thus good general arguments for development agencies to focus their attention upon small farms, but what actually needs to be done? What specific forms of assistance and intervention are known to be effective in strengthening small farms? Little is apt to be learned by turning to the agricultural research system in countries like the US or UK where average farm size in the twentieth century has been relatively large.[10] Parts of Central Europe, on the other hand, are potentially much more informative since average farm size there *c.*1900 was often about four hectares. Moreover, several of these regions had state plant-breeding stations which were explicitly aimed at the needs of small farmers and which proved to be very successful.

3 Peasant-friendly plant breeding and Europe's Green Revolution

The current development industry, however, will have learned nothing from these institutions' experience because they are almost entirely unknown. I have so far found but a single reference to them in the development literature, and no one in development studies with whom I have corresponded seems to have heard of them. This is unfortunate, to say the least, since their history offers illuminating parallels (as well as contrasts) with attempts since 1945 to support peasant farming in the global South.

To be sure, these institutions are not well known even among historians of agriculture and plant breeding since the literature on them is quite small (Kochendörfer 1996; Schneider 2002; Moser 2003; Wieland 2004; von Kittlitz 2005). But as I will show in the chapters that follow, those stations about which we do know something generally undertook several kinds of work. All of them, for example, devoted considerable effort to testing available plant varieties in order to advise the small farmers in their region which of these were suitable for the prevailing growing conditions, both ecological and economic. Many of the stations also undertook the development of new varieties, seeking to breed crop plants that were better adapted to local conditions. Those that were successful were turned over to local farmers at little or no cost for testing, multiplying and distribution. In addition, stations took responsibility for improving those crops of local importance that had been neglected by commercial breeders. Several stations also offered courses in plant breeding so that interested local farmers could learn how to carry out basic selection procedures. And at least one station devoted a great deal of effort to helping small farmers organize themselves into local crop improvement associations so that they could more effectively market their produce and convey their needs to state institutions.

Perhaps because of the stereotype of peasant agriculture as 'backward', some have tended to assume that any technology developed for smallholders must necessarily be 'primitive' or elementary rather than based upon advanced science and technology. As we shall see in Chapter 3, however, this was not the case. The directors of the peasant-friendly stations, for example, usually had doctorates, and some were members of the German Genetics Society and published occasionally in journals of genetics. To be sure, the breeding methods deployed at the stations (and taught to interested farmers) included an older and simpler method – mass selection – but the most modern of methods at that time – hybridization – was also extensively used. Peasant-friendly plant breeding, therefore, was no less 'scientific' than was commercial breeding or that conducted at the universities; it was simply designed with a different group of users in mind (cf. Levins 1986; Bray 1998).[11]

State-funded institutions of this kind emerged in a range of countries within a relatively short period of time: Switzerland (1898, 1907), Austria (1902), Alsace (1905), and in the German states of Bavaria (1902), Hessen (1904), Württemberg (1905), Baden (1908) and Saxony (1908). Moreover, there are signs that similar institutions conducting peasant-oriented breeding also existed elsewhere in Europe at that time.[12] Because the founders of these institutions were often aware of (and sometimes influenced by) similar developments elsewhere, one can speak of a wider European 'movement' for peasant-friendly plant breeding around the turn of the century.

This movement constituted one element within a wider transformation of European agriculture from the late nineteenth century (cf. Chapter 1), a transformation that bore a marked resemblance to the development programmes later known as the Green Revolution. Because of this resemblance, I have chosen to refer to the transformation as 'Europe's Green Revolution'. Some historians are reluctant to accept this equation, rightly pointing to various differences between the European transformation and the Green Revolution.[13] For one thing, the scale of poverty in the global South was and is much greater than in Europe *c.*1900. For another, the Green Revolution was initiated, and to a large extent carried out, in the global South by development agencies and agricultural experts from the North while the European transformation was an indigenous phenomenon. Furthermore, patterns of land ownership were different: 90 per cent of south German peasant farmers (for example) owned their own land while a much smaller proportion of peasants enjoyed such independence in the developing world. And lastly, the cultural and institutional contexts were very different; Europe's agricultural experts, for example, could rely upon much higher levels of literacy among the peasantry and much better developed extension services than could their GR counterparts.

Nevertheless, there are also striking similarities between Europe's Green Revolution and that in the South after 1945 (see Figures I.1 and I.2). To begin with, in both contexts agriculture provided a large proportion of the employment. At the moment the proportion of the population engaged in agriculture in East and Southeast Asia is about 40 per cent while that in Latin America is roughly 20 per cent (IFAD 2001: 41). In south Germany the proportions for Baden and Bavaria around 1900 were both over 40 per cent (Hecht 1898: 676; Haushofer 1975: 56). And although the proportion of the population living from agriculture in southern Africa today is as high as 65–80 per cent (Zerbe 2001:

Figure I.1 Planting potatoes, early twentieth century Germany (courtesy: Schiffer-Archiv, LVR-Institut für Landeskunde u. Regionalgeschichte, Bonn).

Figure I.2 Ploughing, late twentieth century Asia (credit: John and Penny Hubley, Wellcome Images, image no. WOO 26553).

658), the proportion in several Eastern European countries *c.*1930 was also 75–80 per cent (Tiltman 1934: 16; cf. Chang 2009: tab. 1 and 2).

Second, agriculture in both contexts is/was dominated by 'peasant farms'. There is little consensus on the meaning of 'peasant' among rural sociologists or agricultural historians, partly because of the diverse usages of the term in nineteenth and twentieth century Europe. But the secondary literature does identify several features that are common to farms in both contexts (Shanin 1987a; Baumann 1993: 20–25). First of all, the average farms in both contexts are small. Consider the global South. In Africa, for example, 80 per cent of the farms are on average 1.6 hectares (ha) in size (Wiggins 2009: 2) while the vast majority in Zimbabwe are between two and four hectares (Eicher 1995: 808). Subsistence farms in the central highlands of Mexico are about 6 hectares (Wellhausen 1976: 136) while the average farm in Pakistan, Thailand, Turkey or Colombia is 3–4 hectares (Hazell *et al.* 2010: fig. 2), and estimates of the average in India vary between one and two hectares (e.g. Nair 1979: tab. 4.6). Less well known, however, is that most farms in Central Europe were also small; in south Germany at the start of the twentieth century, for example, they averaged about 4–5 hectares (Stiefel 1977: 1718; Hecht 1912: 478) while farms in most Eastern European countries were even smaller (Tiltman 1934: 18). In Germany two hectares was generally regarded as the minimum size necessary to support a family (Hecht 1898: 678).[14] According to administrative terminology of the time, 'peasant farms' (*Bauernhöfe*) there were classified as 'small' (2–5 hectares), 'medium' (5–20) or 'large' (20–100). Using these categories, 60–80 per cent of the farms in south Germany were classed as 'small'. Anything larger than 100 hectares was designated an 'estate' (*Gut*). Since in south Germany 95 per cent or more of farms were in the small- or medium-sized peasant category and a small fraction of 1 per cent were estates, agricultural economists today would probably regard the distribution of land ownership there as 'unimodal'.

Moreover, the nature of the household economy on these farms is similar in both contexts. The labour requirements, for example, are provided mainly or entirely by family members. Crops are grown mainly for subsistence, though surpluses are sold on the local market. On most of these farms a shortage of cash or credit means that little use is made of improved technology; in both contexts, for example, small farmers cannot afford chemical fertilizer.[15] As a result, the principle aim of peasant-friendly breeding in Europe as well as in many GR programmes from the 1970s (cf. Chapter 7) was that the peasant-farm sector should not be left behind in the larger agricultural transformations that were under way. Finally, in both south Germany *c.*1900 and in the Green Revolution of the 1950s and 1960s a major aim of bringing new technology to these small farms, as we shall see, was to undermine peasant radicalism.

But the similarities between the two contexts are not confined to general features of the agricultural economy. If we focus specifically upon the technology central to both revolutions, the same is true. In both contexts, for example, the basic vision of technological 'modernization' was based upon high-yielding plant varieties and chemical fertilizer. The focus of breeding activity in both contexts was upon cereal crops which were important both for subsistence and for sale on the market. In both revolutions the breeding work that was aimed at peasant needs sought to increase the reliability or stability of yield rather than merely trying to maximize yield. In Europe, as in the global South, breeders developed short-stemmed varieties in an attempt to achieve higher yields (e.g. Strampelli 1933). And lastly, in view of the similar technology deployed, it is perhaps not surprising that some of the social consequences for which the Green Revolution was criticized had already been encountered in Europe much earlier. In Italy in the 1920s, for example, new high-yielding wheat varieties were adopted more rapidly by large landowners in the Po Valley rather than by smallholders in the hilly regions (Tiago Saraiva, pers. comm.).

For those concerned with development today, therefore, there is much to be learned from looking more closely at 'Europe's Green Revolution'. My point in using this term, however, is not to emphasize similarity while downplaying difference but rather to open up the possibility of comparative analysis. For it is precisely this combination of similarity *and* difference between two contexts which gives comparison its heuristic payoff. And one of the central problems of the book is to understand why, despite the various similarities between peasant-friendly breeding in Europe *c.*1900 and GR programmes after 1945, they nonetheless had quite different outcomes.

Because comparative history is necessarily demanding, relatively few take it up. Historians of the transformation of European or North American agriculture, for example, rarely consider comparable phenomena in either colonial empires or the Green Revolution. Conversely, the literature on the history of development which has emerged over the last decade (reviewed in Cullather 2000; Frey and Kunkel 2011; Unger 2010b; Cooper 2010) has so far focused entirely on the global South without reflecting upon parallel processes in Europe. But those who nonetheless have opted for comparative analysis have shown how powerful it can be. For example, Michael Kopsidis (2006) noticed the similarities between agricultural revolutions in Europe *c.*1600–1880 and those in the developing world during the twentieth century, above all in the behaviour of peasant farmers. This encouraged him to apply theories from development economics (which were based upon data from the latter context) to earlier European revolutions, and he found that some of them worked very well. Similarly, Francesca Bray's study of the transformation of a wide range of Asian rice economies over many centuries demonstrates how much comparative analysis can contribute to our understanding of development (Bray 1986).

The challenge of comparison, of course, is even greater when the historian ventures onto very different disciplinary terrain (namely development studies) since the chances of screwing up are so much greater. This is no doubt one of the reasons why academic 'tribalism' is so common, discouraging scholars from different disciplines from talking to one another, even when they have closely related interests. But the risks of disciplinary trespassing are worth taking simply because of the potential intellectual gains. Although in this book I am primarily concerned with what history can offer to development studies, it bears saying that the kind of trespassing I am recommending is decidedly of mutual benefit. That historians have much to *gain* from development studies is evident, for example, from the work of Kopsidis just cited, but I have experienced it myself. When circumstances prompted me several years ago to read the secondary literature on the Green Revolution, the parallels with European agricultural development *c.*1900 were so striking that I completely reconceived the book which I had at the time been planning to write. Since then, reading the development literature has taught me to think about European plant breeding from new angles. The concept of 'elite capture', for example, is useful in making sense of the differences between the organization of breeding in north versus south Germany (Chapter 2). Analyses of the Green Revolution alerted me to the possibility that establishing agricultural research institutions can be driven, not just by economic but also by political aims (Chapter 1). Criticisms of the Green Revolution prompted me to look more carefully at the class background and social attitudes of staff at the European plant-breeding institutions in order to better understand their success (Chapter 3). More examples could be cited.

The dialogue between history and development studies is thus a win–win game. That said, in this book I can only make a partial case for its value. For the chapters that follow are in large part concerned to show that development studies may have something to gain *from* history. And as we shall see, that means not just from 'history' as an academic discipline but perhaps more importantly, from the development industry examining its own past practices. Moreover, for those historians who wish to reach beyond the ivory tower, this dialogue has other virtues. By engaging scholars from development studies, it also makes history more accessible to development *practitioners* so that it becomes available as a resource for policy.

4 History and policy

But can history really contribute anything of use to development policy? Some members of the development industry certainly think so (Compton 1989a; Biggs and Farrington 1991: 76–77; Chambers 2005):

> We believe that the failure of a great many development projects to achieve even their most fundamental objectives is due to a reluctance on the part of development practitioners to appreciate the significance of history.... Past lessons are seldom examined.... Planners ignore the relevant literature.
>
> (Porter *et al.* 1991: xv and 131)

I have heard similar views expressed by several people in development studies, and the literature in this field is littered with attempts to extract 'lessons' from history (e.g. McIntyre *et al.* 2009a: 116ff.; McCalla and Brown 2000; Gow and Morss 1988). Some authors have pointed to promising models drawn from history (e.g. Johnston and Kilby 1975; Tendler 1993), and occasionally historical perspectives inform research in development studies.[16] Others have cited illustrations of how *not* to proceed, the most common example of which is perhaps the East African groundnut scheme of the late 1940s (e.g. Wiggins 2009; Johnson and Ruttan 1994). And recently the World Bank has remarked that although 'lessons from the past may not always apply to the future', nevertheless 'lessons from experience ... offer useful guidance' (World Bank 2007: 226, 243).[17]

Nevertheless, the 'history' that development experts usually invoke is rarely based on systematic reflection upon the past (Richards 1985 is a notable exception) and is more often a hastily cobbled together and highly selective version of the past which can be used to justify a particular course of action (Pritchett 2011). Moreover such 'history' is almost always of a very restricted kind, often extending back only to the 1950s (or even just to the 1970s) as though nothing of interest occurred before the Second World War (Staatz and Eicher 1990; Dorward *et al.* 2004; McIntyre *et al.* 2009b). A number of observers, for example, attribute this truncated perspective to a reluctance to confront the colonial origins of the development industry (e.g. Kothari 2011). In Jonathan Crush's view, the development industry seems to have an

> almost overwhelming need to reinvent or erase the past. Most [national development] plans contain a formulaic bow to the previous plan period, a technocratic assessment of its failings designed as a prelude to the conclusion that this time 'it'll go much better'. But prior histories of the object of development – the people, country, region, ... – are deemed irrelevant, best left to the ivory tower academic who has, by definition, no contribution to make to today's problems and tomorrow's solutions. Because development is prospective, forward-looking, ... there seems little point in looking back.... Occasionally [the past] might have 'lessons to teach', but not very often.
>
> (Crush 1995: 9)

Making a case for history's relevance to development policy, Woolcock *et al.* (2011) have recently argued that policy-makers would do well to pay more attention to historians, their analyses and their methods. Their argument is persuasive as far as it goes, but the problem identified above is unlikely to be remedied by requiring experts and agency officials to read books on the history of development. For it appears to reside at a more fundamental level. Judging from what those in the industry themselves say, many experts appear unaware of the success or failure of approaches pursued within their own region only a few years previously (Porter *et al.* 1991: xv). As one economist comments, 'Looking over the fence to other past and parallel projects and seeing what they are doing and learning from them, and perhaps collaborating with them is often *not* encouraged' (Biggs 2007: 277, emphasis in original; cf. Biggs and Gauchan 2001).[18] An unfortunate consequence of such myopia is that 'old and discredited policies are offered as new recipes for rural development' (Williams 1987: 433).[19]

So what can history actually offer? One thing that history (as a discipline) can do is to provide policy-makers with a kind of cross-check. Some have argued plausibly that policy-makers, like all of us taking decisions in everyday life, routinely draw upon 'historical memory', relying upon assumptions about the past which are widely shared within their milieu. But such history is often 'naive, simplistic and implicit [as well as] derived from unconscious assumptions or vague memories' and thus tends to be selective and 'largely unverified' (Woolcock *et al.* 2011: 75; cf. Hall-Matthews 2011). If so, professional historians may be able to provide policy-makers with rigorous and critical feedback.

Feedback of that kind is certainly useful, but can history offer more than this, namely guidance on what will work and what won't? Clearly there is no easy link between history and policy. If one accepts, as most historians do, that history does not mechanically repeat itself, then one cannot simply extract solutions from the historical record and apply them in the present. As we have seen, for example, there were substantial differences between the Central European context *c.*1900 and the developing world after 1945, and many of these persist today. But the utility of history does not depend upon the past and present being identical. What the past offers is not a recipe but a list of issues to watch out for, a wider range of options from which to choose, a set of tools for thinking.[20] In 1990, for example, as the communist bloc in Eastern Europe was beginning to fall apart, the British Foreign Office, presumably realising that they were going to have to deal with a new set of political problems, is reported to have organized a conference to which they invited, among others, historians of the Austro-Hungarian Empire. The rationale was evidently that an awareness of what happened after the collapse of that empire in 1918 might be helpful in anticipating what the dynamics of a reconfigured Eastern Europe after 1990 might be. Since the global financial crisis of 2008, similarly, numerous politicians and commentators seem to be taking John Maynard Keynes' remark seriously: that to be an effective economist one 'must study the present in the light of the past for the purposes of the future' (cited in Skidelsky 2009: 57). And at British officer-training academies like Sandhurst and Dartmouth, I am told, military history is taught, not to confer cultural polish but to provide perspective for cadets who are learning to think analytically about battle strategy. One historian of colonialism sums it up nicely:

> The point of historical analysis is not to commend one form of politics or condemn another but to spell out the range of possibilities, the different conseqences that could ensue from each, and the possibilities of different trajectories following upon particular combinations of actions.
>
> (Cooper 2005: 232)

So much for the general case. What are the specific 'lessons' of Europe's Green Revolution for policy, and what can be learned from studying the history of peasant-friendly plant breeding? Given the scale of rural poverty in the developing world, the key question for policy is: what kinds of technical and institutional arrangements are most likely to be useful for peasant farmers and the landless poor? Despite the development industry's preoccupation over the last 30 years with 'market-led solutions', the book's main conclusion is that state-funded research and development not only can be successful but is probably essential. As we shall see, many of the policies supporting small-farm agriculture that development agencies failed to implement after 1945 but which are currently being promoted were first introduced a century ago in Central Europe (and elsewhere) to good effect. There is something to be learned, therefore, from 'looking back'.

5 The book's structure

The first five chapters deal with the emergence of the south German plant-breeding stations as well as their work, impact and decline. Chapter 1 outlines the economic and political context within which nineteenth century German agricultural authorities decided to establish peasant-oriented institutions for extension and research. In the second chapter the focus shifts to south Germany during the 1890s where a conjuncture of movements and concerns prompted agricultural ministers to see state-funded plant breeding as both economically and politically desirable. In Chapter 3 I recount the early history of the south German stations up to 1914, noting their organization, growth and activity in varietal testing, breeding and extension work. Chapter 4 moves to the interwar period, by which time the stations had begun to make a definite economic impact but also became immersed in controversy. The main focus of Chapter 5 is the period of National Socialism when, despite the regime's oft-declared support for peasant agriculture, the stations came under considerable pressure to abandon their peasant-friendly mission in order not to compete with private-sector breeding.

The last three chapters trace the history of the Green Revolution from the 1940s into the present from a comparative perspective based on the European experience. In Chapter 6, after introducing the Green Revolution and its critics, I examine the attempts by development experts from the 1970s to establish why the revolution had largely failed to benefit smallholders. Their diagnoses are then 'tested' against a particular case: the early history of the Green Revolution in Mexico. Chapter 7 asks to what extent this period of criticism and reflection gave rise to peasant-friendly reforms of the Green Revolution, beginning with new approaches from the 1970s (agroecology, farming systems research, participatory plant breeding). It concludes with an assessment of the status of peasant-oriented approaches today and an examination of the recent claims that biotechnology offers the most promising foundation for a 'second Green Revolution'. In the final chapter I place the history of the Green Revolution in a wider political-economic context, summarize the argument that state-funded agricultural research offers a viable development strategy for peasant farming, but concede that such a strategy is unlikely to make much headway so long as the development industry remains structured as it is.

Since the book is aimed at a variety of audiences, readers may find some chapters of more interest than others. *Agricultural historians* are likely to find Chapters 1 through 5 most relevant to their concerns since these trace the rise and fall of peasant-friendly breeding before 1945. Readers interested in questions of *science and public policy*, on the other hand, may wish to focus upon Chapters 4 and 8 since they challenge a mantra often heard since the 1980s in Britain and the US: that public-sector institutions should concentrate upon fundamental research, leaving applied work to the private sector since the latter does it more efficiently. Academics and practitioners concerned with *development* will probably find Chapters 6, 7 and 8 of most interest since these document the remarkable extent to which green revolutionaries appear to have ignored the success or failure of earlier development programmes. And by juxtaposing the European plant-breeding stations *c.*1900 with the current enthusiasm for biotechnology, I argue that the latter's potential will never be realized as long as the development industry continues to restrict the roles which public-sector institutions can play.

1 The origins of peasant-friendly research in Germany

Where did European peasant-friendly plant breeding come from? More particularly, what were the circumstances which made possible the establishment of the plant-breeding stations which are central to my story? Part of the answer, as we shall see in the next chapter, is to be found in conditions specific to south Germany as a region. But the stations' emergence was also facilitated by more far-reaching developments, both economic and political, throughout many of the German states in the late nineteenth century. And it is here that we begin.

As is well known, public-sector institutions for research and extension emerged in many western countries in the late nineteenth century and have played a significant role in the transformation of agriculture during the twentieth (e.g. Rossiter 1979; Brassley 2000). The question I address here is: what prompted states to establish them? The usual answer encountered in the literature is an economic one: that these institutions were intended to help farmers adapt to rapidly changing markets as international competition forced down prices from the 1870s. In Europe countries responded with several different strategies. One was to redirect production toward less competitive markets; another was to reorganize farmers so as to increase efficiency; and a third was to intensify production so as to reduce costs. In each case farmers were thought to need expert help, whether it was economic advice on farm management or technical advice on the best cultivation methods. And that meant that state-funded institutions providing testing, experiment and extension had an important role to play.

Occasionally, to be sure, there are hints in the literature that economic imperatives may not have been the whole story. Some historians have suggested, for example, that the reason why the British government allocated funding for agricultural research around 1900 was in order to quell farmers' anger at the fact that, unlike France and Germany, Britain was not going to introduce protective tariffs (cf. literature cited in Palladino 1990). And in Belgium from the 1880s agricultural research and education were part of a wide-ranging programme of measures designed to secure the peasantry as a 'bulwark' against socialism (Van Molle 2008). More recently, as we shall see in Chapter 6, historians of the Green Revolution have shown that one of the principal aims behind the US government's promotion of agricultural development in the global South after 1945 was the hope that it would lure the peasantry away from a 'red revolution'. Nevertheless, systematic attempts to place the late nineteenth century emergence of research institutions within a political context are as yet few and far between.

In the case of Germany the existing historiography has been part of the problem. For as several authors have pointed out, there has been an unfortunate division of intellectual labour which separated the agricultural history of Germany from its social and political history (Dipper 1987; Farr 1986; Moeller 1986b). Into the 1980s, for example, social historians working on rural Germany during the period 1871 to 1933 tended to focus upon the political behaviour of estate owners and peasants, driven by a concern with these classes' eventual role in the rise of National Socialism. They were not particularly interested in the social, economic or technical aspects of agriculture.[1] On the other hand, until the 1980s most historians of German agriculture have been largely concerned with the economic and technical dimensions of the eighteenth and nineteenth century agricultural transformation while paying little attention to social and political conflict, as though these were irrelevant to what was essentially an economic process (e.g. Haushofer 1963; Klein 1973; Henning 1978). And although these historians did recognize the importance of the state, their focus was almost invariably upon trade policy such as tariffs and other protectionist mechanisms. Surprisingly little attention was paid to *development* policy, such as supporting credit facilities, cooperatives, land reform, education or research. Over the last decade or so, however, historians have begun to call for more attention to be paid to these areas (Reif 1994; Aldenhoff-Hübinger 2002; Hanisch 2002).[2] And since the 1980s, to be sure, there has been a substantial shift in the historiography of German agriculture, giving long overdue attention to social history (Finlay 2001; Zimmermann 1998; Friedeburg 2004), but as Ulrich Kluge (2000) has observed, very little of this work has been concerned with production (and thus with the associated issues of research, extension or education). As a result, the wide gap between studies of production and analyses of political context persists, at least for the nineteenth and twentieth centuries.[3]

In this chapter, therefore, I draw upon the political history of the period in order to illuminate agricultural historians' accounts of the latter nineteenth century agricultural transformation, knitting the two together where possible with primary sources. The picture that emerges is a dynamic and complicated one, characterized by rapid economic change, a proliferation of new farmers' organizations, and political conflict. In the first half of the century, as industrialization increased the demand for food, German agricultural officials responded with measures to promote technical modernization, and farmers began to organize themselves so as to take better advantage of improved methods. By the 1890s, however, economic crisis had wreaked havoc. Farmers frustrated by government agricultural policy cast about in search of organizations that would defend their interests – be they those of estate owners or of smallholders – while state officials looked for ways to contain the situation. The challenge the latter faced was twofold. On the one hand, noticing the economic viability of peasant farming, officials sought to direct more resources toward the technical modernization of small farms. On the other, they also wanted to quell political unrest in the countryside by creating a new, attractive and more representative farmers' organization, hoping that this would drain support away from militant lobby groups. One of the measures that they devised embodied both of these aims. The 'chambers of agriculture' were a new kind of institution designed to improve the provision of research and extension for peasant farmers while also giving them a larger voice in agricultural policy.

1 Industrialization stimulates growth: the 1830s to 1860s

Although the transformation I am calling 'Europe's Green Revolution' occurred from the late nineteenth century, important forms of technical modernization in the German states also took place earlier. Population growth, for example, prompted late eighteenth century governments to encourage the use of some innovations previously developed in England, such as the use of mineral fertilizers, stall-feeding, the enclosure of common land and especially improved systems of crop rotation (Klein 1973; Bittermann 1956: 109–110; Rösener 1997: 91–94; Teuteberg 1977). In Prussia, however, the state took only limited interest in agriculture until the nineteenth century.[4] After the Napoleonic Wars, in an attempt to reinvigorate the Prussian economy, the state introduced measures aimed at furthering the technical improvement of agriculture. The most important of these was land reform which gave peasants the right to purchase land from estate owners that they had previously farmed (Schissler 1978; Goltz 1903). Henceforth, decisions over crops and cultivation method could be taken by individual peasants rather than collectively. In 1811, too, the Prussian king released an 'edict on agriculture', outlining various measures which might in future be taken by government as well as dispensing technical advice. Estate owners, for example, were urged to form agricultural societies where farmers could widen their experience through lectures and exhibitions in a spirit of self-help. Over the next few years, however, only a small number of these 'Agricultural Associations' (*landwirtschaftliche Vereine*) were set up, and their impact upon farming practice was slight (Altrock 1917; Pruns 1979; Achilles 1993).

During the 1820s, however, the collapse of agricultural prices prompted state officials to think again, but little was done until the late 1830s when estate owners called for more state support (Pruns 1979). At issue was the growing demand for food due to an industrializing economy. The growing population in industrial areas, higher average incomes, and a rising birth rate increased the demand for food at the same time that the number of people engaged in agriculture was declining.[5] Some of the demand was met by increased imports (Klein 1973: 122), and the growth of the rail network made it possible for rural areas to meet some of the urban demand. But the circumstances prompted a more fundamental question: whether the state could and should play a larger coordinating role.

Although the local Agricultural Associations had been growing throughout the 1830s,[6] they had worked for the most part in isolation from each other. By

1840 officials were concerned that Prussian agriculture – and especially its peasant farms – appeared to be among the most backward in Europe. Regarding the Associations as the key to promoting technical advance, they concluded that the Associations' work ought to be better coordinated and that they should be better integrated with the formulation and execution of state policy. To achieve this, local Associations were linked together to form new regional Associations (*Zentralvereine*), and the latter were represented on a new body at state level: the State Agricultural Commission (*Landes-Oekonomie-Kollegium*, est. 1842) (Altrock 1917; Lang 1971).[7] As the official state body representing the farming community, the Commission was charged with advising the Interior Minister on agricultural matters (there was as yet no agriculture ministry) as well as implementing agricultural policies as directed by the Ministry. Although the Commission's initial budget was quite modest (compared to state subsidies for commerce), a fourfold increase during the 1840s allowed it to channel funding to the regional Associations which in turn were able over the following decade to establish 15 agricultural secondary schools, several agricultural colleges and experiment stations, and over 60 model farms (Altrock 1917; Finlay 1992; Grantham 1984; Schmiel 1987 and 1991). With the creation of these institutions, the Associations' promotion of technical innovation was no longer confined to lectures, agricultural shows, prizes, exhibitions of agricultural machinery, and publications. The price of state subsidy, however, was a loss of independence. Although the Associations had originally been founded as voluntary associations funded largely through members' subscriptions, they increasingly became dependent upon state funding, and government officials usually played a prominent role, at least in the regional Associations.[8]

But technical modernization was not the only reason why states took an increasing interest in agriculture. In Prussia both the Interior Ministry and the State Agricultural Commission were agreed that since farmers constituted the 'foundations of the state', subsidies were essential for securing contentment (Pruns 1979). In Baden food riots in the aftermath of poor harvests in the mid 1840s prompted officials to begin thinking about the consequences of an unreliable food supply (Borscheid 1976). Matters then came to a head during the revolutions of 1848. A 'Congress of Agriculture' presented a programme of reforms which called, among other things, for improved credit facilities and agricultural education for all farmers. Significantly, the Congress also called attention to the need for a national organization which would represent farmers' interests to both government and parliament (Haushofer 1963: 118–123). Although the revolutions collapsed, pressure from peasants succeeded in eliminating remnants of their various feudal obligations to estate owners, and officials were forced to rethink agricultural policy. As a result, in 1848 Prussia became the first state to establish a separate Ministry for Agriculture; Saxony and Baden created state agricultural commissions;[9] and Bavaria established a 'Ministry of Commerce and Public Works' whose responsibilities included agriculture (Pruns 1979: 151). In Saxony the minister responsible took a keen interest in intensive cultivation and increased budgets for agriculture; and the state helped to fund its first agricultural experiment station (Finlay 1992). In Baden officials rapidly expanded the budget for agricultural education, introduced reforms to the agricultural secondary curriculum, and established an agricultural lecture series at the University of Heidelberg designed for school teachers who would be teaching peasants' sons (Borscheid 1976).

Officials were aware, of course, that 'the farming community' was not a homogeneous group; the powers, interests and needs of estate owners and peasants were quite different. At the start of the century support had been directed primarily at estate owners, perhaps because in Prussia and elsewhere in Europe large farmers had been the earliest adopters of new technology (Koning 1994; Ballwanz 1978; Teuteberg 1977). And that is not surprising, given that investing in machinery, mineral fertilizer or improved seed required capital or access to cheap credit which estate owners were much more likely to possess than were small farmers. Moreover, peasant farmers were routinely branded in the nineteenth century as 'stubborn' and 'conservative'. While peasants may not have been especially quick off the mark to embrace new technology, however, they observed which of the innovations tried by the local estate owner were successful and adopted those which were appropriate to their own circumstances. Many were interested in experimenting with intensive methods whenever presented with the opportunity, and already in the early nineteenth century they were taking advantage of the changing demand for food due to industrialization (Moeller 1981; Mooser 1992; Zimmermann 1998; Mütter 1988).[10]

By the late 1830s Prussian agricultural officials were concluding that smallholders were going to require assistance and had earmarked funding for model peasant farms (Pruns 1979). In 1845 the Minister went further, instructing the State Agricultural Commission to give preference to peasant farming, leading to increased funding for agricultural secondary schools for peasants and the awarding of prizes. Of course, promoting new cultivation practices among small farmers was also one of the roles foreseen for the Agricultural Associations, but by mid century it had become evident that this was not working.[11] Peasants were not finding the Associations very helpful, in part because the membership was dominated by estate owners, local professionals, and regional agricultural officials (*Landräte*), all of whom stood well above the peasantry in the social hierarchy, but also because the Associations' status as quasi-state organizations prevented them from acting as a farmers' lobby (Achilles 1993: 339–343; Blackbourn 1984; Crone-Münzebrock 1912).[12]

If smallholders were to advance their interests, therefore, some other form of organization would be necessary. In the 1860s a more promising candidate took shape: the Peasants' Associations (*Bauernvereine*). Unlike the Agricultural Associations, these were not only independent of the state but also self-consciously political organizations which sought to take their members' concerns to government. The first of these was established in Westphalia in 1862, grew very quickly – its membership (mostly peasants with medium and large farms) increasing by tenfold from 1870 to 1880 – and was widely imitated elsewhere. In addition to providing practical services (inexpensive property insurance, credit facilities, advice on taxes), some Associations also ran agricultural secondary schools and experiment stations, reporting the results in their own newspapers which were distributed to members (Moeller 1986a; Fahlbusch and Hartwig 1984; Pyta 1991).[13] Initially, the Peasants' Associations met with mistrust from Prussian agricultural officials who saw them as suspiciously 'political' and resented the fact that officials were not offered membership (Albers 1999; Ullmann 1988: 36ff.; cf. Muth 1975).

Despite the help that peasants could get from the Peasants' Associations, however, they still faced financial obstacles if they wanted to intensify production using commercially available inputs because they could not obtain credit on as favourable terms as estate owners enjoyed (Schissler 1978). The situation eased somewhat from 1851 when Prussia – apparently with the aim of securing peasants' political loyalty (Blackbourn 1997) – established rural state landbanks. These, however, could not meet the scale of the demand, so peasants continued to be dependent upon the private sector for credit. And this was the context in which Friedrich Wilhelm Raiffeisen set up rural cooperatives from the 1860s in order to provide credit to smallholders on more favourable terms than they could get in the marketplace. The new cooperatives found a warm welcome from the Peasants' Associations, and the Prussian state was also keen, passing a law in 1867 which gave cooperatives legal protection. Although most cooperatives by the end of the century had been organized to provide cheap credit, others were set up to facilitate bulk purchasing of inputs, shared use of machinery, or processing (e.g. butter) (Merl 1994).[14]

From the 1830s through the 1860s, therefore, a growing demand for food enabled farmers to enjoy reasonable prosperity, and the Prussian state had begun to target small farmers for special assistance. But continued prosperity relied to a considerable extent upon buoyant cereals prices, and this was not to last.

2 Impending crisis: the 1870s and 1880s

From the 1870s German agriculture had to contend with major changes in the international economic context which, in turn, unleashed a certain degree of domestic political turbulence. A contributory factor was the emergence and diffusion of new technologies which had begun to shift the international balance of agricultural power. Steam shipping and railway systems, in Europe as well as abroad, opened European markets to foreign producers while the development of refrigeration and canning methods made it possible for overseas cattle and dairy farmers to sell on European markets. In places like the US and Argentina, for example, great expanses of inexpensive and uncultivated land meant that farmers had no need to invest in mineral fertilizer; once their land became exhausted, they could simply move on. In this way they were able to produce very large amounts of grain cheaply. International competition began to affect German producers, most significantly in cereals where prices in world markets began a long-term decline that lasted until the mid 1890s.

Declining prices affected Western European agriculture in diverse ways. Where governments refused to introduce protective tariffs, farmers had to adjust to the market. In the Netherlands, for example, farmers tended to shift from growing cereals to vegetables while in Denmark (as well as in Britain and to some extent in Germany) they moved into animal husbandry since imported grain (for animal feed) had become so cheap (Abel 1962). In Germany where tariffs were not imposed until the late 1870s, the collapse of grain prices had serious consequences for cereals growers, above all for estate owners since they relied heavily upon cereals (unlike smallholders who generally pursued mixed farming). Grain from eastern Prussian estates, for example, now had to compete in British (and German) markets with cheaper grain from Russia and the United States.[15]

In an attempt to ease their situation, farmers looked to agricultural organizations for support. And in view of their different interests, large and small farmers opted for different kinds of organization. Many estate owners, for example, joined Agricultural Associations; by 1895 the number of Associations in Prussia had increased nearly threefold (Ullmann 1988: 86). But some joined a new organization: the German Agricultural Society (*Deutsche Landwirtschafts-Gesellschaft*), established in 1885. Like the Agricultural Associations, the Society declared itself to be a non-political organization which would focus solely upon promoting technical improvement (Hansen and Fischer 1936; Haushofer 1960). But the Society's founder, Max Eyth, was very critical of existing agricultural organizations; the Agricultural Associations he felt to be too local, too poor and too conservative to promote technical innovation on the necessary scale. And state and national bodies such as the Prussian State Agricultural Commission or the new German Agricultural Council (*Deutscher Landwirtschaftsrat*)[16] had only advisory powers. Although the Council was dominated by estate owners and their concerns (Steinbeck 1997), it met only once or twice a year, prompting Eyth and other members of the Agricultural Society to dismiss it as little more than a talking shop (Hansen and Fischer 1936). By contrast, Eyth's vision for the Society was of an independent organization whose high annual subscription would allow it to undertake major projects while at the same time making sure that members would be keen to secure technical advantages in return. One of its first steps on behalf of members, for example, was to negotiate lower prices for commercial fertilizer. The arrangements were handled by a 'Fertilizer Division' whose tasks were not only to provide fertilizer at a discount but also to conduct experiments on fertilizer use and to inform members of knowledge relevant thereto. Among the other divisions created, as we shall see in Chapter 2, was one for plant breeding which functioned in much the same way.

The Society grew quickly, its membership rising from about 2000 initially to some 12,000 members a decade later (Haushofer 1963: 204). In many states, however, only a minority of estate owners were members (Anon. 1892: x), and in the nationwide farming community of some five million, the Society's membership represented a drop in the bucket. Nor was the membership in any way representative of the community as a whole. By setting subscriptions at what was then regarded as a very high level, the Society effectively excluded the owners of small- or medium-sized peasant farms (Achilles 1993: 359; Herrmann 1994: 214). In any event, Eyth and other founders conceived the Society as an elite organization made up of 'educated farmers' who would set an example for peasant farmers (thought to be backward and conservative). But the latter were to be kept at arm's length; no major role within the Society was foreseen for them (Föhr 1999). Smallholders thus had to look elsewhere for support.

To be sure, for the most part peasant farmers were more buffered from the fall in cereals prices because although many sold grain, the majority of their income came from animal husbandry (especially pigs). This meant that they were well placed to take advantage of the increasing demand for meat (Berghahn 1987: tab. 19) while feeding their animals with cheap imported grain. As a result, many peasants began to pursue animal husbandry more intensively (Webb 1982; Hunt 1974; Moeller 1981), especially in northwest Germany where the proximity to northern ports gave farmers good access to imported grain while nearby industrial cities provided large markets.

But shifting more heavily into animal husbandry meant acquiring more animals. And with cereals prices falling, peasant farmers turned to cooperatives in search of credit, such that the number of these in Germany trebled during the 1880s (Haushofer 1963: 218; cf. 249) and trebled again between 1890 and 1896 (Goltz 1903: 378–379). But they were also obliged to rely upon the private sector for credit, leading to numerous complaints from the 1870s about loan sharks. So peasants continued to lobby for favourable legislation via the Peasants' Associations, and the latter enjoyed some success in securing better credit facilities (Pyta 1991: 28ff.). Since the Associations were local and regional organizations, however, peasant farmers still lacked political representation at national level (Puhle 1975, 58–60; Flemming 1978, 48–53).

Faced with international competition as well as growing discontent among farmers, Prussian and Reich governments were under pressure to respond. The policies that they enacted during the 1870s and 1880s suggest that they were pursuing both a long-term economic strategy as well as a short-term political one. The best-known short-term strategy was the Reich government's imposition of tariffs in 1879. As in several other European countries, the tariffs were placed on the import of cereals as well as cattle, meat and dairy products and were increased several fold over the next decade (Gerschenkron 1966; Aldenhoff-Hübinger 2002). Since the Reich chancellor, Bismarck, came from the East Prussian land-owning nobility (the Junkers) whose members were regarded as an indispensable bulwark of the political order, most historians regard the tariff decision as an attempt to shore up this Junker elite (e.g. Henning 1996: 889ff.). Indeed, tariffs did alleviate the pressure upon estate owners because a high proportion of their income came from cereals. But one must also recognize that even peasants supported the tariffs, partly because they offered protection against meat imports (Hunt 1974), though also because cereals were a source of animal feed and straw and in some regions contributed 10–20 per cent of small peasant farms' income (Moeller 1981; cf. Ballwanz 1978).

Over the longer term, governments sought to increase 'competitiveness' in several ways. Noticing smallholders' enthusiasm for cooperatives, for example, the Reich provided subsidies for new ones (Altrock 1917; Peal 1988) but also introduced a Reich law in 1889 that loosened the legal requirements for establishing cooperatives (Aldenhoff 1996; Merl 1994). In addition, governments established educational and research institutions designed to foster the modernization of agriculture. Although the first university departments of agriculture had emerged during the 1860s, most of the rest were created during the 1870s (Harwood 2005), as were several new experiment stations and a large number of agricultural secondary schools (Achilles 1993: 372–373; Mütter 1988; Hudde 1965). During the 1880s and 1890s the Prussian state took over responsibility for experimental facilities dedicated to branches of the food, brewing and distilling industries (Lundgreen *et al.* 1986; Klemm 1992), and in 1898 the first institute for agricultural research at Reich level was founded (the *Biologische Abteilung für Land- und Forstwirtschaft beim Kaiserlichen Gesundheitsamt*) (Sucker 1998).

In 1886 the Prussian government also introduced a policy of 'internal colonization' which was evidently intended to appeal to both estate owners and peasant farmers. The new policy allowed the state to buy up East Prussian estates, break them up into medium-sized peasant farms (10–20 hectares), and issue mortgages to smallholders or agricultural workers on favourable terms (Aldenhoff 1996; Frauendorfer 1957). An official justification for the policy was to attract small farmers to East Prussia in order to counter the 'flight from the land' whereby agricultural workers in the eastern provinces were migrating to the industrializing areas in Germany in search of better wages, a shift which was perceived to have undesirable biological and social consequences. Another aim of the policy was to shift the balance of ethnic power in the eastern provinces in order to undermine the nationalist movement in the Polish population who constituted the region's majority (Hagen 1980; Rolfes 1976). Peasant Associations were predictably keen on internal colonization, and it seems fairly certain that an unstated aim of the policy was to lure peasants and agricultural workers away from the Social Democratic Party (e.g. Dade 1913: 555; cf. Moeller 1986b; Schreiner 1974: 39; Jatzlauk 1994). Some estate owners were also enthusiastic since, in one historian's view, the policy effectively served as a 'relief agency' for heavily indebted Junkers (Hagen 1980: 177, 156).[17]

Nonetheless, tariffs failed to halt the continuing fall in cereal prices during the 1880s, and most of the other policies could only be expected to have an impact over the long term. Rural discontent grew, and it was evident to contemporaries that the problems of German agriculture were more fundamental. Throughout the 1880s declarations of 'crisis' were commonplace: among officials in agricultural ministries, in the daily press, in agricultural newspapers, and in both Prussian and Reich parliaments (Steinbeck 1997: 27–28).

3 Rural unrest in the 1890s

The escalating agricultural crisis reached its climax in the 1890s. The catalytic event was the Reich government's decision in 1891 to reduce tariffs on both cereals and meat imports. Its aim was to bring down food prices and avoid a trade war in an attempt to lift industry out of recession (Barkin 1970). Following the reduction of tariffs, however, cereals prices fell by a further 25–30 per cent; the scale of estate owners' indebtedness rose yet further; and the ensuing protest from farmers forced the Reich chancellor's dismissal in 1894. Significantly, however, none of his successors during the 1890s chose to restore the tariffs (Abel 1962). Government policy had taken a new turn.

The tariff decision, however, had wider repercussions. For the Junker elite, accustomed to being favoured by governments as a pillar of the social order, the decision was an alarming sign of preferential support for industry over agriculture and was seen to pose a major threat to their traditional social role. In their anger, both estate owners and peasants became politicized as never before. There were big increases in electoral participation, and rural politics acquired a volatility previously unknown (Blackbourn and Eley 1984: 265; Lerman 2001).

The single most important political response to the new tariff policy was the formation in 1893 of the Agrarian League (*Bund der Landwirte*). In contrast with the Agricultural Associations' declared neutrality, the League was a self-consciously political organization of a new kind which sought allies among conservative and centre parties in order to lobby on behalf of farmers' interests, above all with the aim of restoring tariffs. It quickly attracted a large membership, reaching 180,000 within its first year (nearly as many as the Agricultural Associations), rising to 300,000 by 1906 (Achilles 1993). With a huge central office in Berlin, an extensive regional network of committees, a publishing house, several newspapers and numerous agricultural magazines and journals, it was well placed to voice its demands (Puhle 1966; Ullmann 1988; Fricke and Hartwig 1984). In doing so, the League lobbied vigorously and soon managed to get its leading figures onto all of the important public agricultural bodies. But it also drove a hard bargain with members of parliament, pledging its considerable organizational support at election time only if candidates toed its line. As a result, parties that had previously defended free-trade succumbed to pressure during the 1890s and began to advocate the restoration of tariffs (Blackbourn 1984: 64–65). By the late 1890s the League had induced nearly one-third of Reichstag deputies to endorse its programme (Ullmann 1988: 93), enabling it to secure its greatest victory in 1902: the restoration of tariffs. Given its power, of course, the League also quickly attracted enemies. Economic liberals in the Reichstag were highly critical, and various political opponents decried the League's forceful campaigning practices as 'demagogic' and 'terrorist' (Vascik 1991). The Prussian Agriculture Ministry found the League an awkward customer to deal with, and the German Agricultural Council saw the League as a threat to its own position as the official representative of German farmers (Gottwald 1984).

Just which sector of the farming community the League represented has been a matter of some dispute among historians. On the one hand, its spokesmen took every opportunity to insist that it represented the entire farming community, and they could point to the fact that only 1 per cent of its members were Junkers while about three-quarters were small peasants (Puhle 1986). But some historians have pointed out that almost all of the League's leadership were estate owners, concluding that this elite, even though a tiny minority of the membership, nevertheless dominated the League to the detriment of peasant interests (Puhle 1966). More recently, however, other historians have insisted that this interpretation is too simple because peasants' decision to join the League was actually rational (Achilles 1993; cf. Tirrell 1951: 171). For one thing, peasant farmers also earned some income from the sale of grain, thus gaining from tariffs; for another, the tariff reduction opened the door to meat imports that hurt them (Hunt 1974). Finally, it is probably the case that peasants had little alternative but to back the League in the early 1890s since neither of the parties traditionally dominant in the countryside – the Conservatives and the Catholic Centre – had much to offer farmers. Moreover the one organization then dedicated to defending peasant interests – the Peasants' Associations – was not a centralized organization which could act effectively on the national level nor did it possess the resources that made the League so powerful. But this is not to say that peasants perceived their interests to be *identical* to the League's. That this was not the case is noticeable from the persistently ambivalent and sometimes tense relations between the Peasants' Associations and the League (Moeller 1986a: 29–30; Flemming 1978, 48–53; Fahlbusch and Hartwig 1984; Mütter 1988: 656). The conclusion, therefore, must be that peasant farmers' support for the League was a tactical move, undertaken for lack of a better alternative.

Even though estate owners commanded disproportionate political clout within both government and state, however, their numbers were very small (less than 1 per cent of farmers). In terms of parliamentary politics, therefore, it was peasant voters (over five million of them *c.*1900) whose radicalization during the 1890s made them the force to be reckoned with. Although peasants in Protestant regions had traditionally voted for conservative parties (and for the Catholic Centre Party in the south and west), they became more critical of these parties from the 1880s as it became apparent that national policy – above all, tariffs – could substantially affect their economic well-being (Nipperdey 1990: 222). And this dissatisfaction opened the door to the emergence of a variety of radical movements which began to bid for peasant support (Massing 1967; Blackbourn 1984; Moeller 1986a).

One of these was the (socialist) Social Democratic Party which began from 1890 to focus upon the peasantry and agricultural labourers in their campaign to achieve a majority in the Reichstag. With the radicalization of rural voters after 1891, the prospects of wooing peasant voters away from the Conservative and Centre parties must have looked quite good (Lehmann 1970: 263–278). While selling socialism to agricultural workers may have seemed straightforward, however, recruiting peasant farmers was going to be tricky. For a recurrent theme in Marxist writings on agriculture since the 1870s had been that small- and medium-sized farms were doomed. Since they could not compete with large, heavily capitalized ones, there was little point in trying to save them. In any event, tariffs were thought to be out of the question since they pushed up the price of food for urban workers. Since peasants would inevitably be driven into the agricultural proletariat, therefore, their ultimate salvation was thought to lie in collective farms made possible by nationalizing land (Hussain and Tribe 1981: 26–27; Lehmann 1970: 16–17). Armed with this rather unpromising sales pitch, Party members nevertheless headed for the countryside in 1890–1891 with a programme of agitation. Things did not go well. For one thing, proselytizing industrial workers who arrived in villages loaded with pamphlets were met with suspicion at best (and violence at worst). For another, the Agrarian League was soon recruiting peasants with rather more success (Maehl 1980).[18] As a result, in the 1893 national election the Social Democrats increased their peasant vote but still fell far short of their target, prompting a decision at the Party conference to give prominence to 'the agrarian question' in future. At the Party conference the following year, accordingly, there was vigorous debate over which policy changes might be necessary in order to attract more peasants to socialism. One pragmatic and reformist line, advocated by the head of the Bavarian branch of the Party (Georg Vollmar), was particularly controversial. Under his leadership the Bavarian party, presenting itself as a party of workers *and* peasants, had developed a programme of peasant-friendly policies which proved very attractive to rural voters (Hussain and Tribe 1981: 84, 95). Moreover, Vollmar argued, social democratic parties in Denmark, Hungary and elsewhere had done the same thing with success. If the German peasantry were not to go over to enemy parties, he concluded, it was essential that the party reconsider its policy on the nationalization of land. Unsurprisingly, his proposal drew criticism from Engels and Party leaders, but the conference nonetheless voted to consider Vollmar's and other reform proposals formally at the next year's conference (Hussain and Tribe 1981: 96–97). In the event, Vollmar's proposal was subsequently defeated, but the Party's apparent willingness in 1894 to radically reconsider its policy toward the peasantry was, as we shall see, a matter of some concern to the Prussian government.

4 Growing state interest in the peasantry

Overall there is general agreement that the League was the most important and influential agricultural pressure group from the 1890s until 1933 (Klein 1973; Flemming 1978). But its rapid growth, its claim to represent all farmers, and its effectiveness in modifying the agricultural policies of centre and right parties placed the Prussian government in an awkward situation. From the early 1890s, accordingly, agricultural officials began to rethink their policies in order, as they put it, to 'calm the situation' (Steinbeck 1997: 26–27). The problems were both economic and political.

Some years ago Niek Koning's (1994) comparative analysis of agricultural transformations from the late nineteenth century in the US and Western Europe drew attention to shifts of government policy in several countries, prompted by widespread perceptions that smallholder agriculture was gaining a competitive advantage. The evidence on Germany from the 1890s clearly confirms this view. For one thing, there was alarm at the scale of indebtedness. As we saw, estate owners, above all in East Prussia, were especially hard hit by the fall in cereals prices. In an attempt to restore their income, however, relatively few had chosen to abandon cereals for more profitable sectors. Instead they sought to expand production by buying more land, taking out loans in order to do so. When cereals prices plummeted in the 1870s, however, paying off loans became increasingly difficult, and levels of indebtedness began to climb (Henning 1996; Jatzlauk 1994; Tirrell 1951: 195–196). By the early 1890s, an estimated 40 per cent of estate owners were thought to be bankrupt (Steinbeck 1997: 33ff.).[19]

Consequently, in 1894 the Prussian Agriculture Ministry decided to call a conference of estate owners, officials and academic political economists in order to discuss the situation and explore possible solutions. Significantly, no peasant representatives were invited, and the press was excluded; this was evidently to be a confidential discussion of sensitive matters among members of the agricultural elite. As the Minister told the participants, he hoped the meeting would yield information about such matters as levels of debt and the availability of credit facilities. Asked to outline their financial circumstances, estate owners replied, denying the truth of newspaper reports that many were living above their means. Debt was a serious problem right across the farming community. Their own debt, however, had been aggravated by the decline in tariff protection, along with high taxes and the costs of the newly introduced welfare measures for workers. Moreover, debt reduced the amount that they could afford to spend on technical modernization. Not everyone at the meeting agreed with this analysis. One economist doubted that increased taxes and welfare costs actually accounted for much of the debt, and several called for proper statistics on relevant variables such as levels of debt in different sized farms, the frequency of sales, and the reasons for them in order to pin down the exact nature of the problem (Anon. 1894, *Die Agrarkonferenz*).

One economist offered a different explanation. Theodor von der Goltz, an eminent professor of farm management and a member of the Prussian State Agricultural Commission, argued that estate owners had paid too much attention to increasing production and not enough to profitability. As a result, by the early 1870s they were paying too much for land in relation to profit margins and were thus becoming heavily indebted (Goltz 1903: 390–414; 1894).[20] The reason for this behaviour, he suggested, was economic illiteracy. A study of farm management by the German Agricultural Society, he noted, had indicated that most estate owners lacked even an elementary understanding of the economic and management issues involved in running a farm. It was quite common, for example, to become an estate owner after a career as an officer or top civil servant without any background in agriculture. In addition, estate owners were often distracted by work in parliamentary bodies or with their responsibilities in various agricultural organizations. Furthermore, once their estates went into debt, estate owners failed to adjust their lifestyles accordingly, continuing instead to spend substantial amounts on maintaining the lifestyle to which they were accustomed.[21] Loans and price subsidies, Goltz concluded, were just keeping badly run estates going.

Another economist at the conference, Johannes Conrad, went further, stressing that it was important not to treat all sections of the farming community in the same way. In particular, estate owners and peasant farmers might require different kinds of assistance. Credit arrangements for the former seemed to be quite good, for example, but they did not provide adequate credit for the latter, so the state needed to create better facilities specifically for peasants. Goltz agreed, pointing out that the level of debt among peasant farms (i.e. as a proportion of the value of the farm) was much lower. The reason, he argued, was that unlike many estate owners, peasants were well versed in agriculture, quicker to respond to the market by moving to intensive methods, and prepared to reduce their standard of living in times of scarcity. And entering labour-intensive sectors (vegetables, sugar beets, animal husbandry), for example, was an easier option for smallholders because they didn't have to pay high labour costs; they could simply work harder or draw more heavily upon family members. And that was why they were less hard-hit by the crisis of the 1890s than were estate owners.[22] While estate owners at the meeting objected to this diagnosis, the Finance Minister, Johannes Miquel, agreed with the need for better statistics, accepted that debt among estate owners and peasants might need to be treated separately, and subsequently accepted that economic incompetence and lavish lifestyles were partly responsible for the high level of estate owners' debt.[23]

But peasant farming's advantage was not confined to levels of debt. The rapid growth of cooperatives during the 1890s, for example, indicated that among peasants there was considerable demand for organizations and credit facilities that would assist them in adjusting to changing markets.[24] Nor were peasant farmers being driven out of business. Evidence from farm censuses of the period showed no decline in the number of small farms (Frauendorfer 1957: 403ff.), and mid-sized peasant farms were actually thriving.[25] From the 1880s to the turn of the century numerous studies of profitability on farms of different sizes were published (Frauendorfer 1957: 386ff.). An analysis of survey data from 1882 and 1907, for example, indicates that peasant farms were shifting more rapidly toward animal husbandry than were estate owners (Achilles 1994: 201–202; Kluge 2005).[26] In addition, yields were often higher in regions where small farms predominated than in regions where estates were common (Ballwanz 1978; cf. Schreiner 1974: 39).[27] Despite their ability to adapt quickly to changing market conditions, however, peasant farmers lacked formal education, von der Goltz argued, and thus found it harder to get the information that they needed. It was therefore essential that the state made technical innovation more accessible to them (Goltz 1899; cf. von Rümker 1900).

The growth of interest in peasant farming was evident in various spheres. In the 1890s, for example, books on 'bookkeeping for small- and medium-sized farms' began to appear,[28] and over the next decade or two a number of important works on the economics and sociology of peasant farming were published in German (Shanin 1987b). Moreover, by the turn of the century the German Agricultural Society's Committee on Accounting was beginning to develop simplified bookkeeping forms for peasant farmers (Haushofer 1960: 133ff.).

But if policy in future were to focus more heavily upon supporting technical improvement among peasant farmers, the state would have to find some way to outflank the all-too-vocal Agrarian League whose demands, if met, would have diverted funding toward estate owners and slowed the pace of restructuring.[29] And that meant that the government's problem was also a political one. As we have seen, during the early 1890s ministers were disturbed by the growing strength of the Social Democratic Party and especially the prospect of the Party attracting the peasant vote. Facing a similar problem a decade earlier, Bismarck's strategy for undermining the growth of socialism had included a series of social insurance measures designed to win the loyalty of workers (Wehler 1983: 100ff., 137). Comparable measures to counter socialist agitation in the rural areas began in the winter of 1890–1891 when both government and the Catholic Church organized propaganda, warning peasants that the socialists planned to eliminate private property (Lehmann 1970: 59–60, 103). The policy moved up a gear in 1893 when the Prussian Finance Minister, Miquel, was sufficiently concerned to embark upon a formal policy of offering legislative inducements to various sections of the electorate in order to lure them away from the Social Democratic Party.[30] Given the very large number of peasant voters, as well as the then oft-voiced idea that the peasants constituted a 'bulwark' against collectivism (Lehmann 1970: 17–18), they became a prime target of this policy. In proceeding in this way, the government reckoned it could also count on support, not just from conservatives but also from other groups nervous about the rise of socialism.[31]

Accordingly, Prussian expenditure on agriculture nearly doubled during the 1890s (Goltz 1899: 287). Miquel's strategy emerges particularly clearly in a letter written to the Kaiser in 1894. Drawing attention to speeches on the agrarian question by Georg Vollmar – and describing him as one of the smartest, most moderate and thus 'most dangerous' social democrats – he argued that the government's agricultural reforms were essential to head off the socialist threat. There were many things that could be done, he added, including maintenance and consolidation of small- and medium-sized farms, improvement of credit facilities for peasants, practical agricultural instruction aimed at small farmers, and the promotion of animal breeding:

> these and many other issues have started to be addressed and should, in my view, be solved in future decades. Then this revolutionary troublemaking will fail, confronted with the thick skull of the German peasant.[32]

Over the next few years several institutions were established whose character was consistent with this strategy. One of these was the creation in 1895 of a state agency which would promote cooperatives by making cheap credit centrally available (*Preussische Zentralgenossenschaftskasse*) (Aldenhoff 1996: 44). Significantly, this institution was not a response to demands from cooperatives themselves but rather a case of *state* initiative. Since the 1880s Prussian officials had been steering support toward those cooperatives that were deemed to be politically neutral and away from those with a left-liberal orientation. In the 1890s, similarly, Miquel and co. saw the new credit agency as a way to mobilize support from smallholders who might otherwise be attracted to socialism (Merl 1994; Peal 1988).[33]

Another new institution is particularly relevant to my concerns here: the Prussian 'chambers of agriculture'. Established in 1894, the chambers were a novel kind of institution, combining a wide array of functions which since the early twentieth century have usually been delegated to specialized organizations. For the chambers were not only responsible for providing research, extension services and agricultural secondary education; they were also the officially recognized spokesman for the farming community in their region, authorized to convey the needs and wishes of farmers to the Ministry of Agriculture. The law making it possible to establish a chamber in each of the Prussian provinces was passed in 1894, and by 1914 most other German states had followed suit (Altrock 1925).[34]

The chambers provided the Prussian government with a neat solution to both of its agrarian problems. As far as promoting technical modernization was concerned, for example, the Ministry reckoned that the Agricultural Associations were not up to the task (Steinbeck 1997: 34–35). They were dominated by estate owners and civil servants whose commitment to advancing peasant farming was doubtful; they had few peasant members; and their budgets were too small. Membership of the chambers, by contrast, was compulsory for all farms beyond a given size (including nearly all peasant farms), and the chambers were far better funded than the Associations.[35] Endowed with a healthy income, chambers could employ agricultural scientists, technical assistants and farm labourers as well as administrators and clerical staff.[36] In addition, they usually possessed their own experiment stations which conducted research and carried out tests on soil, commercial fertilizers, seed and sometimes machinery. A chamber's programme of work was generally proposed initially by its expert committees (in areas such as agronomy, animal husbandry, economics, fertilizer, credit, cooperatives, education) and then considered by the elected members of the chamber before being implemented by the appropriate department. Active in extension, the chambers provided both economic and technical advice, implemented government guidelines on best practice, monitored the uptake of new practices, published agricultural magazines and flyers on particular technical questions, and arranged courses on specialist topics. On the educational front, they ran 'winter schools' for small farmers during the off-season, and their teachers toured the province's more isolated regions during the growing season to hold short courses there. Moreover, the chambers were legally obliged – as the Agricultural Associations were not – to provide technical and financial support to both Peasants' Associations and cooperatives.

But the chambers also provided the government with an antidote to rural unrest. Neither of the existing national agricultural organizations, for example, were effective mouthpieces for peasant concerns. Since membership of the Agricultural Associations was voluntary and peasant members few and far between, they could hardly claim to speak for the entire farming community.[37] For its part, despite its peasant membership, the Agrarian League was equally unrepresentative since it was far stronger in northern Protestant regions than in the Catholic regions of the south or west where the Peasants' Associations predominated. Moreover, as an organization independent of the state and well financed to boot, the League was difficult to harness to state policy. From the state's perspective, therefore, what was required was a more representative farmers' organization which was entitled – as the Agricultural Associations were not – to lobby government on behalf of farmers' interests, just as the League had done so successfully at national level and the Peasants' Associations, in the regions. A well-funded organization which could provide peasant farmers with technical support while offering them a way to voice their grievances might wean them away from the League as well as immunize them against a socialist infection. The chambers were designed to do both.

Thus the chambers served as both pressure group and state agency. Unlike the Agrarian League – which was free to lobby both parliaments and the public – the chambers were expected, as state institutions, to convey farmers' demands solely to the Ministry of Agriculture, conveniently allowing the Minister to deal with such demands out of the public view.[38] And as agencies of the state, the chambers were charged with advising officials on agricultural matters and cooperating in the introduction of government measures. State control of their activities was assured by the Ministry's financial contribution to them, its right to approve their budgets, and its power to dissolve them.[39]

Finally, the way in which the new institution was received by the farming community suggests that the chambers represented a substantial shift of power toward smallholders. When the draft law on chambers was presented to the Prussian legislature early in 1894, it unleashed a prolonged and heated debate which dragged on over five months. The Agricultural Associations expressed their resentment at being displaced as the voice of the farming community while Junkers argued that the bill's electoral procedures (which allowed for direct elections) let too many peasant members into the chamber, ignoring the Junkers' traditional role as leaders of the agricultural community (Gothein 1910/1911: 496ff.; Wygodzinski 1916; Steinbeck 1997: 36ff.; cf. Lichter 1994: 54ff.).[40] The evidence on peasant perception of the chambers is mixed, but in some regions at least they were favourably regarded.[41] To be sure, the way in which the chambers were organized gave disproportionate influence to estate owners (Harwood 2008), such that they were not ideal from the smallholder's standpoint. But faced with the alternatives on the national scene – the Agricultural Associations or the Agrarian League – the chambers represented an offer that peasant farmers could not refuse.

Conclusion

A dual policy of promoting modernization while pacifying the peasantry was not new in the 1890s. As we have seen, during the 1840s several German states were so inclined, prompted on the one hand by the beginnings of industrialization and the weakness of agriculture in their region, and on the other by the connections between food shortage and rural unrest. What was novel about the 1890s, however, was that these dual concerns came to shape the formation of institutions of research and extension which, in some form, still exist. Moreover, as we shall see in the final three chapters, this reliance upon technology in order to solve political problems has been a recurring feature of agricultural development schemes throughout the twentieth century, whether in European colonial empires before 1945 or in Green Revolutions thereafter.

2 The movement for peasant-friendly plant breeding, 1880–1905

As we have seen, from the 1890s the Prussian government, faced with the apparent viability of smallholder agriculture as well as the political dangers posed by militant peasants, responded *inter alia* by establishing peasant-friendly research institutions. Against this general backdrop, let us now shift the focus toward Central Europe in order to find out why agricultural officials at this time chose to create state plant-breeding stations. Like earlier public institutions charged with quality control (experiment stations, seed-control stations), these new stations conducted testing – in order to provide the region's farmers with better information on the choice of crop variety – but they were also designed to promote the development of *new* varieties better suited to their region. What made these stations distinctive, however, is that they were explicitly designed to help peasant farmers.

The story is one of confluence. In Germany and elsewhere, a wide range of largely independent developments during the 1890s – new kinds of research institution, new cultivation technologies, worsened market conditions, intensified government concern with the peasantry – shifted perceptions, among leading figures in the agricultural community as well as in government, as to what was both possible and desirable. Significantly, however, state involvement came relatively late. During the 1880s those who took the initiative in improving plant varieties via breeding were either estate owners or agricultural scientists, followed in the early 1890s by a few cooperatives. Quasi-official organizations representing farmers, on the other hand, played a largely passive role at this time. Only from the late 1890s did various Central European states begin to explore the possibility of promoting varietal testing or plant breeding. And in Germany when officials eventually decided to embark upon state-funded breeding, their aims were both economic and political, just as with the Prussian chambers of agriculture a few years earlier.[1]

1 The emergence of commercial plant breeding

Given its economic importance, it is remarkable how little attention historians have yet paid to commercial plant breeding in Germany.[2] German attempts to improve potato varieties began in the 1840s (Hillmann 1909; Scheibe 1961), but the earliest major successes in commercial breeding were with sugar beets. The first sugar factories emerged around 1800 and began to proliferate from the 1830s. Concentrated in Lower Silesia and central Germany – above all in the Prussian province of Saxony which possessed both the fertile soils well suited to beet cultivation and the cheap coal necessary for power – the factories were typically built on estates whose owners grew beets but also extracted the sugar (Perkins 1984 and 1990; Mende 1997; Becker-Dillingen 1931). Attempts to increase the sugar content of beets began at several horticultural breeding firms in Saxony during the 1850s and by the 1890s had succeeded in roughly doubling sugar content (von Rümker 1894; Fehse 1980).[3] In addition to the breeders' success, the fact that sugar beets were unusually responsive to mineral fertilizers as well as to more careful soil preparation (plus the use of mechanization for the latter purpose) helped to make Germany the largest sugar producer in the world from the 1870s to the First World War (Perkins, J. A. 1990) (see Figure 2.1).

As a result, sugar-beet cultivation came to be seen in Germany as a model of what could be achieved through intensive cultivation practices. It comes as no surprise, therefore, that the province of Saxony was one of the earliest centres of intensive agriculture (Müller 1989) as well as the centre of the commercial plant-breeding industry. Given high grain prices into the 1860s, cereal growers were one group which took a particular interest in the methods that had proved so successful with sugar beets. Extending the model to cereals, however, was not so straightforward. For when one applied large amounts of mineral fertilizer to the local German varieties of wheat or rye, yields did not increase substantially, and the plants tended to topple over in response to wind or rain (von Rümker 1912–1913; Krzymowski 1951: 270–275). Innovation-minded farmers were aware, however, of the existence of improved varieties elsewhere that responded

Figure 2.1 Selecting beets (source: Hillmann 1910a).

well to mineral fertilizers and produced yields much higher than did German local varieties. During the 1860s and 1870s, therefore, cereal growers – especially in the sugar beet-growing regions – began to import such varieties, initially from Denmark, France and Britain and later from Sweden, Russia and North America. The best known of these was 'Squarehead', a high-yielding British wheat variety which was brought to Germany in the 1870s from Denmark where it was proving very successful (von Rümker 1928b; Schindler 1909: 211). German growers soon found, however, that while the imported varieties produced twice the yields of German varieties, they did not cope well with severe continental winters and generally produced flour of poor baking quality (Steglich 1896; Kiessling 1914).[4]

Some farmers wondered, therefore, whether it might be possible to get the best of both worlds. After all, horticultural breeders in Britain had been using hybridization since the beginning of the nineteenth century; German potato breeders had been using it successfully since the 1840s to construct disease-resistant varieties; and by the early 1860s Patrick Shirreff had managed to produce hybrid cereals varieties (Olby 2000; Schikorra 1914b; von Rümker 1895). Encouraged by these precedents, a few German cereals growers set out to develop hybrid varieties which would combine the strengths of both imported and German varieties.[5] The first German breeder to explore this method systematically appears to have been Wilhelm Rimpau (1842–1903) whose father had taken up sugar-beet breeding in the 1840s (Thiel 1904; Meyer 1970; Meinel 2008). The younger Rimpau turned in 1867 to rye, initially using mass selection to improve a local variety, but in 1875 he decided to use hybridization with wheat, aiming to combine the high yield of Squarehead with the winter hardiness of other varieties. By 1889 he had produced 'Rimpau's early hybrid' which was one of the leading German wheat varieties for several decades (Hillmann 1909; Reitemeier 1904: 84–85).[6]

In the 1870s and 1880s a number of other German estate owners – many of them based in and around the province of Saxony – also took up the challenge and began to try to improve the best of the existing cereal (and sometimes potato) varieties. As Thomas Wieland has pointed out, this first generation of breeders – among them Ferdinand Heine, Gustav Bestehorn, Otto Beseler, Otto Steiger, Friedrich Strube, Otto Cimbal and Ferdinand von Lochow as well as Rimpau – shared a number of significant characteristics (Wieland 2004: 35–36). Unlike the traditional stratum of aristocratic estate owners, almost all of these men were of bourgeois background. All had some experience of agricultural higher education, were keen on technical innovation, and ran their farms for a profit. Accordingly, they all turned to plant breeding originally in order to make their own farms more productive, only later focusing upon the sale of their improved varieties to other farmers.[7]

By 1914 almost all varietal development in Germany was using hybridization (Edler 1913). Those trying to improve wheat often crossed Squarehead with one or another of the German local varieties, but the methods used were not restricted to hybridization. In addition to mass selection, some breeders isolated high-yielding or otherwise distinctive lines out of heterogeneous local varieties, as had Shirreff from 1819, while others adopted the pedigree method which had been pioneered by Vilmorin in the 1850s and Hallet somewhat later (Reitemeier 1904: 23–29; Beseler 1909; von Arnim 1925; Hillmann 1910b; Gayon and Zallen 1998). Perhaps the most successful of these selectionists was Ferdinand von Lochow (1849–1924). His winter-rye variety 'Petkus' won first prize in the German Agricultural Society's varietal tests every year between 1891 and 1914, and by 1930 was estimated to cover 75 per cent of the German rye acreage (Aufhammer 1970; Merkel 1919; von Rümker 1925b; Peters 1986).[8]

Between about 1880 and the First World War the yield of various cereals and potatoes in Germany increased by between 60 and 90 per cent (von Lochow 1913; Steglich 1907; Kiessling 1914), and experts reckoned that about one-third of that increase was attributable to breeding (Ramm 1925: ix.; Lucas 1931: 83–84; Nathusius 1955: 37). Since cereal prices from the late nineteenth century were low, many growers attempted to compensate by expanding production through planting high-yielding varieties. Demand for them was thus buoyant, and this in turn encouraged more individuals to take up breeding. The number of sugar-beet breeders, for example, increased from 10 in 1867 to over 40 by 1895 while the number of cereals breeders increased about tenfold between the 1880s and the 1910s (von Rumker 1895 and 1928a; Kühle 1926). By 1914 there were between 200 and 300 commercial breeders in Germany, far more than in the US, Britain, Sweden, Denmark or the Netherlands at that time (Hillmann 1910a; Schikorra 1914b; Bohne 1928).

2 'The varietal question': what to plant?

The growth of commercial breeding created a huge number of new crop varieties. By the 1890s one observer estimated that there were some 500 varieties of wheat alone on the German market (Steglich 1896: 2; cf. Gisevius 1912: 10) For a few examples see Figure 2.2. As a result, a new item came onto the agricultural policy agenda: 'the varietal question' (*die Sortenfrage*) as it was commonly known. Progressively minded farmers were now confronted with the problem of choice, not least because these varieties were of very uneven quality (Kraus 1900; Reitemeier 1904: 95). Which commercial breeders could be relied upon to produce high-quality seed? When similar questions concerning the quality of commercial fertilizers (and later, of commercial animal feed) had arisen in the 1850s, farmers had turned to the voluntary sector for help; the first experiment stations were established by the regional Agricultural Associations. From the early 1870s similar experiment stations – seed-control stations (*Samenkontrollstationen*) – dedicated to assessing the quality of commercial seed (e.g. its purity as well as its germination rate) began to emerge in several parts of German-speaking Europe. Subsequently both kinds of station were taken over by the state,[9] but that still left the question unanswered: which *varieties* were best for particular growing conditions?

Throughout the nineteenth century various individuals had conducted comparative tests of varieties, but these had been isolated and uncoordinated efforts

Figure 2.2 Varieties of wheat (source: Koernicke 1885).

(cf. Popplow 2008). In Baden, for example, until 1900 the weekly newspaper published by the state's Agricultural Association published only an occasional article on the question of varietal choice. In Württemberg local varieties from outside the state were sometimes imported where they were thought to be well adapted to harsh climates (such as the state's uplands), but hardly anyone went to the trouble of testing the imports' performance against that of native varieties (Fruwirth 1907b).[10] That the varietal question was still relatively new even at the end of the century can be seen from one high profile discussion of how the state might aid the spread of technical improvements. While devoting three pages to the identification of the best animal breeds and their improvement, the author's remarks on improving agronomic practices are restricted to manuring, drainage and seed testing; not a word was said about varietal testing or improvement (Goltz 1899: 237–240).[11]

During the 1880s about the only place where varieties were being systematically studied was at the newly established German Agricultural Society (est. 1885). One of the Society's first steps was to create a Seed Breeding Division which brought together breeders and academics in order to obtain and distribute information about new varieties (von Lochow 1904; Hillmann 1910b; Hansen and Fischer 1936). One mechanism by which it sought to promote the use of improved varieties was its Seed Office (*Saatstelle*, est. 1888) which was intended to serve as an honest broker between commercial breeders and farmer-members. The Office required breeders to provide key information about their varieties: for example, yield, the type of soil and climate in which it had been bred and multiplied. Seed also had to meet the Office's requirements of purity, germination rate and reasonable price. Seed that passed these tests was then sold by the Office to members of the Society, and the Office served as referee in cases of dispute between buyer and seller.

The Division also introduced several measures designed to help farmers make informed choices about the varieties they planted. In 1887–1888, for example, it set up Germany's first systematic programme of varietal testing in nine regions across the country, based on seed samples supplied by those breeders willing to take part. The initial focus was on cereals though from the mid 1890s other crops were included, and the results were published in the Society's journals.[12] These tests provided useful information for any farmer who was willing to wade through tables of data before choosing the most appropriate variety. But in 1897 the Division also became allegedly the first organization in Germany to 'certify' varieties whose production met its standards of excellence (Hillmann 1910b; Merkel 1910; Fruwirth 1911a).[13] When a farm where a variety was bred (or multiplied) applied for certification, members of the Division would visit it in order to establish that what was growing in the breeder's fields was in fact the same variety as in the seed sample the Division had received, that it was free from disease, that it was growing in visually uniform stands, that the breeder's claims for the variety (e.g. that it was 'new' or 'improved') were justified, and that proper breeding procedures and facilities were in place. Once certified, the variety was placed on the Society's published 'Seedlist'. By 1905 the Division had certified over 200 varieties (Hansen and Fischer 1936: 197, 203).

Certification not only gave buyers an assurance of quality; it also gave competent breeders favourable publicity while helping to reduce fraud and misleading advertising by the unscrupulous (Liebscher 1893; Wohltmann 1895; von

Rümker 1907a).[14] The Division also took other steps to single out first-rate breeders. Already in 1887 it began to award prizes to the best breeding establishments (Saatzuchtstelle der DLG 1905b), and in 1905 the prizes were superseded by an exclusive list of only the most outstanding varieties: the 'Hybred Varieties Register' (*Hochzuchtregister*).[15] And like certification, the Register offered farmers a hallmark of seed quality, but inclusion of a variety also entitled the breeder to use a protected trademark in promoting it.

While the Seed Breeding Division's activities were undoubtedly an important step in promoting the use of improved varieties among farmers, they did not serve all sectors of the farming community equally well. As various experts pointed out, the Division's varietal-testing programme was just too restricted to do justice to the enormous diversity of growing conditions across the country (Lochow 1904: von Rümker 1898 and 1899; Böhmer 1914).[16] One reason for this was that almost all of the varieties tested were those developed by the commercial sector. And this was a serious limitation because by the 1880s it was becoming clear that these varieties often fared poorly in regions like south Germany. In today's language, here was a straightforward case of 'market failure'. (As we shall see in Chapter 6, after 1945 the Green Revolution's high-yielding varieties also proved ill suited to many regions of the developing world for virtually the same reasons.)

Part of the problem with commercial varieties was ecological. Since they had been bred by firms in central and north Germany, they were ill adapted to south Germany where soil types and climatic conditions were not only very different but also enormously diverse.[17] But there were also economic obstacles. One of these was that commercial varieties were bred to do well under intensive cultivation, i.e. to produce their maximum yield when weeding was thorough and large amounts of mineral fertilizer were applied. But this made them too 'fussy' for most south German farms where intensive cultivation was much less common (Kryzymowski 1913; Lang 1914b; Böhmer 1914; Wacker 1911).[18] Thus, the Society's varietal tests – which compared commercial varieties when grown under intensive cultivation – were doubly unhelpful to most southern farmers. The other economic obstacle was cost. Commercial varieties were substantially more expensive than good quality local varieties and thus unaffordable for most farmers (Dix 1911; Holdefleiss 1908: 14–15; Fruwirth 1919a).[19]

Faced with these disadvantages, most small farmers – like their counterparts in the US during the nineteenth century and into the twentieth (Olmstead and Rhode 2008: 188) – chose to save seed from one harvest to use for the next planting. But for those smallholders who were nonetheless on the lookout for improved seed there were two options. One was to buy 'derivative' varieties (*Absaat*). These were first, second or third generation offspring of the 'original' version of a variety released by a commercial breeder which were cheaper than the original because they often yielded somewhat less than the parent (Liebscher 1892). Sometimes, however, derivatives performed even better than the parent, though it was not easy for the farmer to find this out since the Division made little attempt to test derivatives (Fruwirth 1907b; Hansen and Fischer 1936: 195).

The other option was to look around for a better 'local variety' (often known today as 'traditional' or 'farmers'' varieties). These were varieties which had been grown in a region for generations and were thus necessarily well adapted to local soil and climate conditions through the process of natural selection. They were also adapted to local cultivation practices since for generations they had survived farmers' annual decisions as to which plants should provide the seed for the next season.[20] And that meant that they were also 'unfussy': they could produce a decent yield even in regions where growing conditions were relatively poor. In addition, local varieties offered another advantage. Since they were almost always genetically heterogeneous populations, local varieties were not as vulnerable to shifts of weather or pest populations as were genetically uniform commercial varieties. Thus a local variety's yield, though modest, was relatively stable from one season to the next, and this was an important consideration for smallholders who relied on their crop for subsistence (Schindler 1907a). Finally, local varieties were often highly regarded by brewers and millers who reckoned them to be of higher quality than many commercial varieties. Despite their lower yields, therefore, local varieties offered a number of advantages to smallholders. For the farmer wanting to find out which local variety might be best for his region, however, the Society's varietal-testing programme was of limited use since only the most famous of local varieties seem to have been tested.[21]

In addition to the Seed Breeding Division's preoccupation with commercial varieties, however, it displayed a marked tendency to 'look after its own'. For one thing, the varietal-testing programme relied upon farmers volunteering to plant the test varieties. A prerequisite for this, however, was that a farm possessed sufficient acreage and staff who could be allocated to such work. This meant that realistically only large farmers could afford to participate, and there were few of these in regions like south Germany. For another, the Division – mainly the brainchild of the breeder, Wilhelm Rimpau – was dominated for the first 15 years or so by commercial breeders along with a few estate owners and like-minded academics.[22] Representatives of peasant farmers or of south German interests only began to appear in small numbers around 1900 and never amounted to more than a tiny minority of the Division's main committee before the First World War. And finally, gaining access to the results of varietal testing was awkward since for the first 15 years or so the results were published only in journals of the Society (and occasionally in the weekly magazines published by regional Agricultural Associations). These were readily accessible to members of the Society or of an Association or to those with access to good libraries, but none of these conditions applied to the average farmer. Moreover certified seed was available only via the Seed Office, and the Office sold only to members of the Society (Hillmann 1910b).[23]

In the same vein, one might ask whose interests the Seed Breeding Division actually served. Its activities certainly strengthened the position of commercial breeders in several ways: by creating a market for their varieties (through the Seed Office), by publicizing their varieties' characteristics (via varietal testing), and by providing limited protection for intellectual property (through certification and the

Hybred Varieties Register). By contrast, it is rather less obvious that the Division's procedures met the needs of farmers (cf. Böhmer 1914).[24] For example the Division was not prepared to admit all varieties, no matter how good, to its certification procedure. Applications from foreign breeders were considered only where those countries were open to the import of German varieties (Hansen and Fischer 1936). The same bias is evident in the Division's definition of what was to qualify as a 'hybred' variety (and thus which varieties were to be considered for inclusion in the Hybred Varieties Register). The decisive criterion was not to be a variety's outstanding *performance* in comparative tests, but rather the type of *method* used to breed it. Only varieties bred using the most 'demanding' of methods – which meant pedigree selection, with or without hybridization – were to qualify for the Register. The Seed Breeding Division's administrator could not deny that mass selection had produced some useful varieties (Hillmann 1909). But the Division's rule nonetheless meant that local varieties which had been improved by mass selection, no matter how well they performed, were not eligible for inclusion (Saatzuchtstelle 1905a: 410). As Michael Flitner (1995) pointed out several years ago, the Division's definition of a 'hybred variety' was thus self-serving.[25]

That the Seed Breeding Division's activities should have been of little help to the small farmer was predictable. For as we saw in the previous chapter, the German Agricultural Society was established by a group of estate owners keen on technical innovation in order to serve a limited constituency of like-minded peers. Regarding itself as an elite organization, the Society made no attempt to recruit peasant farmers, one consequence of which was that south Germany was weakly represented in both the Society's membership and its executive committee (Föhr 1999: 48).

In many respects, therefore, small farmers from south Germany were stuck. The virtual absence of private-sector breeders in their region (Hillmann 1910a)[26] meant that they had to turn to commercial breeders from central and north Germany whose high-yielding varieties were poorly suited to southern conditions. In addition, the Agricultural Society had largely turned its back on smallholders. Under these circumstances, how were peasant farmers supposed to identify which of the available varieties were most suitable, and who was going to produce new and better ones for them?

3 The beginnings of peasant-friendly breeding

During the nineteenth century, as we have seen, German agricultural officials were not indifferent to productivity. Along with quasi-official bodies like the Agricultural Associations, they had encouraged farmers to intensify crop production through crop rotation and improved fertilization. But into the 1890s, as some experts complained (von Rümker 1898), state agencies had paid scarcely any attention to the varietal question. And this was a serious omission since it was reckoned that big gains were to be had merely from planting the best available variety in each locality (Stoll 1905).

In Bavaria deliberations of the State Agricultural Commission in the early 1890s had little to say about crop production. In 1894, to be sure, one member called for government support of various measures to improve cereals cultivation and marketing, but the annual reports for the next three years had nothing more to say about seed quality (*Jb. General-Comites* 1890–1894; *Jb. Bay. Ldw-Rats*, 1895–1901). In its 1897 survey of Bavarian agricultural progress, the only varietal testing to which the Interior Ministry could point was being conducted at the occasional experiment station, a few branches of the Agricultural Association in one region, and at a few agricultural schools and colleges (Bay. Staatsmin. Innern 1897: 74).

In Württemberg, as French and British improved varieties began to arrive at markets in the 1880s, the State Agricultural Commission (*Zentralstelle für Landwirtschaft*) encouraged farmers to take up varietal testing of potatoes and cereals in order to establish which of the imports were really suited to local growing conditions. Into the 1890s, however, few farmers bothered to do so. As in Bavaria the only places where testing was conducted were the agricultural college at Hohenheim,[27] a few agricultural secondary schools, and a few large farms. During the early 1890s some interest in high-quality seed began to emerge, as seen in the creation of annual seed markets (Hillmann 1910a: 482), but the Agricultural Association's weekly newspaper printed hardly any advice on the choice of seed or the importance of varietal testing, nor does the issue seem to have been discussed in the State Agricultural Commission.[28] Plant breeding also found no mention in this newspaper during the 1890s, despite the fact that as in Bavaria, work on cereals breeding was in fact going on at the agricultural college (Fruwirth 1907b).[29]

Faced with the state's lack of interest in the varietal question and the Agricultural Society's neglect of smallholder interests, during the 1880s and into the 1890s small farmers themselves, with the help of an occasional estate owner or agricultural scientist, began to take matters into their own hands. The institutional arrangements they devised – whether for varietal testing, certification, producing seed or developing new varieties – were diverse.

One such novel institution was the seed-growing association (*Saatbauverein*). This was a kind of cooperative, often consisting of small farmers, whose aim was to supply local smallholders with high-quality seed at low cost (Bischoff 1922; Holtmeier-Schomburg 1908). They did this by buying original varieties from commercial breeders and testing them alongside local varieties in order to identify the best ones for local growing conditions.[30] Having done so, they then needed to grow up large quantities of the variety in question. The easiest place to do so would have been on large estates, but as we have seen, in south Germany these were few and far between. The only alternative was to multiply seed on small farms, but this was not ideal. For one thing, since small farms' agriculture was generally diversified, the risks of contamination from another crop grown on the farm were much higher than on a specialized large estate. The existence of many other small farms nearby posed the same problem. Furthermore, small farms could rarely afford the equipment required to clean and purify seed after harvest. In order to get around these limitations, therefore, small farmers set up cooperatives. A larger plot of land, drawn from adjacent farms, was dedicated to the production of that variety judged best for the locality. Measures to reduce contamination were agreed, and equipment was purchased collectively and shared. The material advantages of this arrangement were substantial. By buying and selling in bulk, the members could obtain better terms from suppliers and purchasers than they could individually. Local smallholders also gained since the associations were able to sell their seed at substantially lower prices than the original variety (von Rümker 1911b; Fruwirth 1911a).

In some regions the best variety turned out to be a local one. The earliest cooperative of this kind was formed during the 1880s in the Probstei region of Schleswig-Holstein. There local varieties of oats and rye had been grown for seed and sold in other German states as well as in Eastern Europe since the mid nineteenth century (Kochendörfer 1996; Hillmann 1910a: 581–582; Holtmeier-Schomburg 1908) (see Figure 2.3). Indeed, the local rye variety's excellent qualities made it the starting point for many of the German commercial rye varieties bred before the First World War (Funke 2008). Another such case was the Pirna region of Saxony where a local variety of rye possessed properties which had made it attractive as seed-rye in other parts of Germany for many years and as a starting point for the extraordinarily successful

Bild 332. Schutzmarke.

Figure 2.3 Trademark of the Probstei seed-growing cooperative (source: Hillmann 1910a).

commercial variety, Petkus. In the Fichtel Mountains region of Bavaria, too, a local variety of oats, grown and sold by several cooperatives formed during the 1880s, was widely planted in south Germany as well as in upland areas elsewhere and won prizes at World Fairs in 1900 and 1904 (Raum 1912; Kraus 1899a; Fruwirth 1907b).

Seed-growing associations proliferated rapidly from about 1900, tending to emerge in regions that were ill served, for whatever reason, by the German Agricultural Society's varietal-testing programme. Occasionally they were set up in northern or eastern regions which possessed unusual growing conditions and for that reason were not included in the Society's programme.[31] But associations were especially common in south Germany and in similar regions where farming was dominated by small farms. As a result, by 1914 there were small numbers of seed-growing associations in most states and eight in Prussia, but several hundred in Bavaria. (Merkel 1914; Bischoff 1922).[32]

Seed-growing associations offered a solution to small farmers faced with the problem of obtaining good quality seed cheaply. But as we have seen, the varieties available commonly possessed one or other deficiency. Commercial varieties were often poorly adapted to south German conditions, and local varieties, though well adapted and of high quality, produced only modest yields. The solution, some agronomists suggested, was to try to improve local varieties through breeding. In a paper to the International Congress of Agriculture in 1890, for example, Emanuel von Proskowetz and Franz Schindler[33] argued that in view of local varieties' valuable properties, it would be worth trying to increase their yield through breeding at specialized experiment stations (Pribyl *et al.* 1890: 348). Proskowetz (1849–1944) was a Moravian estate owner who in the early 1870s had noticed the decline of cereal yields there as well as in Bohemia. At that time the varieties favoured there were imported British or German varieties. In 1873, therefore, he began to collect local barley varieties from the Hana region and test them against the commercial ones. Finding that one of the Hana varieties was consistently superior, not only in quality and early ripening but also in yield, he began to improve it by mass selection, and in 1884 it was introduced to the market as the 'Proskowetz Original Hana Barley'. It proved very popular with brewers, and by 1900 commercial breeders had begun to use the Hana as their starting material (Proskowetz 1937; Tschermak 1958: 81–88).

But during the 1890s (and for decades thereafter) the suggestion that local varieties should be improved was controversial for it represented a fundamentally different approach to breeding than had been pursued by commercial breeders (Kraus 1914; Pammer 1907).[34] Indeed, the same tension – between what I will call a 'local' and a 'cosmopolitan' breeding strategy – continues to divide breeders today in their attempts to develop improved varieties for the global South (Chapter 7). The basic aim of the cosmopolitan strategy, as its name implies, was to develop one outstanding variety – then referred to as a 'universal variety' (*Universalsorte*) – which would produce good results practically everywhere. Since intensive cultivation – i.e. careful soil preparation, liberal use of mineral fertilizer, and thorough weeding and pest control – was known to enhance yield, the technical task for the breeder following this strategy was to develop a variety which could thrive under these conditions. The starting point for breeding was generally a high-yielding variety which might come from almost anywhere. (Where its place of origin was very different from the breeder's location, it was often crossed with a local variety in order to ensure that high yield and adaptedness would be combined in a single plant.) This plant was then selected further to ensure that it gave high yield under intensive cultivation. Because such varieties tend to perform well in a range of locations, contemporary GR breeders often emphasize their 'wide adaptation'. As Ceccarelli (1989) has pointed out, however, 'wide' in this usage means that these varieties will thrive under intensive cultivation in any environment which resembles that of the location *where they were bred* (and this means they can potentially occupy a geographically wide space). It does not mean that these varieties will perform well in a wide range of *different* environments or where cultivation is not intensive. (This is why, as we have seen, commercial varieties bred in central Germany did not do well in south Germany.)

The local strategy – recommended by Proskowetz and Schindler and eventually institutionalized at the south German stations – started from very different premises. Its aim was not to develop one 'universal variety' but instead a multitude of improved varieties, each well adapted to a particular growing condition. The starting point for breeding, therefore, was not one outstanding high-yielding variety but rather many local varieties, each chosen because it offered high quality and was well adapted to a particular locale. The breeder's task was to increase the yield of these plants. And the fact that a local variety generally consisted of many genetically distinct lines, often varying in yield by as much as 50 per cent (Lang 1911a and 1914b), meant that this could be done within a few generations by selection. If the vision underlying the cosmopolitan strategy was a physiological one – in the sense that the breeder sought to develop a single plant design which was maximally efficient – the local strategy was rooted in an ecological perspective. Starting with an awareness of the enormous variability of growing conditions which needed an improved variety, the breeder judged that it would be easier to increase the yield of a well-adapted plant than the adaptedness of a high-yielding one.

The cosmopolitan strategy was the one that had been pursued in the late nineteenth century by commercial breeders located in the favourable growing conditions of central and north Germany. Many agricultural experts and innovation-minded farmers at the time endorsed this view. Believing that the only realistic way to increase production was via intensive cultivation and knowing that local varieties did not respond well to mineral fertilizer and were anyway low yielding, they regarded any attempt to improve local varieties as a backward step (Kiessling 1912; Schindler 1928; Kulisch 1913; Broili 1899: 218).[35]

Nevertheless, Proskowetz and Schindler's recommendation found a definite resonance. For by the 1890s even the most sought-after local varieties were beginning to face competition from higher yielding commercial varieties and thus were perceived to be in need of improvement. Some scientists in the state sector were sympathetic to the local strategy. At Dresden's new experiment station (est. 1890), for example, one of Bruno Steglich's (1857–1929) first aims was to develop varietal testing throughout Saxony as a complement to the German Agricultural Society's work (Steglich 1891). He also believed that the local Pirna variety would be able to continue to compete only if it were improved through breeding. But who was going to take on this task? The commercial breeders of central Germany, as we have seen, were not interested in improving local varieties, and in the south there were very few commercial breeders who might have been persuaded to do so.[36] In Steglich's view the best way to proceed, therefore, was for a group of interested local farmers to organize themselves into a cooperative which, under his technical guidance, would use a simple method (mass selection) to improve the Pirna variety (Steglich 1893 and 1896).[37] Within two years a group of local farmers had decided to follow his advice, and in 1896 a breeding and sales cooperative was formally set up. The results were dramatic. The improved Pirna rye quickly won prizes from the German Agricultural Society (1898) and at the Paris world's fair (1900), attracting attention from scientists and state agricultural officials in other states who saw in it a potential model of how small farmers could secure high-quality seed at moderate cost through self-help (Steglich 1910; Edler 1898; Schindler 1907b).[38] Members of the seed-growing cooperative in the Fichtel Mountains region of Bavaria, for example, were interested in the Pirna model since, as in Pirna, by the late 1890s competition from commercial oat varieties had begun to eat into the markets for their local variety. Accordingly, in 1900 with financial support from the Bavarian Interior Ministry the Fichtel Mountains cooperative also embarked upon a breeding programme to improve their local oats variety, and as at Pirna and the Probstei, the cooperative relied upon technical advice from an agronomist, Prof. Carl Kraus of the agricultural college at Weihenstephan.[39] Although these three breeding cooperatives were the best known, by the turn of the century a handful of similar organizations had been set up in other German-speaking states (Kraus 1899a and 1900).

By the end of the 1890s the possibility of improving local varieties through systematic plant breeding was starting to get discussed in various German-speaking agricultural organizations and publications (Kiessling 1922b; Anon. 'Ueber den Anbauwerth' 1900; cf. Fruwirth 1896). In addition, several experts were suggesting that since existing arrangements for meeting the needs of small farmers were insufficient, new kinds of institution might be required. In 1898, a high profile proposal of this kind was made in a lecture to the German Agricultural Society by the professor of agronomy at the University of Breslau, Kurt von Rümker (see Figure 2.4) (von Rümker 1898; cf. von Rümker 1900). Acknowledging that commercial varieties were probably too expensive for the small farmer, he suggested that the way forward could be through the proliferation of breeding cooperatives such as those recently established in the Probstei

Figure 2.4 Kurt von Rümker (1859–1940) (courtesy: Niedersächsische Staats- und Universitätsbibliothek Göttingen).

and Pirna regions. There was no reason, he insisted, why small farmers should not be encouraged to undertake basic plant breeding themselves. While some crops were unsuited for technical reasons and some breeding methods were too complicated for laymen, cereals breeding using simple methods would certainly be feasible on small farms. A specialized state institution, however, he believed was not necessary since technical support could be provided by teachers at local agricultural secondary schools or the regional chamber of agriculture (cf. von Rümker 1889 and 1905: 28–31).

A year later a Bavarian estate owner, Josef Broili, made a similar appeal, calling upon the German Agricultural Society to make a greater effort to address the interests of small farmers. His lecture to the Seed Breeding Division ended with a formal proposal that the Society should encourage the improvement of south German local varieties, and the state should lend a helping hand (Broili 1899). One of the experts who spoke up in support of Broili in the ensuing discussion was Carl Kraus, director of the agricultural college at Weihenstephan who had recently made many of the same points (Kraus 1897) and who had been doing varietal testing at the college since 1892 (Kraus 1903 and 1909b). Despite some individuals' scepticism, he argued, local varieties could be successfully improved via selection, and it was important that more work of this kind be undertaken. As to how such proposals could be implemented, Kraus also endorsed breeding cooperatives as a useful way to meet the needs of small farmers, at least in some circumstances. But unlike von Rümker, he suggested that state plant-breeding stations could also be valuable in some cases. The latter could be useful in overseeing a state-wide programme of varietal testing which would include monitoring the quality of the available commercial varieties as well as studying the range of local varieties in order to identify which might be worth improving. And while such a station could not attempt to breed varieties for an entire state, it could at least provide the technical support for small farmers who wished to take up breeding (Kraus 1899a, 1899b, 1900; cf. Liebscher 1896).

Proposals of this kind are likely to have fallen on receptive ears. As we saw in the last chapter, during the 1890s German experts and officials were beginning to conclude that small farms were potentially economically viable as long as they had the necessary technical support. The proposals by von Rümker, Kraus and others were almost certainly noted by agricultural officials in south Germany because the region was poorly served by both commercial plant breeding and the German Agricultural Society's Seed Breeding Division. Accordingly, in 1897 and 1898 the Bavarian Interior Ministry asked the State Agricultural Commission to consider whether a more systematic programme of state funding for varietal testing might make sense. By the start of 1900 the minister was exploring whether the state should also initiate a programme of plant breeding (Bay. Staatsmin. Innern 1905: 109; Wieland 2004: 84–90).[40] By this time, however, there were several models available as to how the state might support such work. One was to support the formation of a breeding cooperative which would be advised by a technical expert from a local experiment station, as at Pirna. Alternatively the expertise could be provided by a local chamber of agriculture or by creating a section for plant breeding within an existing experiment station.[41] But the institutional model most widely cited in the German-language literature around the turn of the century was probably the plant-breeding station established in 1886 at Svalöf in southern Sweden (Kraus 1900; Pammer 1905; Steglich 1905; Elofson 1905; Stutzer and Gisevius 1904; Ulander 1906; DLG 1907). As one German observer pointed out, although Svalöf was a public-sector institution, its varieties were competing successfully against those of older and more famous European private breeders (Arnim-Schlagenthin 1901).[42] Given such interest, the Prussian Agriculture Ministry commissioned an expert to visit Svalöf about 1903 in order to report back on the way in which breeding work there was organized as well as the kind of work undertaken (Holtmeier-Schomburg 1908).

By 1900, therefore, many of the conditions conducive to the establishment of peasant-friendly state-supported plant breeding were in place. Yet by then all of these factors had existed for a decade or more. The value of improved varieties was evident from at least the 1880s. The economic potential of small farms and the rationale for promoting the use of quality seed among peasant farmers was clear by about 1890. And several institutional models were available *c.* 1890, showing how states might support the improvement of local varieties. Indeed, already in the winter of 1888–1889 the Prussian State Agricultural Commission had unanimously recommended to the Minister that the state should provide funding so that provincial agricultural associations could award prizes to outstanding cereal breeders and organize varietal testing of cereals which would be overseen by staff from the experiment stations. Moreover, it was 'desirable', they noted, that 'smaller farmers' should be involved with such tests.[43] Yet nothing happened on this front for another decade. And then within the space of just a few years peasant-oriented plant-breeding institutions were established in many places. Why the delay followed by a burst of activity? The answer, I suggest, is to be found in the political events of the 1890s.

4 The political context of institutional innovation

The 1890s were just as turbulent in south Germany as in Prussia and elsewhere, but the political landscape was different. Given the largely agrarian nature of the economy, farmers were a major sector of the electorate, but the predominance of small farms and near absence of estate owners in south Germany meant that the peasantry – rather than the Agrarian League – was the force to be reckoned with. (For convenience, I will focus in this section upon events in Bavaria.)

With the collapse of tariff protection in 1891, as we have seen, farmers throughout Germany sought to change trade policy by lobbying the Reich government through organizations like the Agrarian League. Other forms of assistance, however, could only be secured at state level since there was no Reich ministry of agriculture. The Bavarian government and other state-level bodies, therefore, became major targets of peasant dissatisfaction. Neither of the organizations to whom peasant farmers had traditionally turned for support, however, now seemed up to the task. One of these was the Bavarian Agricultural Association, recognized by the state as the quasi-official representative of the farming community, but it was of little use. As elsewhere in Germany, its membership was dominated by estate owners, civil servants and professionals, and only about 10 per cent of farmers bothered to join (Bay. Staatsmin. Innern 1890: 785). Moreover, the facts that its statutes largely forbade 'political' activity and its leading members were state officials made it very unlikely that the Association would be prepared to fight on farmers' behalf (Schlögl 1954; Haushofer 1986). During the 1880s a Bavarian farmers' organization had called for reform of the Agricultural Association but got nowhere. With growing rural unrest in the 1890s and the creation of chambers of agriculture in Prussia, radical critics of the Association called for the establishment of a Bavarian chamber since the latter was perceived as potentially more representative and more 'political' (Hundhammer 1926; Bergmann 1986: 80–88). In response the Bavarian Agriculture Ministry asked the State Agricultural Commission to consider whether existing arrangements for the representation of farmers' interests were satisfactory. The Commission was in no doubt about the depth of farmers' anger (*Jb. General-Comites* 1893: 1–2), but it nonetheless recommended retaining the Agricultural Association as the farming community's representative while making a few reforms in order to meet critics' objections.[44] By 1901 even the German Agricultural Commission – not exactly a radical voice – was calling upon the Bavarian government to establish a chamber-like organization, but the request was again declined (Hundhammer 1926: 109ff.). Having conceded so little, both the government and the Association were henceforth under pressure to defend their record in looking after peasant farmers' interests.[45]

The situation in the Bavarian parliament was no better. The party for whom most of the rural population had traditionally voted – the (Catholic) Centre Party, the largest party in the Bavarian parliament – no longer enjoyed their confidence since it had endorsed tariff reduction and was seen as having done nothing for agriculture. An alternative from its inception in 1893 might have been the Agrarian League, but Bavarian smallholders perceived the League as a lobby group for estate owners and kept their distance (Schlögl 1954: 567). Instead peasants began to organize themselves, forming a number of local 'peasant leagues' in order to contest local elections against the Centre Party. This strategy succeeded, and in 1895 a number of local leagues merged to form the Bavarian Peasants' League (*Bayerischer Bauernbund*), the only peasants' organization in Bavaria to operate as a political party at both state and Reich level. In the Bavarian parliament the League called *inter alia* for the restoration of tariffs but also for the establishment of agricultural advice bureaus and more practically oriented agricultural education. This platform regularly secured the League about 10 per cent of the votes in the Bavarian parliament up to 1914 (Hochberger 1991; Bergmann 1986).

The Peasants' League's rapid rise was cause for alarm in the Centre Party which soon regarded the former as its main opponent. Furthermore, the League's support for various radical measures (e.g. opposing increases in military spending) placed it closer to left-liberals and Social Democrats in the Bavarian parliament. For the government this raised the awkward prospect of an alliance between the League and other opposition parties. That prospect seemed especially threatening since, as we saw in the last chapter, the Bavarian wing of the Social Democratic Party was actively courting peasant members. Its leader, Georg Vollmar, was a gifted orator who spoke in upper-Bavarian dialect and in parliament repeatedly challenged the Centre Party to say what they had done to help peasant farmers. Moreover he was open to forming coalitions with the Peasants' League and other oppositional factions at election time (Lehmann 1970: 64–76). All of this put pressure on the Centre Party to address peasant concerns more energetically (Möckl 1972).

Nervous about the League's advances, the government kept a close eye on its activities via informants' reports on meetings and dossiers on leaders, watching for signs of libel which could lead to prosecutions. But more importantly, the League's success forced the government to reconsider its agricultural policies. Sensitive to the League's criticisms of the Agricultural Association, the government urged the Association to engage more actively with peasant demands. Accordingly the Bavarian Agricultural Commission's annual reports during the 1890s emphasized how much effort was being expended on farmers' behalf. As one report declared, the government, legislature and agricultural officials had stood by the farming community, and one could be grateful for the funding they had allocated toward the establishment of public institutions, especially of cooperatives. Here was 'cause for optimism'.[46] Indeed, after 1893 there was a burst of government legislation on rural matters (Farr 1978). State appropriations for agriculture increased nearly fourfold between 1880 and 1904 (Bay. Staatsmin. Innern 1905: tab. XI), some of it going toward the provision of cheap credit, agricultural secondary education, and the promotion of cooperatives.[47]

Another indication of the government's concern that its efforts should be properly recognized was its publication in 1897 of a survey on the state of Bavarian agriculture (following similar surveys in 1862, 1872 and 1890)[48]. The volume served two functions. Hoping for a little *Vorsprung durch Technik*, one aim was to stimulate technical improvement in those regions thought to be lagging behind by promoting best practice. In the covering letter sent with the volume to all district governments, for example, the Ministry urged that more attention be given to encouraging studies of cultivation methods and to the organization of seed markets.[49] Its other aim was to chart the progress in agriculture which had been made since 1890. 'Progress', however, was very narrowly defined: the volume was supposed to focus only upon those measures favourable to agriculture which had been introduced by government, state legislature, and the Agricultural Association. While there is little doubt that measures taken by a variety of other organizations – for example, cooperatives, peasant associations – had also contributed to agricultural progress, these were not to be included, suggesting that the volume was conceived less as a survey of progress per se than as an exercise in public relations.[50] In the introduction to the volume the

Ministry conceded that while it had not been able to secure an increase in agricultural prices since 1890, success had been achieved through improvements in production and sales of commodities, better organization of credit arrangements, increased protection of farmers against animal disease and hailstorms, and a reduction in the tax burden. Over the following 350 pages each area of policy relevant to agriculture was discussed in turn, noting which measures the state had taken, how much they had cost, and with what success. In the case of plant production, for example, the volume explained that the Agricultural Association had actively promoted the use of the best seed through funding for seed markets, encouraging collective purchase of seed, varietal testing, and better procedures for cleaning and sorting seed (Bay. Staatsmin. Innern 1897).

On publication the volume was distributed to provincial governments, other Bavarian ministries, the State Agricultural Commission, and cooperatives, among others.[51] Determined to make sure that the book's message escaped no one, a government spokesman in the state legislature devoted a lengthy speech to its contents. The legislature had often been criticized recently for having neglected agriculture, he noted, but the newly published volume clearly demonstrated just how much the government had done for the farming community since 1890. Bavaria, he claimed, was the envy of other German states in several respects including its support for cooperatives. And not wishing to forego the party-political opportunities which the volume presented, he rounded off his address with an attack upon the behaviour of individual members of the Bavarian Peasants' League as well as a dig at the Social Democratic Party.[52] The book's political intent did not go unnoticed. As word of its forthcoming publication leaked out, one Munich newspaper declared its suspicion that the undertaking was basically designed to demonstrate the government's role in securing 'improvements' for agriculture.[53]

As the new century opened, agricultural officials took every opportunity to declare their support. Following Carl Kraus's 1900 talk to a farmers' organization, recommending state support for plant breeding, for example, a top agricultural civil servant present at the meeting endorsed the recommendation, noting that although 'much had been done in Bavaria to improve agriculture', further measures could still be introduced to foster the use of better seed. The Ministry, he reassured his audience, was 'warmly in favour of this development and keen to support it.' Considerable sums had already been spent for this purpose, and 'I can assure you that the government will carefully consider all such applications [for funding]' (Kraus 1900: 88). But the pressure did not let up. In the autumn of 1901 rural unrest reached a crescendo when a large meeting of farmers took place in Regensburg with the aim – for the first time – of unifying the Bavarian farming community. Several speakers called for state intervention to protect farmers against foreign imports while one spokesman for the cooperative movement (Georg Heim) went further. It was 'noteworthy', he said, 'how unresponsive our [Bavarian] government has been to the peasantry's demands' (cited in Schlögl 1954: 571).

A year after Heim's complaint, the new Bavarian plant-breeding station was opened. To my knowledge, the Bavarian Peasants' League, though keen on practically oriented education as well as extension services, had never demanded peasant-friendly experiment stations. And it is difficult to know what peasant farmers in general may have thought about such stations. What is clear, however, is that the plant-breeding station was just one of several Bavarian experiment stations which were all established (or taken over by the state) at about the same time: a research station for moorland agriculture in 1900 (*Bayerische Moorkulturanstalt*), a fisheries research station in 1900 (*Bayerische Biologische Versuchsanstalt für Fischerei*), and an experiment station for agricultural botany (*Agrikulturbotanische Anstalt*) in 1902. Moreover, the government and allied agricultural organizations seized every chance in various publications to draw attention to these stations and the state expenditure which made them possible, the work at some of them on varietal testing, and the favourable impression they had made on foreign visitors (Bay. Staatsmin. Innern 1897: 326–327 and 1905: 16–24, 102ff.; *Jb. Bay. Ldw-Rats* 1898). Whatever *farmers'* perception of the stations may have been, therefore, it seems clear that agricultural officials themselves were keen to promote the stations as an important contribution to the well-being of the farming community.

In trying to account for the creation of peasant-friendly stations in south Germany, I have focused so far upon the conditions in that region. But the economic and political explanation I have given is also borne out if we widen the analytical focus to include north Germany. Since I have argued that peasant radicalism combined with a dearth of commercial plant breeding were crucial in prompting the formation of the south German stations, one should not expect to find such peasant-friendly institutions in north Germany where these conditions were absent. For one thing, the north possessed a strong private-sector breeding industry. For another, the large number of estate owners there meant that rural protest was dominated, not by peasant organizations, but by the elite-led Agrarian League. If we look at the character of non-commercial breeding institutions in north Germany, accordingly, the picture is very different from that in the south. First, although almost all German states had seed-growing cooperatives by 1914, there were far more of these in Bavaria than in Prussia (even though the latter had five times as many farms).[54] Second, although several public-sector plant-breeding institutions were established in the north in the years before the First World War, none of them was designed to serve peasant farmers. All of these institutions' research and development work was intended either to support the private sector or to complement it rather than providing farmers with an *alternative* to it. Before the war the institutes of plant breeding at the Universities of Breslau and Halle, for example, did not attempt to develop new varieties but instead concentrated upon fundamental research of use to breeders: for example, the inheritance of commercially important traits or improving breeding methods (Stephani 1911; von Rümker 1912). To be sure, by the 1920s Halle's institute under Theodor Roemer and Erwin Baur's department of genetics at the agricultural college in Berlin were both developing new varieties, but promising new ones were not handed over to regional farmers or cooperatives but instead sold to commercial breeders in central Germany (cf. Chapter 3). Perhaps closest in character to the south German stations was the plant-breeding sub-department of the Kaiser Wilhelms-Institut für Landwirtschaft at Bromberg in East Prussia. It, too, was charged with breeding new varieties well adapted to the particular growing conditions of the eastern provinces, advising the region's farmers on suitable varieties, and providing technical support to local farmers who wished to take up breeding. In the event, however, for the first few years of the sub-department's existence there is no evidence of any work in plant breeding being conducted there.[55] And when varietal development finally got under way a few years later, the new varieties actually developed were confined largely to crops in which the private sector was not interested: grasses, clovers and other forage plants (Gerlach 1908).[56] As one breeder there declared, 'the state should stay out of areas where the private sector is active so as not to compete' (Schikorra 1914a). Thus the quite different kinds of breeding institutions established in north Germany before 1914 are consistent with the argument that the establishment of peasant-friendly stations in south Germany soon after 1900 derived from specific economic and political circumstances in the region.

Conclusion

Although the Bavarian station was obviously intended to promote agricultural development, that was almost certainly not the government's sole aim in creating it. For it is clear that the Bavarian state was under considerable pressure from the early 1890s to demonstrate to farmers that it, along with the quasi-official Agricultural Association, was taking the agricultural crisis seriously. And after 1900 both bodies pointed to the Bavarian station *inter alia* as evidence that they were doing so.

Although I have focused in this chapter upon south Germany and especially upon Bavaria, the institutional developments there were not unique (as we saw with the similar work by Steglich at Dresden from the 1890s). Soon after the establishment of the Bavarian station (1902) seed-growing associations began to be set up in other regions dominated by small farms (e.g. Hessen, the Rhineland) which carried out both varietal testing and plant breeding under the technical supervision of a nearby academic institute (Hillmann 1910a: 417ff. and 459; Remy 1910; Anon. 'Förderung' 1911a). Chambers of agriculture were also quick to promote these activities. By 1914 even in Prussia nearly all of the chambers had made some kind of provision for plant breeding,[57] though some observers – thinking perhaps of the new stations in south Germany – felt that Prussia still had some way to go in its provision of research for smallholders (von Rümker 1911a and 1914). Nor was peasant-friendly breeding confined to Germany. Elsewhere in Central Europe agricultural officials were taking similar decisions, Lausanne being a case in point, as we have seen. Another example was to be found in Vienna where in 1898 Gustav Pammer took up the challenge posed earlier by Proskowetz and Schindler, began to improve a local variety of rye, and quickly achieved substantial yield increases. Inspired by a visit to Svalöf three years later as well as by von Rümker's endorsement of seed-breeding cooperatives, he succeeded in persuading the Austrian government in 1902 to establish a department of plant breeding for him at the Royal Seed Control Station in Vienna which would oversee varietal testing in cereals as well as the improvement of local varieties by regional cooperatives (Pammer 1902, 1903, 1904, 1905).[58] A similar development took place in Alsace-Lorraine where agricultural conditions were much like south Germany's. Soil and climate varied greatly over short distances; farms were overwhelmingly small; imported commercial varieties performed poorly; and regional plant breeding was non-existent. From 1904, accordingly, the state-funded agricultural experiment station at Colmar organized a programme of varietal testing in cereals and began to improve local wheat varieties via mass selection, handing over promising ones to a regional seed-growing cooperative (Kulisch 1913; Hillmann 1910a: 471). The south German stations, therefore, were just one part of a wider movement around 1900 which aimed to help small farmers find a solution to the varietal question.

3 Research, development and extension at the south German stations

The single most important feature of the south German institutions is that although they bore the name of 'plant-breeding stations' (*Saatzuchtanstalten*), their work encompassed far more than breeding. As we shall see in this chapter, they were in fact responsible for improving their region's agriculture through a variety of means, including varietal testing and certification, continuing education, and extension. And this broadly conceived role almost certainly contributed to their popularity with farmers and politicians as well as their success in stimulating the regional economy.

Since the Bavarian station was the largest and most effective of the three, and the archival and printed sources on it are especially rich, it provides the best vehicle for introducing the main issues. After discussing the organization and work of the station, I turn to the stations in Baden and Württemberg for comparative purposes.

1 Organization

The Bavarian station was established on the grounds of the Agricultural College at Weihenstephan and was formally attached to the College.[1] Accordingly, it was initially funded by, and accountable to, the division of the Interior Ministry then responsible for higher education.[2] In order that it would not lose sight of agricultural concerns, the station was assigned an advisory committee (*Beirat*) whose members included representatives from the Bavarian Agricultural Council, the district branches of the Agricultural Association, and the Ministry (*Jb Bayern 1903*).

The first director of the station was Carl Kraus (1851–1918) (see Figure 3.1). Trained as a plant physiologist, he started his career at an agricultural secondary school, devoting much of his research to barley and hops. In 1892 he became director of the agricultural college at Weihenstephan, moving in 1901 to the Technical College in Munich as professor of agronomy (Kiessling 1918a). Since Kraus's administrative and teaching responsibilities at the Technical College tied him much of the time to Munich (40 km away), it made sense for the station to have a deputy director on-site who could oversee day-to-day activities. The person appointed was Ludwig Kiessling (1875–1942) (see Figure 3.2). Educated at the agricultural college

Figure 3.1 Carl Kraus (1851–1918) (source: Kiessling 1918b).

in Weihenstephan and the Technical College in Munich, he became an assistant (*Assistent*) at Weihenstephan in 1898 and Kraus's deputy at the newly founded station in 1902. On Kraus's retirement in 1910, Kiessling took over as director of the station (and was appointed professor of plant breeding at the agricultural college). While the original concept for the station and the negotiations that led to its founding were Kraus's achievements, its remarkable expansion and eventual

Figure 3.2 Ludwig Kiessling (1875–1942) (source: Hanns Kreutz 1942).

impact upon Bavarian agriculture must be credited to Kiessling's broad vision, skill as an organizer and sheer dedication (Raum 1959; Scheibe 1942; Gassner 1927).[3]

Apart from Kraus and Kiessling, the station's initial staff consisted of two technicians (*wissenschaftliche Hilfsarbeiter*) and a janitor. Its facilities were equally modest: a laboratory each for botany and chemistry and three additional rooms in the College's main building, a corner of the botanical garden for breeding experiments, and 4 hectares with a barn on the nearby state farm as an experimental field (*Jb Bayern 1903*; Kiessling and Gassner 1927: 6). As we shall see, all of this was to change radically over the next two decades.

2 Developing varieties for small farms

The Bavarian station's statutes gave it responsibility for promoting the development of plant breeding within the state. This task was to be achieved by giving courses for those farmers who wished to take up breeding (see Figure 3.3), providing advice and technical support to those who went on to set up breeding enterprises, giving lectures on breeding, publishing popular articles, and conducting practically oriented breeding work in-house.[4]

As we saw in the last chapter, the breeding strategy pursued at the stations was diametrically opposed to that used by commercial breeders. In contrast with the latter's cosmopolitan strategy, the station's strategy was local in two respects. On the one hand, the starting materials for its breeding work were local varieties. In 1906, for example, about 90 per cent of the breeding work overseen by the station used local varieties as raw material; the remainder used varieties of non-Bavarian origin (*Jb Bayern 1906*: 94). On the other, the process had to be decentralized in order to take account of the much higher degree of geographical variation in soil and climate conditions in Bavaria (as in south Germany more generally) compared to central or northern Germany.[5] The aim was thus to improve the performance of a local variety in its own locale.

One way to achieve decentralized breeding would have been to establish branches of the station in all of the major agro-ecological zones of the state. Since this was too expensive, the station opted to train breeders who came from all parts of the state and monitor the quality of their work (Kraus 1907). The first step, therefore, was to hold introductory courses on breeding for interested persons (at no charge). The first such course took place in 1902 with 20 participants. Demand appears to have been buoyant since courses were held annually over the next few years.[6] Promising 'students' who wished to proceed were given the station's support over the initial few years to get them started. Would-be breeders were also chosen with an eye to their geographical location; the aim was to develop 'breeding centres' (*Zuchtstellen*) in all of the major growing conditions across the state who would improve their region's local varieties for local farmers. The role of the station was to advise the centres which of their local varieties offered the most promising starting material for breeding (based on varietal testing at the station), monitor the progress of breeding, assist with more complicated procedures, and assess the quality of the end products, certifying them where appropriate. For the most part these services were provided free.[7]

The number of these centres grew quickly. Within three years there were already about 50 of them, rising by 1922 to well over a hundred (*Jb Bayern 1906*; Kiessling and Gassner 1927; Kraus 1905b). Overseeing such a large number of centres scattered all across the state was too burdensome for the limited number of staff based at Weihenstephan, so outside help was sought. In the first few years the station arranged for about half of the centres to be advised by a teacher from their nearby agricultural secondary school (*Jb Bayern 1906*: 98). But gradually a network was built up in which each of the major growing regions in the state was assigned a specialist in plant breeding (called an 'inspector': *Saatzuchtinspektor*) who took responsibility for all of the centres in his region. These inspectors served as extended arms of the station, providing advice, technical support and quality control to breeders but in addition conducting regional variety testing, advising farmers on choice of variety, maintaining standards at grain stores and seed-production farms, and more generally promoting best cultivation practice. The first of these inspectors was employed in 1910, and by 1927 there were eight more.[8]

Those who took up breeding under the station's supervision were generally of three kinds: individual farmers, staff at public-sector institutions (e.g. agricultural secondary schools, state farms), or cooperatives and similar kinds of farmer organization.[9] The clientele that the station served changed significantly over time. In the early years, for example, individual farmers constituted less than half of the breeders.[10] This proportion rose steadily, and by about 1920 they made up a large majority. The interesting issue, however, is what proportion of these farmers were actually small- or medium-sized peasant farmers. Rather inconveniently, given the station's declared mission, in the first year of its operation virtually all of the individual breeders that the station supported either owned or leased estates. Large farmers were also heavily overrepresented among those who attended the station's introductory courses on plant breeding held over the first five years. Moreover, there are signs that in the decade before the First World War the station discontinued its support for a substantial number of small breeders (Fruwirth 1919a: 37–38; *Jb Bayern 1914–1918*: 95). By 1907, however, the proportion of estate owners among the individual breeders had fallen to about two-thirds and by 1913, to a little under one-half. After the war the proportion who were estate owners dropped still further, but even then they still constituted a far higher proportion than their presence within the Bavarian farming community as a whole (*Jb Bayern* 1903 to 1919–1921).

This disproportion posed a rather awkward public-relations problem for the station. In an attempt to grasp the nettle, its director asserted right at the outset that for the station to assist only peasant farmers and cooperatives would be 'one-sided'.[11] That he should have felt obliged to strive for such 'balance' perhaps had something to do with the fact that two-thirds of the station's advisory council at that time were estate owners.[12] But the problem didn't go away. Given the dominance of Bavarian agriculture by small-holders, there were soon complaints to the Ministry that the station was serving 'only' big farmers.[13] The station was quick to reply, emphasizing that of 52 breeding centres, 'only 11' were located on big farms and that this was in accord with the distribution of land in Bavaria; all of the rest were associated with small- to medium-sized farms.[14] As we shall see in the next chapter, this tension between small and large breeders was to increase sharply during the 1920s.

Initially most of the breeding centres devoted their attention to improving barley and oats, the former as a cash crop and the latter as animal feed largely for on-farm use. By the First World War, however, they had widened their scope to include all of the major cereals as well as potatoes (Kiessling 1913: 8; Kraus 1905b; Gassner 1927). By the late 1920s the range of crops bred had broadened yet further, but barley and wheat commanded the most attention, the former due to the large brewing industry in the state and the latter because it fetched the highest prices of any cereal (Scharnagel 1929; Kiessling and Gassner 1927: 17).[15]

The idea that small farmers could be taught to do plant breeding undoubtedly struck some observers as overly optimistic or even dangerous. But once one recalls that many farmers had, in effect, been applying the simplest of breeding methods – mass selection – to their crops for centuries, it is not surprising that various public-sector breeders committed to the improvement of local varieties saw no problems in peasant farmers carrying out mass selection (Fruwirth 1919b; Schindler 1907a; Bischoff 1922). Accordingly, this was thus the method with which the station's new centres began their work. But since Kiessling felt that the limits of mass selection were soon reached, some centres were able to graduate within a few years to individual (pedigree) selection. And by 1909 some of the larger centres were thought sufficiently competent (as well as endowed with the necessary resources) to take up hybridization (Kiessling 1906, 1912; Pfetten n.d.; Bauer 1993; *Jb Bayern 1909*: 62).

But the station did not restrict itself to training breeders and advising them on procedure. It also provided other kinds of support. One form which this took was the conduct of what was called 'scientific' research (the kind of work which in science policy circles would now be called 'strategic' or 'fundamental' research). This consisted of relatively basic studies which were of potential commercial importance over the medium term but not of immediate utility. This included studies of the inheritance of important traits (e.g. ear form in cereals, nitrogen content, kernel size, disease resistance), the physiological foundations of the growth process in particular crops, the causes of lodging in cereals and degeneration in potatoes, the efficacy of selection in pure lines, and the best ways to design varietal tests (Kiessling 1910a; Kraus 1905a). In addition to these kinds of inquiry, the station also tested commercially available devices used in breeding as well as devising new kinds of apparatus (Kiessling and Gassner 1927; Kiessling 1916). As noted above with respect to hybridization, the station also carried out tasks which were too complex or expensive for the smaller breeding centres. But even large centres lacked the facilities for certain kinds of breeding – for example, breeding for disease resistance or for baking quality in wheat – so the station filled this gap as well.[16]

Lastly, the station also provided the breeding centres with organizational support. As the number of centres grew rapidly, Kiessling felt that they needed not only a trade organization to represent their commercial interests, but also a vehicle through which the station could provide uniform scientific, technical and economic support to all breeders in the state. For that purpose he joined with two of the largest and most successful Bavarian breeders in 1911 to establish the Association of Bavarian Plant Breeders (*Bayerischer Saatzuchtverein*) which had a technical advisory office based at the station. All of the breeders and most of the seed-growing organizations were members, and by 1922 the Association had well over 200 members. An important function of the Association was also to serve as a sales office for members' seed (Kiessling 1913, 1922a, 1938; Müller 1927).[17]

In addition to such forms of support, of course, the station also conducted its own in-house breeding. That this work was designed to assist small farmers is evident in part from the choice of crops to be improved. Cereals received a lot of attention, reflecting their importance to small farmers: barley and wheat were cash crops while rye or oats were produced on-farm as animal feed. But considerable attention was also given to improving what are now called 'orphan' or 'minor' crops: those in which private-sector breeders were not interested, either because the market for them was too small or because they were technically difficult (thus expensive) to breed. One group of crops in this category were pulses, important as a protein-rich animal feed as well as for crop rotation. Other such groups were oil-bearing plants (e.g. lupins, rape, maize, poppy), soybeans and fibrous plants (e.g. flax, hemp), all of which were unattractive to private breeders on both counts but had proved strategically important during the war and so were bred at the station from 1917 (Kiessling and Gassner 1927; *Jb Bayern 1914–1918*: 38).

Another particularly good example of how the Station sought to compensate for market failure was its work on clover and grasses. With industrialization pushing up the demand for meat and dairy products and large expanses of south German agricultural land devoted to pasture and meadow, some agricultural authorities began to think that the traditional approach to grassland – via extensive cultivation methods – was no longer good enough. If production was to be intensified, however, farmers needed advice on the measures available as well as improved grass and clover seed. But around the turn of the century the demand for seed was being met largely through imports, especially from Denmark and the US (Weller 1928), and by 1914 Germany was the largest importer of animal feedstuffs in the world (Nipperdey 1990: 193). The trouble with imported varieties, however, was that although cheaper than native varieties, they were more poorly adapted to German growing conditions and thus much lower yielding.[18]

The prospects for developing improved native varieties were good since farmers in some parts of south Germany had very positive experience with local grasses.[19] But few commercial breeders wanted to get involved; individual selection with grasses was very laborious because the seed was tiny and hard to manipulate. Moreover since most of these species are naturally out-breeding, much effort would be needed to avoid cross-contamination. In addition, the commercial risks were high. If a cereals breeder produced a surplus and could not sell it as seed, it could still be sold for consumption; not so the grass breeder (Spennemann n.d.: 16; Schikorra 1914a).[20] In an attempt to fill this gap, therefore, Kiessling began in 1907 to encourage the formation of seed-growing associations for alfalfa (lucerne). A few years later breeding work in this area took off at the station, fostered by the construction of a dedicated greenhouse (1910) as well as the creation of a department for grasses and clover (1913), and by 1914 the station's programme for promoting the production of clover and grass seed won the approval of the Bavarian Agricultural Council and the Ministry.[21] During the ensuing war it became all too evident how dependent German farmers had become upon the import of animal feed. The result was a huge expansion of fodder-plant production nationwide after 1918 (Uekötter 2010: 193ff.). In Bavaria the number of grass- and clover-seed producers more than doubled between 1919 and 1927, and by 1926 about a dozen breeding centres were developing improved varieties of these crops.[22] At first finished varieties were bred in-house and released to seed-growing farmers for multiplication and sale (*Jb Bayern 1919–1921*: 8–13), but within a few years unfinished varieties could be handed over to the larger or more skilled centres to complete the breeding process themselves.

That the station's breeding was tailored to the needs of small farmers can also be inferred from its choice of breeding goals, i.e. the particular traits which were selected for. Since most Bavarian smallholders sold at least part of their crops on the market, breeders sought to increase the yield and disease resistance of barley and wheat as well as improving brewing quality and baking quality (Scharnagel 1914; Weller 1951). But these breeding goals probably did not distinguish the station from public-sector breeders elsewhere. Other goals, however, were more revealing of the station's intended beneficiaries. By crossing wheat with spelt (a near-relative), for example, breeders sought not only to achieve adequate yield and good quality but also stability (reliability) of yield. As staff realized, faced with the choice between high-yielding varieties whose yield fluctuated greatly from one harvest to the next and lower-yielding ones which were more consistent from one year to the next, small farmers invariably opted for the latter. Estate owners may have been able to tolerate an occasional poor harvest, but the risks for a small farmer were simply too high (Kiessling and Gassner 1927; Kraus 1914). Another goal in some cereals was increasing the yield of straw, for it was well known that one limitation of commercial varieties was that they did not produce enough straw to meet the needs of small farms (Scharnagel 1953). Unlike more specialized large farms, smallholders engaged in mixed farming and thus required cereal varieties that yielded decent amounts of grain for sale/consumption but also of straw for animal husbandry. Finally, under the station's supervision breeders also aimed to develop varieties that would flourish under relatively poor growing conditions, in keeping with the circumstances of most small farmers in the state (Kraus 1914; cf. Kiessling 1912).[23] That was one reason why local varieties were chosen as the starting material for most Bavarian breeding; they were 'unfussy'.[24]

Finally, it is instructive to take a look at the particular breeding methods deployed in the station's in-house work. There were basically three. When beginning to improve a local variety, mass selection was initially used to obtain a more uniform population. Thereafter individual selection was used in order to isolate promising constituent lines from this population (e.g. *Jb Bayern 1906*: 33). Before the war this was how most of the varietal development conducted at the station proceeded (Kiessling 1922b). While these methods proved very productive, however, from about 1908 staff began to turn to hybridization. (In keeping with the station's breeding strategy, one of the parents in such crosses was almost always a local variety, chosen because of its desirable characteristics.) At that time hybridization was regarded in some breeding circles as 'modern' and was much championed by academic proponents for the new Mendelism (Harwood 1997). Kiessling's enthusiasm for the method was more restrained, but by the 1920s it had become the predominant method at the station (*Jb Bayern 1912–1913*: 19ff.; *Jb Bayern 1919–1921*: 7; Kiessling 1914).

The way in which the station handled the hybridization process is particularly interesting because it bears a certain resemblance to what is today called 'participatory plant breeding' (cf. Chapter 7). Following hybridization, the station often did not attempt to complete the breeding process (through several generations of selection in order to obtain a recombinant which was stable). For one thing, 'finishing off' a variety in this way would have required time and space that the station did not have, but it was also the kind of relatively simple process which almost any breeding centre could do for itself. So whenever staff felt that the results of a cross looked potentially useful for a particular set of growing conditions, they did not attempt to finish it but instead handed it over to a centre in the appropriate region as a mixture of promising recombinant lines. The centre could then complete the breeding process by selecting for those lines that were best for its own particular growing conditions.[25] This division of breeding labour was effective because it coupled the station's generalized scientific and technical skills with the specific knowledge and experience of the local breeders, but it also made economic sense in the Bavarian context. Since hybridization was expensive, by carrying it out in-house, the station would reduce breeding centres' costs so that their varieties could then be sold more cheaply to farmers (Ackermann 1929: 71; Scharnagel 1930; 1953: 178). Almost all of the centres remained dependent upon the station for support of this kind, but by the 1920s several of the largest ones had hired their own breeding directors and no longer needed the station's assistance. This gradual divergence of the interests of large and small breeders, as we shall see in the next chapter, was at the heart of a noisy political row from the late 1920s.

Despite the station's overall shift to hybridization, however, it is important to note that its breeders did not abandon the older and simpler technique of mass selection. It continued to be used, for example, in order to improve local varieties of oats or barley because increases in yield of 20 per cent and more could be obtained at low cost (Lang 1913; Glaser 1902). Although this method would not have produced genetically homogeneous populations (which was generally the aim of those using individual selection or hybridization), that was no bad thing for small farmers. Indeed, it was the heterogeneity of local varieties that accounted for their yield stability despite fluctuating weather conditions or pest invasions. And it was this same principle that prompted the station – at farmers' request – to construct *mixtures* of pure lines, some of which brought good results and were released onto the market. 'Purity' was thus a mixed blessing (Kiessling 1912; cf. Lang 1910c).

Once the station had developed promising material, however, how was it actually transferred to breeders or seed producers? A straightforward sale to farmers was ruled out by the station's statutes. What was permitted, however, was a 'handover' (*Uebergabe*) of the material, either to the state farm or to individual Bavarian seed growers for planting and multiplication. Precisely what 'handover' meant is not easy to establish, given the dearth of records on this procedure. Nevertheless, the director outlined the general nature of the station's practice,[26] claiming that for the most part 'handover' meant one of two things:

(a) Where the station released a *finished* variety that had been bred in-house, the recipient was obliged to pay a fee (said to be 'very modest'). Before the First World War, for example, a number of varieties were turned over to the state farm on this basis, which in turn multiplied the seed and sold it to local farmers. The cost to farmers would have been low since the farm was a public-sector institution and in addition had itself not been obliged to pay the market price to acquire the variety. The station's concern to keep the cost of seed as low as possible is also reflected in Kraus's undertaking that in those cases where it looked as though a commercial variety might be suitable for cultivation in Bavaria, the station itself would buy it, breed it further, and then distribute the improved version of it at low cost to farmers.[27] The other suitable recipients for finished varieties, in Kraus's view, were cooperatives, growers' associations or district committees of the Bavarian Agricultural Society. Their suitability presumably derived from the fact that all of them were organizations that were expected to act in the collective interest.[28]

(b) The other form that handover took (as we have just seen in the case of hybridization) concerned *unfinished* varieties. These were released free to breeding centres for further work, multiplication and sale.[29] In fact, however, the situation was a bit more complicated, for it was not just unfinished varieties which were released free. In the case of grasses, even *finished* varieties bred in-house were also handed over at no charge to seed producers for multiplication and sale (a practice that eventually got the station into trouble; cf. Chapter 4). Both forms of handover were obviously intended to subsidize the nascent breeding industry as well as seed producers in a way that would make improved varieties more accessible to small farmers.

3 Promoting the use of improved seed

Although the Bavarian station's original statutes charged it with promoting the development of plant breeding within the state, its range of activity soon expanded beyond this.[30] Above all, given the size of regional farms, extension work was essential if the south German stations were to achieve their goal of improving cultivation methods. As Kiessling observed, extension was not really a problem for large farmers; they adopted new technology and methods relatively quickly without requiring much assistance. It was small farmers who needed help, both technical and organizational, and this was where state agencies had a role to play (Kiessling 1922b).

Cultivation improvement at national level had already been promoted in various ways since the 1880s by the German Agricultural Society, among others through varietal testing. During the early years of the station's existence, overall responsibility for testing lay with the state's Institute for Agricultural Botany (*Agrikulturbotanisches Anstalt*). The only testing which the station itself conducted was designed to establish which varieties offered the best starting material for breeding. Within a few years, however, Kiessling had come to the conclusion that this division of labour did not make sense; the payoff in terms of changing users' practices was limited because the station was only one of several varietal-testing institutions in the state, and they did not always act in concert (Gassner 1927). By 1913, however, he had managed to persuade the Ministry that responsibility for varietal testing needed to rest exclusively with the station.[31]

But this entailed a great deal of work. All the varieties thought to be potentially of value in Bavaria had to be tested, including commercial ones from outside the state. Those that looked promising in the station's preliminary tests were then sent to the appropriate growers' associations for local testing (since testing, like breeding, had to be decentralized). Barley was the top priority, given the state's large brewing industry. Since the German Agricultural Society's varietal-testing programme had devoted little attention to barley (Hansen and Fischer 1936: 190–214), the Bavarian Brewers' Federation (*Bayerischer Brauerbund)* asked the Agricultural Council in 1903 to set up a system for testing barley varieties. And in 1904 the station was given formal responsibility, not just for varietal testing but more generally for raising the standards of barley-cultivation throughout the state.[32] One of the first tasks the station undertook was to compare local varieties of barley with commercial varieties in order to check the sometimes extravagant claims for the latter made in advertisements. The outcome, in keeping with tests elsewhere, was that local varieties were equal to or better than their competitors (*Jb Bayern 1906*). Testing soon extended to other crops and increased rapidly in scale so that by 1905 the station was testing hundreds of cereals varieties alone (Scharnagel 1929; Steuert 1905: 227). This, of course, gave it leverage for requesting more resources. In 1914, for example, Kraus argued that the station needed an experimental baking and milling facility in order to widen the range of traits in wheat and rye varieties which it could test. And from 1916 the station took over responsibility for the advancement of potato cultivation, setting up testing stations throughout the state.[33]

As we saw in the last chapter, however, farmers needed to know not only which variety was most suitable, but which suppliers produced seed of that variety in high-quality form. The German Agricultural Society's certification procedure had been introduced to provide that assurance for seed produced mainly by commercial breeders in the north, but as Bavarian farmers began to turn to seed produced in Bavaria or elsewhere in south Germany, the Society's procedures were increasingly irrelevant. For that reason in 1905 the Bavarian Agricultural Council introduced its own certification process (designed by Carl Kraus) under the technical direction of the station at Weihenstephan. Only seed that was suited to Bavarian growing conditions was certified.[34] Station staff were aware, however, that the Bavarian system would not be able to adopt the Agricultural Society's system unaltered. Given the dearth of large farms in Bavaria, Bavarian-bred seed would have to be multiplied on cooperatives consisting of small farms which would have difficulties in meeting all of the Society's requirements for certification.[35] For this reason in 1909 the station devised special procedures for certifying cooperative seed producers, and the number of Bavarian farms producing certified seed rose very quickly.[36] To encourage the purchase of certified seed, each winter the station published a list of certified varieties in the region's weekly agricultural newspaper.[37]

But the station's extension activities went well beyond testing and certification. Providing advice to farmers also meant dealing with their inquiries, publishing – in farmers' magazines or flyers – useful information about new varieties or tips on collecting seed from valuable crop plants as well as holding courses on varietal testing for teachers at agricultural schools.[38] Kiessling and Kraus also gave lectures on crop improvement to local agricultural societies and hosted visits to the station by numerous agricultural organizations, individual farmers, and foreign breeders.[39] In order to persuade farmers to adopt improved varieties, however, the station reckoned that crop shows were much more effective than lectures (*Jb Bayern 1904*: 51; cf. Lang 1912a, 1912b). These were organized at local level according to guidelines provided by the station, and the best varieties exhibited were awarded prizes by committees consisting of farmers, brewers or millers, and experts from the regional branch of the Agricultural Society. Field visits with prize-giving were also organized, and the results were published in local papers. Both kinds of occasion served not only to diffuse information on best practice but also to provide staff with feedback on the station's impact.[40]

Probably the most important extension activity in Bavaria, however, was the promotion of crop-improvement associations (*Ackerbauvereine*). The idea behind these organizations was twofold. First, they would bring benefits of scale

Figure 3.3 Field demonstration at the Bavarian Plant Breeding Station (courtesy: Bayerisches Hauptstaatsarchiv, Munich).

to small farmers. This advantage was particularly marked in Bavaria, partly because of the predominance of small farms but also because of the importance of barley as a cash crop. By the end of the nineteenth century Munich's rapidly growing brewing industry found that they could not get enough good quality barley from Germany and thus were having to import a large proportion from Central Europe (Teich 2000: 129–132). As Kraus remarked at the time, however, the brewers in fact had a high opinion of Bavaria's own local barley varieties, but they were only willing to pay premium prices for large batches of uniform and well-cleaned grain. And small farmers, often growing different varieties from their neighbours and lacking access to grain-cleaning machinery, could not meet those requirements. The solution, he argued, was to form local crop-improvement associations who would agree on which high-quality variety all members would plant, would arrange collectively for cleaning of the grain, and would pool their harvests for sale in large batches. The idea found the Ministry's backing.[41]

The other rationale for the improvement associations was that they would also much simplify the process of extension. (As one member of the station's staff put it, the associations were the ideal link between breeders and farmers (Scharnagel 1914)). For it was clear to the directors that breeding new varieties would be pointless unless there were reliable arrangements for multiplying the new seed and getting it to farmers.[42] Simply making low-cost improved seed available for inspection at crop shows was certainly necessary, but it would not suffice to persuade the mass of farmers who tended to be cautious about new practices. The key lay in showing farmers that it was in their own interest to organize themselves into improvement associations (Kiessling 1913).

The central figure behind this movement was Kiessling, and his vision of the associations' role was a broad one.[43] Their aim, as he saw it, was not just to produce larger batches of barley nor to promote the use of improved seed but to raise the level of cultivation practices overall. Brewing quality in barley, for example, was affected not just by the variety planted but by cultivation, harvest and storage practices as well. And the means whereby the associations would be effective agents of extension, as one observer put it, was social pressure. The prospect of an association receiving good prices for a high-quality product put pressure on individual members to bring their cultivation practices up to the required level. Experiments, demonstrations and varietal tests organized for members by a local advisor would display the advantages of improved cultivation methods or apparatus, and free or subsidized inputs would allow small farmers to try out these things without risk. But once standards had been agreed by the association, annual visits by the advisor to every member's fields guaranteed that individual members would have to adhere to the rules. Moreover, the existence of an association, backed up by social pressure, meant that *coordinated* improvements could be introduced across the locality such as a collective approach to pest control.[44]

The first step toward organizing such associations got under way in 1903. Finding that he could not get a good price for the relatively small quantities of high-quality barley he was able to produce, Georg Schweiger contacted Kiessling to discuss how the situation of small barley growers might be improved. They agreed that Schweiger should persuade about two dozen of his neighbours to join him in a 'barley-improvement association' whose members would all plant the same variety and adopt uniform cultivation and harvesting practices, following advice provided by the station at Weihenstephan. Schweiger was successful; the Moosburg Barley Improvement Association was formally constituted in 1905, and within a year it was securing top prices for large shipments delivered to a Munich brewer. Details of the new organization were publicized, and its improved barley came to the attention of farmers and brewers alike at a national brewers' exhibition. Since the station had been given responsibility by the Bavarian Agricultural Council in 1904 for raising the standards of barley production, the Moosburg Association soon provided the model for what was to become a state-wide movement, initially dominated by barley but later extending to other cereals, root crops and grasses. By 1910 the number of local associations had grown so quickly that it was decided to form a coordinating organization at state level – the Bavarian Federation of Crop Improvement Associations (*Landesverband Bayerischer Ackerbauvereine*) – which began to publish its own newspaper (*Bayerische Ackerbauzeitung*) two years later. The station provided the Federation with technical advice and an administrative office as well as editing its paper. By 1912 there were about 400 local associations; two years later there were over 600; and by 1927 there were nearly 1500 across the state (Schweiger 1929; Kiessling 1912, 1913; Kiessling and Gassner 1927; Lang 1927; von Rümker 1909; Müller 1927).

Crucial to the successful operation of these associations was that they could draw upon the services of a local expert, usually a teacher at an agricultural secondary school (*Wanderlehrer*) who was employed by the regional branch of the Bavarian Agricultural Society. This person had to be familiar with local economic/ecological conditions and cultivation practices, was responsible for providing advice on technical matters, and was free to consult the station on more specialized matters where need be. The expert was responsible for conducting varietal testing in the region and was thus well placed to advise on the choice of the best variety.[45] The variety chosen might be a local variety (provided it was available in a healthy and relatively uniform condition) or even a commercial variety from outside Bavaria (so long as it was well adapted to the locality), but in most cases it was an improved variety which had been bred under the station's supervision (Kiessling 1913: 18–19). Once the variety had been decided, an association bought seed (as well as fertilizer and equipment) in bulk at modest price, thus saving its members money. The seed was obtained either from a nearby breeding centre or from one of their own members who multiplied improved seed, thus guaranteeing that it would be well adapted to local growing conditions. Cultivation, harvesting and storage procedures were then conducted on the advice of the local expert. The associations also provided a meeting place in which members could learn about best practice through lectures, crop shows, local demonstrations, and tours of outstanding farms.[46] Once the associations were up and running, the local experts constantly urged them to make full use of all available facilities provided by public-sector institutions as well as by the Bavarian Agricultural Society. High usage rates could then be used to lever more resources from the state (Kiessling 1911).[47]

For the first ten years or so Kiessling's efforts to establish improvement associations had run into a fair amount of scepticism, but by 1912 he felt reassured that they had won the support of the Ministry and other bodies.[48] Moreover the speed with which the improvement associations spread, as well as the high proportion of farmers in some regions who became members, suggests that small farmers themselves felt that the associations were meeting their needs.[49] Other south German states soon adopted the Bavarian model. By 1912 crop-improvement associations were being set up in Württemberg, and by 1914 there were 31 across the state (Wacker 1912; *Jb Wurtt 1913–1914*: 69–70).[50] By the war even north German observers were paying tribute to the station's role in promoting the associations (Edler 1913).

Farmers were not the only constituency served by the station. In order to help smallholders gear their production more effectively to demand, the station sought to act as a broker, maintaining close ties with the brewing and milling industries so that the latter's requirements could be channelled to the crop-improvement associations. In the case of barley, both the breeding centres and the station itself addressed issues important to the industry, the former by improving local barley varieties and the latter by studying the cereal's malting properties (Kiessling 1910a). Although brewers were critical during the early years of both the varietal testing and certification of barley varieties conducted at the station, by 1914 they were very satisfied with the quality of the station's new barleys.[51] As for wheat, around the turn of the century studies of baking and milling quality were practically non-existent in Germany, though there was agreement among breeders that it would be desirable. By 1914 the advisory committee at Weihenstephan was concluding that the station ought to extend its varietal testing to this domain, not least because Berlin's institute of flour processing seemed unwilling to analyse Bavarian samples. Unimpressed with imported wheat or rye varieties, Bavarian millers reckoned that the state's local varieties were of better quality. As we have seen, taking note of the millers' requirements, the station's wheat breeder (Theodor Scharnagel) concentrated on developing improved local varieties of high baking quality. Consequently, by the 1930s Bavarian millers, unlike their north German counterparts, had much closer ties to regional breeders and could therefore satisfy their demand for high-quality wheat entirely from Bavarian growers (Kiessling 1914; Scharnagel 1914, 1936/1937, 1953).

In most other German states the creation of chambers of agriculture from the 1890s had relieved the agricultural colleges and other state institutions of responsibilities for extension. Since Bavaria did not acquire a chamber until 1920, however, the station needed to do a lot more than just develop new varieties if it was to have any impact upon the use of good quality seed or of improved cultivation practices more generally. That meant that the station's extension work was particularly important for regional agriculture.

4 The stations in Württemberg and Baden

In most respects the stations in Baden and Württemberg were very similar to that in Bavaria, though there were occasional differences of emphasis. In organizational terms both stations started out about the size of Bavaria's but by 1920 had grown much less. As in Bavaria, Württemberg's station was also established as a part of the state's agricultural college, though it never became independent.[52] Since Baden had no agricultural college, the station was set up on the grounds of an agricultural secondary school and was incorporated into the state chamber of agriculture after the First World War.

All three of the stations were, at least for a time, responsible for varietal testing and certification in addition to breeding. At Württemberg the director, Carl Fruwirth, was happy to promote breeding but regarded varietal testing as the highest priority.[53] Probably due to limitations of staff and facilities, the station in Baden conducted very little in-house varietal development (until it moved to better quarters in 1922) and instead concentrated upon advising its breeding centres and certifying them (Lang 1909c). The director, Hans Lang, argued that big increases in yield could be obtained, not just via breeding but also by reducing disease, cleaning seed, and applying mineral fertilizer.[54]

Both stations shared their Bavarian counterpart's small-farm orientation. In Württemberg Fruwirth insisted that varietal tests must be conducted on small farms using derivatives of original seed (*Absaat*) if the results were to provide information helpful to small farmers (Fruwirth 1907a, 1907b). His successor, Johann Wacker, took the same view (Weiss, 1918) as did Baden's director (Lang 1909e). The same orientation is evident from Fruwirth's view that breeding improved varieties was less urgent than getting advice on the best varieties to the farmer. In Baden similarly, the station believed it important to breed not only for yield but also for yield security (*Jb Baden 1920*, 113–114). This perspective was also reflected in the crops bred at, or with the support of, the stations. Although the large flour mills in the region were mainly interested in wheat, for example, both stations focused instead upon improving its close relative, spelt. From small farmers' point of view, spelt was preferable to wheat because it generated more reliable yields and fared better in unfavourable growing conditions.[55] And since a high proportion of the acreage in both states was pasture or meadow, both stations devoted considerable resources to grass and clover breeding. Baden was particularly quick off the mark in this area, establishing in 1910 an 'advisory centre for fodder plants' which would not only conduct varietal testing and develop improved local varieties but also provide farmers with advice on the most appropriate seed for their growing conditions.[56]

In both states the stations encouraged a certain degree of farmer participation. At Württemberg Fruwirth took the view that although an expert would initially have to show the farmer how to carry out breeding, the farmer should have a say on which traits were to be selected (Fruwirth 1910, 1906: 6–7, 1911b). At Baden one of Lang's first steps was to ask the state's farmers to send in seed samples of local varieties which they found to be especially good[57] and to share their experience with growing fodder plants so that it could inform the station's own experiments in that area (Lang 1914c).

Decentralization of the breeding operation was also characteristic of both stations, each of which maintained external branches where breeding and varietal testing took place under different conditions from those at the main station.[58] And both stations encouraged the establishment of breeding centres across their regions, though the number of these remained far smaller than in Bavaria (Lieber 1932; Wacker 1933).

Although both stations promoted varietal development in the *interests* of the small farmer, the directors at Württemberg had reservations about allowing small farmers themselves to carry out anything more complicated than mass selection (Fruwirth 1906: 6–7, 1910, 1912; Wacker 1912). The result was that by the 1920s the breeding centres supported in Württemberg were exclusively on large farms.[59] Baden's directors were much more comfortable with the idea of small farmers doing breeding or multiplying varieties bred at the station.[60]

Finally, as in Bavaria, both stations used local varieties as starting material for their in-house breeding and had begun to use hybridization on a large scale well before the First World War (Wacker 1925; Roemer 1914; Lang 1914b: 14–15). At both places varieties bred in-house were handed over at no cost, wherever possible, to another public institution – in Württemberg to the state farm, in Baden to the chamber of agriculture – for multiplication and sale at low cost.[61]

Conclusion

On the whole the south German stations were regarded by most contemporaries as a resounding success. One indication of public support was the phenomenal growth of the Bavarian station by the 1920s. In terms of staffing, the number of technical assistants (i.e. those with higher educational qualifications) increased from two in 1903 to 14 by 1926. The first of the regional breeding inspectors was appointed in 1910, rising to a total of nine by 1927. Office staff increased from a single person in 1909 to seven in 1926 (Kiessling and Gassner 1927; *Jb Weih* 1903 to 1922–1926). Correspondingly, the station's budget (running costs plus salaries) rose from 3600 marks in 1902 to 17,600 four years later, reaching 47,000 by 1914–1915.[62] Nor was approval forthcoming only in Bavaria. When the Prussian Agriculture Minister was considering between the wars how Prussian chambers of agriculture might begin to support plant breeding, it was suggested to him more than once that 'the south German stations' would be a suitable model. The Bavarian station in particular was cited as an institution 'which has already released a whole series of very useful varieties to the private sector'.[63] It is thus not altogether surprising that by 1922 Kiessling was claiming that no other region in the world had such extensive (public-sector) facilities for both breeding and seed production as did Bavaria (Kiessling 1922a, 1922b).

But were the south German stations distinctive in their peasant-friendly orientation? To be sure, not all of their activities were unique. In the US during the

1920s, for example, some agricultural colleges offered courses on plant breeding so that farmers could produce their own hybrid maize and then made provision for quality control by certifying the new varieties (Kloppenburg 2004: 106). Similarly, the agricultural experiment station in North Dakota undertook to promote the formation of crop-improvement associations (Danbom 1989). In colonial Java before the First World War Dutch rice breeders used local varieties as their starting material for selection (Maat 2001: 185–186). And in Britain the rationale for regional stations was partly based on the argument that varieties developed at the English station in Cambridge were ill suited to growing conditions in Wales or Scotland (Palladino 1990). But to judge from the existing secondary literature, none of the stations in the US, Sweden, Britain (or France until after 1945) undertook these activities in accord with an overarching mission to serving smallholders (Kloppenburg 2004; Fitzgerald 1990; Äkermann *et al.* 1948; Andersson 1963; Palladino 1990; Bonneuil and Thomas 2009).

From the point of view of current development policy which, as we have seen continues to search for ways of better assisting small farmers, it is obviously important to try and identify *why* the south German stations were so effective. One way to do this is to use the criteria that green revolutionaries themselves arrived at during the 1970s and 1980s as they sought to understand why previous GR programmes had so often failed to reach smallholders. We will look in detail at those criteria and the reasoning behind them in Chapter 6. At this point, however, let us 'jump ahead', noting how the stations measured up against the diagnoses produced by development experts 50 years later. The remarkable conclusion is that many of the later hallmarks of success were already embodied in the stations:

- a decentralized approach to varietal testing and breeding;
- heavy emphasis upon organizing smallholders;
- an overarching concern to improve cultivation practices (backed up by a regional network of extension services) rather than a narrow concentration upon varietal development;
- a concern to develop varieties that did not require ideal growing conditions but were instead well adapted to the circumstances of smallholders;
- strong government support from the states in which they were located.

Lastly, there is the question of staff commitment. In a survey a few years ago of what made for effective development projects in northeast Brazil, Judith Tendler observed that staff at one successful experiment station 'worked to promote the development of their municipality or region in general, and their interest in disseminating improved varieties and practices came out of a larger passion for bringing "development" to where they lived' (Tendler 1993: 1574). How did the stations measure up against this criterion? Many of the stations' directors were of peasant-origin themselves. At the Bavarian station, for example, although Carl Kraus was the son of a Bavarian primary school teacher, his successors, Kiessling and Scharnagel, were both peasants' sons (Kiessling 1918a; Scheibe

1942; Weller 1951), and the same was true of the long-standing director of Württemberg's station, Josef Wacker (Anon. 'Professor' 1934a). Not all directors were of peasant or even rural backgrounds. Carl Fruwirth (first director of Württemberg's station, 1905–1907), Hans Lang (first director of Baden's station), and Paul Kulisch (director of the experiment station in Alsace-Lorraine) all came from urban middle-class backgrounds. But each of them appears to have embraced the stations' service mission. Lang, for example, was a native of Baden and published dozens of short articles in Baden's agricultural newspaper with advice for farmers, some of them peppered with humour and colourful examples so as to appeal to a wide readership (Schulenburg 2002; Lieber 1924; Lang 1909a, 1911b, 1914b: 27–28). And Kulisch's career in Alsace as well as after 1918 (as head of the Agricultural College in Weihenstephan) displayed a consistent commitment to serving small farmers.[64] Thus the stations largely confirm later analyses of successful development projects.

In Chapter 6 we will return to this comparative theme. Meanwhile, however, it is necessary to pursue the stations' history a stage further. The strong support that they were receiving from farmers and agriculture ministers suggests that by the 1920s they were having a favourable impact upon the south German agricultural economy. But as we shall see in the next chapter, it was precisely that impact which made them highly controversial in some quarters.

4 Success breeds trouble

The controversy over public-sector breeding, 1902–1933

As we have seen, the south German stations' statutes cast them in a public service role. They were to serve the regional farming community and promote the development of small plant-breeding enterprises. On the other hand, 'commercial activity' (*Gewerbe*) was explicitly ruled out. What this rather vague term then meant in practice was that where stations developed new varieties in-house – an activity then regarded as reasonable by all concerned – these were not to be sold on the market. The founder of the Bavarian station, Carl Kraus, made a point of stressing that his station would abide by this principle (Kraus 1905b), and his successor, Ludwig Kiessling, regarded this restriction as entirely justified since it was essential, in his view, that stations enjoyed the trust of commercial breeders as organizations that did not compete with them on seed markets.[1]

Before the First World War this issue was little debated since the stations' breeding work was just getting under way, and it was not yet clear in most cases whether their varieties would have any commercial significance. By the 1920s, however, it had become abundantly clear. Public-sector varieties had begun to appear on lists of the best available varieties in Germany, and the stations were beginning to be perceived by private-sector breeders as unwelcome competition. From the mid 1920s through the end of the Weimar Republic, accordingly, a vigorous controversy – in public as well as behind the scenes – centred on which aspects of the stations' breeding work were 'legitimate' and which, in the eyes of the private sector and its allies, constituted 'unfair competition' and should thus be shut down.

Twenty years ago Jack Kloppenburg provided the first substantial account of tension between private and public sectors, focusing upon the development of hybrid maize in the United States and the conflict between the sectors which emerged after the mid 1930s (Kloppenburg 1988). Informed by a Marxist perspective, his account is particularly illuminating on the repeated – and largely successful – attempts by private breeders since the 1930s to displace the public sector from varietal development so as to monopolize control of the seed market. He has rather less to say, however, about the other half of this partnership: for example, how *public* breeders understood their service mission, the extent to which those understandings varied over time and space, or the specific forms of collaboration between public and private sectors prior to the clash. Nor do we learn much about how the state attempted to mediate in this conflict. In the years since Kloppenburg's book appeared, surprisingly little attention has been paid to these issues.[2]

My aim in the chapter is to begin to fill this historiographical gap by looking closely at the changing relationship between private- and public-sector breeders over several decades. Section one traces the initial reception of the stations before the First World War, during which time commercial breeders occasionally expressed concern about them. In section two I show that by the 1920s varieties bred at the stations had begun to make a substantial impact upon the seed market. Section three outlines the character and extent of the economic crisis facing German agriculture from the latter 1920s. Section four surveys the renewed attacks upon the public sector which emerged in consequence, and the final section looks in detail at the form that this controversy took in Bavaria.

By following in detail the evolving nature of the relationship between these sectors over time in a particular context, one can tease out several features which are arguably characteristic of public–private relations more generally. Public research institutions are naturally subject to expectations from various interest groups who seek to shape the institution's research agenda in order to exploit its work for their own advantage. Inevitably, not all groups gain equally; indeed, the south German stations were *designed* to assist a particular section of the public. As a result, they attract considerable criticism from disappointed constituents. Occupying an exposed position in the public arena, every decision such institutions take may be scrutinized, and they can easily become a target for political pressure. Initially regarded as a potential *Milch-Kuh*, such institutions can all too easily end up as a kind of public punching bag.

1 Initial reactions to the south German stations, 1902–1914

The new peasant-friendly stations were welcomed by breeders in both public and private sectors (Edler 1913; Beseler 1911). Perhaps their most prominent academic supporter was Kurt von Rümker, professor of agronomy at the University of Breslau. As we saw in Chapter 2, von Rümker was an early advocate of seed-breeding cooperatives. Interestingly, in his early writings on this subject he didn't think that such cooperatives would pose much threat to commercial breeding since they were likely to develop varieties for small farmers and relatively unfavourable growing conditions while the private sector would produce improved varieties for good conditions. And if some competition did arise, he suggested, it would be a good thing for agriculture as a whole (von Rümker 1898; cf. his 1905 and 1911b). Although he had originally not liked the idea of state plant-breeding stations, by 1909 he had changed his mind. They were still not the right solution, he felt, for north Germany where a strong private sector required instead research and teaching to provide useful kinds of basic genetic knowledge as well as breeding method (von Rümker 1909: 6–7, 33–34, 50ff.; cf. his 1910 and 1912–1913). But he was impressed with the new south German stations and took pride in the fact that they were in part a response to his 1898 proposals. (As we shall see, however, by the 1920s he had changed his mind.)

Nevertheless, despite the fact that the stations' statutes forbade commercial activity, they were controversial right from the start, and the criticism extended as well to other kinds of peasant-friendly breeding institutions such as seed-breeding cooperatives (Bischoff 1922) and chambers of agriculture (Anon. *Ldw-Kammer* 1911b: 128). Unsurprisingly, private breeders were suspicious (Stephani 1911; Kiessling 1912; Pflug 1912).[3] For example, in 1904 the eminent breeder, Ferdinand von Lochow, insisted that breeding was best left to individual initiative and open competition; it was wrong for the state to 'monopolize' breeding (Lochow 1904: 27). Moreover, he noted, the German Agricultural Society took the same view.

The Agricultural Society, however, was soon to be joined by another organization which proved to be an altogether more energetic defender of the private sector's interests. In 1908 an Association of Commercial Breeders was formed (*Gesellschaft zur Förderung deutscher Pflanzenzucht*, henceforth GFP) whose declared aim was to foster ties between 'science and practice' as well as to advance the interests of the private sector (Edler 1908).[4] The GFP welcomed those public-sector institutions that were prepared to cooperate with commercial breeders; namely, university agricultural institutes where what the GFP called 'scientific' work was done on laws of inheritance or methods of breeding but no varietal development. From the start it lobbied the Prussian Agricultural Ministry to establish such an institute, succeeding in 1914 with the creation of an institute of genetics for Erwin Baur at the Agricultural College of Berlin. In addition they supported public-sector breeding in crops in which the private sector was not interested: for example, hops, tobacco, grapes, grasses and clovers (Anon. 1922 'Geschäftlicher Teil'; Anon. 1913 'Eröffnungssitzung'; GFP 1933: 29–31). But the state plant-breeding stations were regarded in a very different light since they undertook varietal development in cereals and other *major* crops. As the chairman of the GFP remarked at the Association's 1914 meeting, state institutions on the Svalöf model were 'entirely inappropriate' for Germany (Frölich 1922: 41).[5]

Members of the GFP tended to portray their organization as representing the interests of all breeders (e.g. von Rümker 1925a; GFP 1933: 27), but in order to understand why the GFP pursued the policies it did, it is important to recognize that it represented only one faction within the private sector. By 1911, for example, its 49 ordinary members (i.e. commercial breeders) constituted only a quarter of the total number of private-sector breeders at that time. By 1924 the membership had climbed to 134, but this was still only about half of the total.[6] Moreover, the membership was dominated by large and well-established firms from central and north Germany, including almost all of the big names of the first generation of late nineteenth century breeders.[7] Breeders from south Germany made up only 5–10 per cent of the membership, and cooperatives were even less common.[8] Indeed, for reasons which will become clear below, most of the small and newer southern breeders and breeding cooperatives were more inclined to see the GFP as enemy than as friend.

The GFP, however, enjoyed support from some sections of the scientific community. Some academic breeders, for example, were equally critical of public-sector breeding, Erwin Baur (see section three) being a case in point (Baur 1914). Another supporter was Paul Gisevius, professor of agronomy at the University of Giessen, who argued that state intervention 'on a foreign model' (he was no doubt thinking of Svalöf) was wrong and would do more harm than good (Gisevius 1912; cf. his 1908). And Theodor Roemer, professor of agronomy and plant breeding at the University of Halle, argued that public-sector institutes would never be as successful as commercial breeders and should therefore leave varietal development to the private sector (Roemer 1914).

In the face of such criticism, the south German stations frequently made a point of stressing that the varietal development which they undertook brought no commercial gain to the station (e.g. *Jb Bayern* 1906: 7–8; 1909: 13; and 1910: 8). Similarly, the heads of some university institutes where work in plant breeding was being conducted also emphasized that this work would be 'purely scientific' and that varietal development was best left to the private sector. When the University of Halle, for example, established a plant-breeding station in 1911, its director underlined the fact that the station's research activity would be restricted to identifying laws of inheritance and improving breeding methods (Wohltmann 1911; cf. von Rümker 1904).

The clash of views over public-sector breeding became especially visible in 1907 when an article by the professor of agronomy at the Agricultural College in Bonn, Theodor Remy, triggered a public debate in the agricultural press. Remy expressed reservations, in part about the sheer numbers of new varieties being generated by commercial breeders, not all of which he felt were of value, but especially about their suitability for a wide range of growing conditions. Part of the solution, he suggested, would be to expand varietal testing at local level in state institutions, such as the chambers of agriculture (Remy 1907b). One eminent breeder, evidently wounded by Remy's critique, denied flooding the market with poor quality varieties and instead declared his commitment to the public good (Richter 1908). And an agricultural official issued a blunt rebuke, defending commercial breeders against the charges and rejecting state intervention out of hand (Sperling 1908). But a public-sector breeder from south Germany endorsed Remy's position and argued that the solution was to create many more state plant-breeding stations, even in north Germany, because state intervention was justified in the interests of the greater good (Broili 1907). Similarly, the director of breeding at a well-established firm agreed that many varieties were of little value, adding that the new south German stations could play an important role in protecting farmers against unsubstantiated claims by unscrupulous breeders. Besides, if stations were established only in regions where commercial breeding did not exist, they would not damage the interests of the private sector (Lang 1907; cf. Sierig 1908).

Two years later a similar exchange took place in an agricultural journal, initiated by an article on the organization of plant breeding by Carl Fruwirth, former director of Württemberg's station. For directors of such stations, he observed, 'trying to find the right stance [on the role of the public sector], given the conflict of interests here, is not always easy', especially in countries which already have a private sector (Fruwirth 1910: 645–646). After explaining the rationale for the stations' various activities, he added that a certain amount of varietal development in-house was unavoidable. Although this tended to attract criticism from the private sector, some of which was justified, there could be no objection where the station bred varieties in which commercial firms were not interested.[9] These remarks prompted a reply from a commercial breeder who remarked in passing that public-sector breeding had been controversial among his colleagues for some time (Pflug 1911). Public-sector breeding, he argued, was unfair since the fact that a variety had been developed by an academic institution would confer upon it additional credibility. And it was not right that a public breeder should hand over a variety to a single multiplier since the latter would get an unfair advantage at public expense. Nor was it fair that a state institution should teach peasant farmers how to breed their own varieties since this would damage the interests of those private breeders – like himself – who sold to peasants. In reply Fruwirth agreed that although stations were potentially competitors for the private sector, this was not a serious problem since (a) they had only been established in regions where commercial breeding did not exist and (b) the small farmers served by the stations were not in a position to buy commercial seed anyway (Fruwirth 1911b). If nonetheless the stations sometimes impinged upon the interests of the private sector, Fruwirth concluded, this could not be helped because as public institutions the stations were obliged to serve *all* farmers.

The debate over public-sector breeding was to become a good deal more heated a decade later, and to understand why, we need to look at the effects that the stations' new varieties were beginning to have by the 1920s.

2 The stations' economic impact between the wars

Varietal improvement did not happen overnight. It took a few years to improve a variety using mass selection and much longer using individual selection or hybridization. So it is not surprising that the number of varieties which the Bavarian station, for example, was able to release before the First World War was small. By the early post-war period, however, this had changed dramatically. Between 1919 and 1921 the station handed over 40 varieties to growers for multiplication and 22 lines to breeding centres for finishing (*Jb Bayern 1919–1921*). One of the stations' critics had predicted in 1908 that they would not be able to compete with private breeders in developing commercially successful new varieties, and this view was to be reiterated by others over the next 20 years (Gisevius 1908). But he was wrong. On the contrary, as we shall see, the intensified criticism coming from the private sector from the mid 1920s was in fact prompted by the stations' *success* in capturing a substantial section of the seed market.

But what does 'success' mean? Ideally an answer to this question would come from the acreages planted in particular varieties, but I have been unable to obtain such data. By 'success', therefore, I mean that varieties, bred either in-house or with the technical assistance of the stations, were highly regarded by breeders (both commercial and academic) and also sold well. Consider, for example, the assessments of public-sector breeders after 1945. Wilhelm Rudorf was especially complimentary about the Bavarian station's work. Weihenstephan's varieties, he reckoned, had had just as much impact upon south German agriculture as Halle's had upon central Germany, and most of the leading cereals varieties in recent decades had emerged from the work of these two institutions. Moreover the fodder plant varieties developed at Weihenstephan had been 'of great economic importance' (Rudorf 1955: 29).[10] Hermann Kuckuck and Rudolf Schick took a similar view. Thanks to its close collaboration with private breeders, Kuckuck argued, Weihenstephan was among the most productive public-sector institutions during the interwar period (Kuckuck 1951: 25, 73; 1978; Schick 1958). Breeders singled out particular crops and varieties from the station for praise. The hybrid winter-wheat variety Tassilo (released in 1930), for example, was widely planted in south Germany, Alsace and Austria into the late 1950s and was often used as a parent for breeding high-quality baking wheats well into the post-war period (Bauer 1993; Scheibe 1987: 20; Wienhues and Giessen 1957; BDP 1987: 53; Bonjean and Angus 2001). The barley variety Isaria came to market in 1924 and was important throughout the central European brewing industry for the next 20 years (Fischbeck 2004 and 1988; Gäde 1993: 209).[11] The station's fodder varieties – grasses, clover, oats – also attracted praise (Fruwirth 1911a; Gäde 1993; BDP 1987: 53).

The stations in Württemberg and Baden, of course, were much smaller institutions which would not be expected to have had such a large impact. Nevertheless, by the late 1920s one-half of the certified varieties on the market in Württemberg had been bred by its station.[12] By the mid 1920s one of the brewing barleys bred with the station's assistance was on the Hybred Varieties Register and had placed first in several competitions (Württ. Landessaatzuchtanstalt n.d.). And by 1953 the station had turned over 38 varieties to Württemberg's breeders, of which half had come to market; the station's fodder-plant varieties in particular were regarded as economically significant (von Kittlitz 2005: 70, 51–52; Rudorf 1955). In Baden by the mid 1920s improved cereal varieties bred under its station's supervision were yielding as much or more than highly regarded varieties from large north German breeders (Lieber 1924). By 1932 about one-quarter of the cereal varieties planted in the state had been bred in Baden (though the proportion was substantially higher in spelt and winter wheat). Maize was a regional speciality. Two varieties bred at the station were listed on the Hybred Varieties Register from 1921 and were widely planted, not only in Germany but also in Poland, Austria and Czechoslovakia (Anon. 1932a; Buss 1925).[13]

Judgements of this kind, of course, are to some extent subjective and in any event indirect. Ideally one wants a more objective indicator of the quality of public varieties. One such measure is inclusion on the German Agricultural Society's Hybred Varieties Register. During the early 1920s south German varieties made up a modest proportion of the varieties on the Register; about 11 per cent were of public origin (Saatzuchtstelle 1925). Things were beginning to change, however, for of the new varieties admitted to the Register in 1924, *all* were from south Germany (Saatzuchtstelle 1924). And by 1933 south German varieties were constituting a substantial minority of the varieties of certain crops; for example, 15 per cent of the oats and winter rye varieties, 35 per cent of the summer barleys, 28 per cent of the winter barleys, and half of the grass and clover varieties (Saatstelle 1932, 1933).

Another measure of a variety's importance by the mid 1920s was its inclusion in published handbooks of high-quality certified varieties. In one such volume (Ramm 1925, 1926) 30 per cent of the barley varieties listed were bred in conjunction with one or another public-sector institution (two-thirds of these were from Weihenstephan).[14] A quarter of the oats varieties had public-sector origins (half of these were from the south German stations). Twenty per cent of the rye varieties listed were of public-sector origin (two-thirds of them from Weihenstephan), and about 20 per cent of the wheat varieties listed were from south German stations. In another handbook of this kind (Leverenz n.d.) the proportions of cereals varieties stemming from the south German stations were rather similar, though as many as 38 per cent of the winter barley varieties listed came from Weihenstephan.[15]

As we shall see in the next chapter, after 1934 Nazi agricultural authorities compiled the first state-sanctioned list of approved varieties, and similar lists were produced after 1945 in both West and East Germany. Inclusion required that a variety be among the very best performing varieties of a given type for particular agro-ecological conditions. On the first such list in 1935 a substantial minority of the varieties in several categories had originated in the public sector, most of them from the south German stations.[16] On the list in 1938 south German varieties constituted 25–30 per cent of the winter rye, winter wheat and summer barley varieties (Anon. *Ratgeber* 1938a), and the proportions in 1940 were roughly similar (Anon. *Sortenliste* 1940b). In 1942 as many as 40 per cent of the summer barleys were of south German derivation (RdP *Ratgeber* 1942). And in 1944 the regime published a handbook of recommended winter wheat varieties, 16 per cent of which were from the south German stations (Voss and Breuninger 1944).

After 1945 public varieties, mostly of south German origin, continued to appear on the state-approved lists maintained by both West and East Germany. Of the breeders listed on the western list in 1949, for example, at least 10 per cent had relied upon public-sector support (Anon. *Sortenliste* 1949–1951).[17] Of the cereal varieties on the list nearly 20 per cent were of public origin, and in grasses, clover varieties and other crops unattractive to the private sector the proportions were predictably higher. Since the varieties developed by the south German stations were bred for growing conditions in their regions (all of which were in the western zone of Germany after 1945), one would not expect to see many of them on a list of varieties approved for the different growing conditions in the East. Nevertheless in 1948 the East German list of approved varieties contained about half a dozen cereals varieties bred in conjunction with the stations, and about 15 per cent of the legume, grasses, alfalfa and clover varieties were also of south German origin (Anon. *Sortenliste* 1948).

In making a judgement of the south German stations' impact upon the seed market, we need to remember that the number of south German stations was actually very small in relation to the number of commercial breeders (*c.*200 in 1910). That the stations should nonetheless have accounted for 20 per cent or more of highly regarded varieties thus indicates that their impact was disproportionately large. Furthermore, when evaluating the sources above, I have been primarily concerned to establish how many of the high-quality varieties emerged from the south German stations. It is important to note, however, that these figures represent an underestimate of the total number of *public* varieties on these lists, partly because additional varieties came from other public-sector institutions such as university institutes of agronomy (e.g. Entrup 2008) or Svalöf as well as from self-provisioning organizations such as breeding cooperatives, but also because public-sector origins were not always evident in a variety's name.[18]

Without figures on acreages planted, it is difficult to know with any precision just how popular different varieties were with farmers. What is quite clear, however, is that from the mid 1920s commercial breeders were worried by the market share commanded by public varieties. GFP officials, for example, were aware that if sales of commercial seed were to be increased substantially, the key market was among the very large numbers of small- and medium-sized peasant farmers rather than the relatively few estate owners (Bohne 1924). South German agriculture, of course, was dominated by farms of that size, but thanks to the efforts of the regional stations, a substantial number of south German breeders had emerged by 1918 who were able to supply the region's farmers with less expensive but high-quality seed. One 1918 survey of breeders in the various German states, for example, indicated that Bavaria was second only to the province of Saxony in the number of its breeders (Evers 1924: 26).[19] In Württemberg by the late 1920s one-half of the state's certified and improved seed had been bred at its station (von Kittlitz 2005: 40). In Baden the proportion of the state's seed requirements that was being met by breeders there increased from only 3 per cent in 1921 to 33 per cent four years later (*Ber. Bad. Ldw-Kammer* 1925: 171; cf. Anon. 'Pflanzenzüchterkurs' 1925; Lieber 1932). As the GFP recognized, the consequence of this public-sector activity was to restrict the markets available to the large commercial breeders of central and north Germany.[20] Though occasionally trying to pretend that public-sector breeding was insignificant (Kühle 1925), by the late 1920s the GFP's chairman was conceding that his members had effectively lost the south German seed market (Kühle 1926; cf. von Rümker 1937: 248). And it was precisely this success that was to make trouble for the stations during the 1920s.

3 The agricultural crisis of the 1920s

Wartime shortages and destruction left agricultural production much reduced in 1918; pre-war levels were not to be achieved for nearly a decade. With demand for food high and imports restricted, commodity prices rose quickly. Runaway inflation over the next few years, however, at least allowed farmers to pay off pre-war debts. With currency reform at the end of 1923, the government encouraged farmers to take out short-term loans, albeit at quite high interest rates, in order to invest in intensive agriculture and satisfy the demand for food. With a stable currency, however, Germany again became an attractive export target for overseas growers, and by 1925 the influx of foodstuffs was driving prices down again. Farmers sought to make up the shortfall in income by increasing production yet further, but this only lowered prices faster. Government attempted to stem the price fall by reintroducing tariffs in 1925 (and continually increasing them in subsequent years), but the prices of grain and animal products continued to fall. With ever less income, farmers were increasingly unable to meet interest payments on their debts, leading to a several-fold increase in foreclosures (Schreiner 1974; Henning 1978; Gessner 1981; Holt 1936). By the time the global depression arrived in 1929, therefore, a farm crisis in Germany was already well under way.

With their finances in a precarious state, most farmers were in no position to pay the premium for commercial seed (Ackermann 1924). In 1925, for example, the seed of original varieties of cereals cost 50 per cent more than ordinary grain (von Rümker 1925a), and by the late 1920s some breeders were charging 100 per cent more. As a result, even farms using intensive cultivation methods were increasingly abandoning original varieties in favour of derivatives or even reusing seed from their own harvests (*Schlipfs Handbuch* 1922: 144; Kappen and Henkelmann 1932). By then many peasant farmers could not even afford certified derivatives.[21] As a result, the sale of commercial seed plummeted, and seed prices were forced down.[22] GFP members complained that with such low prices they could no longer cover the costs of breeding (Stadelmann 1924; Bohne 1924; von Rümker 1925a). And cereal breeders left with large stocks of unsold seed ended up having to sell it as ordinary grain (*Jb Bayern 1922–1926*: 65).

Unfortunately this drastic decline in sales soon led to additional problems. Since the demand for original varieties had collapsed, seed producers began to develop derivatives and other cheap alternatives (Kiessling 1922b). Unscrupulous growers, however, simply bought good quality commercial seed, multiplied it, and sold it under a new name, though it was largely indistinguishable from the original variety.[23] Such ‘plagiarizing breeders’ (or ‘pseudo-breeders’), as they were called, could set the price below that of the original variety since, unlike the original breeder, they did not incur any breeding costs. By the 1920s the seed market was awash with such ‘synonymous varieties’. Of more than 1000 varieties of potato on the market about 1930, for example, only 130 of them were distinct varieties; the rest were copies.[24] From the mid 1920s, therefore, an attempt was made to stem the flood by setting up varietal-sorting commissions (*Sortenregisterkommissionen*) for potatoes and cereals which tested those varieties on the market in order to establish which ones were distinct and then published their results (Appel *et al.* 1932). Although this put pressure on the plagiarizers, however, the law offered no means by which to take them to court.

Faced with the deteriorating market position of its members, the GFP frequently approached the Reich and Prussian governments for support.[25] From 1925 onwards breeders received assistance in the form of mortgages, personal loans, a subsidy to help cover the costs of certification, and a reduced interest rate on loans to farmers who wanted to buy certified seed. One million copies of a flyer were also published, urging farmers to use original seed varieties (Kühle 1926; von Rümker 1928a).[26] The aid received was substantial,[27] prompting a series of seed-breeding and seed-growing cooperatives to complain bitterly that the GFP was receiving a disproportionate amount of government subsidy while their own members' requests remained unfulfilled.[28] In the view of one prominent GFP member, without Reich government support in 1926, most large breeders would have gone bankrupt (Ackermann 1928). Despite the scale of this assistance, the GFP – wanting to have it both ways – did not hesitate to defend the 'free market' whenever it felt its autonomy was under threat (a theme to which I return in the concluding section).[29]

But the GFP's large and established breeders were also dependent upon the public purse in another respect. Given falling income from the sale of their varieties, some breeders sought to reduce the costs of developing new varieties by outsourcing this work to public-sector institutes. These institutes clarified the inheritance patterns for economically important traits, maintained large collections of crop-plant varieties which could provide the parent lines for crosses, and carried out the laborious and expensive process of hybridization. Sometimes they turned over unfinished material to the firm to complete the breeding, but in two cases known to me institutes sold finished varieties to private breeders. One of these was Theodor Roemer's Institute for Agronomy and Plant breeding at the University of Halle which between 1926 and 1933 sold a dozen new cereals varieties to various well-known breeders (Nathusius 1955: 31–32).

The other was an institute directed by Erwin Baur who had enjoyed a close relationship with the GFP right from the start. He was an advisory member of the Society from 1912 and one of only two scientists to be members of the GFP's executive committee (*Hauptvorstand*) in 1929.[30] In addition, as we saw, the GFP's lobbying had been important in the establishment of Baur's institute at the Berlin Agricultural College (Harwood 2002 and forthcoming). After the war, disillusioned with the Prussian government's failure to provide better facilities for practical breeding work at the College, Baur moved in 1928 to a large new institute created for him by the Kaiser-Wilhelm Society: The Kaiser-Wilhelm Institute for Breeding Research (KWIB) outside Berlin. Kaiser-Wilhelm institutes were partly funded by private donors, and in this case the GFP was one of the largest donors (von Rümker 1928a). The GFP published KWIB's house- journal, *Der Züchter* (edited by Baur), and the GFP's large collection of cereals varieties was handed over to the Institute to maintain.[31] The establishment of yet another public-sector breeding institute in the late 1920s might have been a worrying prospect for commercial breeders, but Baur's view of the appropriate relationship between public- and private-sector breeding was quite different from that which prevailed at the south German stations. During discussions in 1927 surrounding the establishment of the KWIB, for example, breeders were assured that the Institute would not become a competitor. It would instead study the genetic foundations of breeding and develop methods for the benefit of breeders. If practical results of economic promise should arise, these would be sold to breeders for finishing and sale (Baur 1933).[32] In the event, the KWIB's organizational structure allowed the GFP to exert a certain amount of influence over the Institute's activities. The chairman of the GFP, Ludwig Kühle (see Figure 4.1), was a member of the Institute's management committee (*Kuratorium)* in 1930 (along with half a dozen other central German breeders) and was also chair of the committee that arranged for the sale of KWIB varieties to private breeders (*Verkaufskommission*).[33]

These ties to Baur's and Roemer's institutes were important. As two of Baur's students later testified, large commercial breeders could not have survived between the wars without support from public-sector research and development (Kuckuck 1951: 25; Stubbe 1982: 13). Despite close ties to these particular institutes, however, the breeders' relationships with the south German stations and other peasant-friendly breeding institutions were very difficult.

4 Interwar conflict between private and public sectors

Pre-war doubts about the propriety of public-sector institutions engaging in varietal development had been largely objections in principle since the stations had yet to make much impact upon the seed market. By the 1920s, however, the situation had changed radically. The critics' earlier suspicions now looked all too justified, and the controversy grew in intensity. Although I focus in this section upon the conflict between the GFP and the south German stations, it is important to recognize that this clash was part of a more general tension during the interwar period between private- and public-sector agricultural organizations. The GFP was not the only trade association to complain of public-sector competition,[34] nor were the stations the only target for the GFP's wrath. In 1926, for example, the GFP filed a complaint with several state agricultural agencies that some chambers of agriculture were undertaking too much varietal development and that this was undermining the position of commercial breeding.[35] In short, the economic crisis brought about an inevitable confrontation between state (and cooperative) institutions committed to public service and commercial organizations in search of profit.

To be sure, during the 1920s the stations continued to have admirers (e.g. Bischoff 1922). One of these reiterated the now familiar warning about the stations competing with the private sector, but he felt nonetheless that such institutions had an important part to play even in north Germany (Evers 1924). Small farmers in the north, he argued, had not had much benefit from the private sector. Thus stations could make a contribution by supporting regional seed-breeding cooperatives and developing varieties in areas neglected by private breeders. Nevertheless, given the dire economic state of commercial plant breeding by the mid 1920s, it was to be expected that private-sector attacks upon the stations would increase.[36]

In 1926 the chairman of the GFP led the way. Seemingly forgetting the enormous diversity of agro-ecological conditions across Germany, he insisted that it was irrelevant where varieties were bred. Seed should be produced where the economic and natural conditions were favourable for producing varieties which would do well everywhere. There was thus no point in each state feeling it had to have its own breeding institution and as many of its own varieties as possible. (Here was a clear statement of the 'cosmopolitan' breeding strategy described in Chapter 2.) Nor was it fair for particular regions to exclude varieties bred elsewhere (Kühle 1926). In order to put across the message more forcefully, the GFP launched a formal campaign, distributing a printed petition to interested parties.[37] Two years later another GFP officer again voiced the allegation (Bohne 1928). While conceding (unlike his predecessor) that there was no single variety which could excel in every region, he nonetheless objected to the large numbers of varieties on the market. Every locality, he complained, seemed to want its own variety, and this had more to do with politics than with plant breeding.[38]

The GFP's campaign found support. Once again academics with close ties to the private sector warned of the dangers of the public sector competing with commercial breeders (von Rümker 1929). And despite Theodor Roemer's own very considerable success in developing new varieties which were eagerly snapped up by commercial breeders, even he found some way to claim in a

Figure 4.1 Ludwig Kühle (1869–?), chairman of the Commercial Breeders' Association (source: Gesellschaft zur Förderung deutscher Pflanzenzucht, 1933).

lecture that 'as in all areas of the economy', public-sector breeding could not be as efficient as that in the private sector.[39] Moreover the GFP could count on substantial support in government circles. At the Association's conference in 1927, for example, a civil servant from the Reich Ministry of Agriculture expressed his conviction that German plant breeding was best conducted in the private sector (Anon. 'Eröffnungsansprache' 1927). And the following year the Reich government set up a committee of inquiry to look into the conditions in which the struggling German economy was having to operate. One of its agricultural sub-committees, responsible for examining the state of German plant breeding, was heavily dominated by GFP members who unsurprisingly took the opportunity to draw attention to the alleged evils of public-sector institutions.[40]

One of the GFP's policies, however, provoked an outcry from several sections of the agricultural community. Since sales of their original varieties were down while the demand for derivative varieties had increased, a committee of the GFP began in 1926 to draft proposals for the introduction of licence fees.[41] This was a fee to be paid to the breeder by any farmer who multiplied one of the breeder's original varieties (i.e. grew it up for sale as seed) and sold it as a derivative under the original's name. In 1928 the GFP duly introduced such fees for cereals varieties bred by its members. The policy did not go down well in south German circles. In Baden, for example, the chamber of agriculture expressed doubts as to whether the GFP's actions were in the interests of south Germany's small farmers or indeed of the small breeders there who relied upon the chamber's assistance. As the head of the chamber remarked, there was resentment at the financial resources possessed by large north German breeders as well as at their skilful politicking. Their actions, he argued, were not in the long-term interests of German agriculture and would hurt peasant breeders.[42]

But there was more to come. Despite the earlier objections, in 1929 the GFP decided to extend licence fees to its members' fodder-plant varieties, prompting one seed-growing cooperative to declare that 'the economic "peace" which was recently agreed between us is now once again suspended'.[43] Believing, however, that fees were a 'healthy and natural' development, the GFP carried on, recommending to the Reich government that licence fees should be incorporated into a bill on breeders' rights which the government was currently drafting.[44] This prospect that licence fees might henceforth become enshrined in law, adding an additional financial burden to already hard-pushed multipliers and smallholders, unleashed a vehement controversy which was reported in both the agricultural and the daily press. Once again the protest came from south Germany, including the Bavarian Agriculture Minister, a Bavarian peasants' association and the Bavarian chamber of agriculture. But governments from other regions where peasant farming predominated also registered objections.[45] By the end of the decade, the GFP was acknowledging that its own interests were in conflict with those of 'consumers' (i.e. farmers).[46] But, they insisted, there could be no justification for taking 'so-called consumer interests' into account in the provisions of the bill on breeders' rights since the 'true interests' of consumers surely lay in maintaining the high standards of German plant breeding.[47]

By this time, however, the GFP was in trouble not just with small farmers. As early as 1925 some observers were reckoning that the GFP was trying to make life as difficult as possible for small breeders (von Rathlef 1925). And it soon became clear that, despite its claims to the contrary, the GFP did not speak for all breeders. There were still, for example, few breeding cooperatives in the GFP.[48] And in 1926 and 1927 a number of them complained to the Prussian and Reich ministries of agriculture that the measures for the protection of intellectual property that the GFP then advocated were not in their interests.[49] Similarly the director of the Bavarian station – with the support of the Bavarian Agriculture Ministry – pointed out that the draft bill on breeders' rights wrongly portrayed the GFP as though it represented all breeders.[50] Indeed, not just cooperatives but many individual breeders were also unhappy with the GFP's policy on licence fees. Over 100 of them who were *not* members of the GFP objected to the policy, and even several GFP members resigned from the Association in protest.[51] By 1932 the volume of criticism coming from breeders themselves, inside and outside the GFP, had grown so loud that it could no longer be ignored. At a meeting of the GFP the chairman felt obliged to declare that the Association was *not* dominated by large breeders, even if its voting rules might seem to suggest this (since they granted more votes to those who paid higher fees) (Kühle 1933a).

By the end of the decade, therefore, the GFP's attempts to rescue its large and established breeder-members from economic decline were running up against considerable opposition, especially in south Germany. Smallholders there resented the effect that licence fees would have upon the cost of derivative varieties, and breeding cooperatives and small individual breeders felt that GFP policies were tailored, not to their needs but to those of the big players. Given its peasant-friendly orientation and its success on the seed market, the Bavarian station found itself at the very centre of the dispute.

5 The clash of interests in Bavaria

The conflict between private- and public-sector breeding was especially marked in Bavaria. And it is here that the different interests of small and large breeders were particularly visible. The latter often referred to themselves during the 1920s as 'independent' breeders, by which they meant that they employed their own trained plant breeder to carry out the breeding process on their (usually large) farms and did not require technical assistance from the station. In 1921, for example, of the 112 breeding centres with links to the station, 10 employed their own breeders (who had often been trained primarily at the station) (*Jb Bayern 1919–1921*: 97). As we shall see, however, such 'independence' was a recently acquired status. For most of those declaring themselves to be 'independents' had in fact relied upon the station's assistance before the First World War when they originally took up breeding.[52]

During the late 1920s a number of independents began to bite the hand that had fed them. From 1928, for example, the Bavarian station came in for criticism for its work in grass breeding. The owner of a Bavarian grass-breeding firm, August von Schmieder,[53] wrote to the Bavarian Ministry of Agriculture, drawing attention to the harmful effects that Weihenstephan's 'commercial activity' was having upon the seed market. The station, he complained, was turning over its own improved grass varieties at little or no cost to farmers who multiplied and sold them at prices below those demanded by the independent breeders such as himself. While it was right that the station should continue its grass-breeding work, he argued, its new varieties ought to be sold at 'reasonable' prices so as not to drag down the market.[54]

The minister forwarded Schmieder's complaint to the director of the station, Theodor Scharnagel, who replied – not a little irritated – at length (in a 16-page letter plus a large number of enclosures), outlining the history of the station's relations with Schmieder. Between 1919 and 1926, he emphasized, the demand for domestic grass varieties was high since Germany no longer possessed enough foreign currency to buy seed from abroad. In Bavaria, however, there were no commercial grass breeders, so the station began to promote the growing of grass seed via the new Bavarian Society for the Promotion of Grassland Agriculture (of which Schmieder was chair). The station developed improved grass varieties and handed them over at low cost to members of the Society who multiplied and sold them as 'Weihenstephan' varieties. By 1927 the station had placed 16 grass and other fodder-plant varieties on the market in this way (Kiessling and Gassner 1927). The aim was to keep the multipliers' costs down; the prices they charged would then be correspondingly low, and Bavarian farmers would acquire good quality seed at reasonable cost.[55] Between 1919 and 1924 Schmieder himself was one of the beneficiaries of this arrangement. In 1923, however, he hired a

Figure 4.2 Experimental fields of the Bavarian commercial grass breeder Dr Karl August von Schmieder (source: Gesellschaft zur Förderung deutscher Pflanzenzucht, 1933).

breeding director – a man who had been trained for several years at the station – and began to do his own breeding in-house (see Figure 4.2). By 1926 this work had advanced far enough that the station could certify Schmieder's new varieties which were then sold under his own name. Such unusually rapid success, Scharnagel insisted, was possible only because of the station's support up to 1924. Until 1928 Schmieder had voiced no objections to the station's involvement in grass breeding, but in that year, unable to sell all of his seed at the price he had set, he became concerned about the competition. By then the station had already advised him on several occasions that his prices were substantially higher than those of other Bavarian breeders, largely because his breeding costs were too high. Other breeders, the station maintained, had managed to reduce their breeding costs in accord with the depressed seed market, and Schmieder ought to do the same. As of 1928 no other grass breeder had complained about the station's role. That it should be Schmieder, of all people, who should object, Scharnagel observed, was ironic, for he was the breeder who had probably benefited most from Weihenstephan's earlier assistance.[56] Although Scharnagel's reply would seem to have undermined Schmieder's complaint, the latter soon gained a partial victory: the station decided to stop doing any multiplication of its own grass varieties 'and is thus no longer a competitor'.[57] A year later Schmieder gained satisfaction yet again when a ministerial decree instructed the station to stop turning over its grass varieties to farmers for multiplication and sale.[58]

Similar criticisms were raised two years later, this time directed at the station's work in cereals breeding. In a public letter to the Bavarian ministry – distributed simultaneously to party leaders in the Bavarian parliament as well as to several agricultural organizations – eight Bavarian breeders (Schmieder among them) complained that the station was competing with the independent breeders.[59] Lead signatory (and the moving force behind the petition, according to top civil servants in the Ministry) was Jakob Ackermann (1873–1938) (see Figure 4.3), a successful cereals breeder, founding member of the GFP, and generally regarded during the 1920s as the leader of Bavaria's estate owners.[60] The station's real aim, Ackermann and co. asserted, was to gain a monopoly over breeding in Bavaria. It was clear to the signatories, however, that a healthy private sector was in the interests of the agricultural community as a whole. The solution, therefore, was for Weihenstephan to abandon varietal development and concentrate upon 'scientific research' relevant to breeding.

As before, the Minister solicited a response from the station's director who reiterated some of the same points from his letter two years previously and rejected a series of 'false allegations'. All of the signatories but one, he observed, were members of a small cartel formed by independent Bavarian breeders in 1927 (the *Interessengemeinschaft Pflanzenzucht*). The station had been objecting to the cartel's actions for several years, not least because its members often set their prices for a given crop at twice the going market price (or more). This had not helped the uptake of improved commercial varieties among Bavaria's peasant farmers. That these independent breeders should now demand that the station stop providing assistance to others was 'grotesque', seeing that every

Figure 4.3 Dr Jakob Ackermann (1873–1938), leading Bavarian commercial breeder (source: Kiessling 1938).

single one of them had had a great deal of help from the station in the past (and in some cases this assistance continued to the present).[61] In addition, as civil servants in the Ministry pointed out, four of the signatories (including Ackermann) had received substantial subsidies from the Ministry in recent years, 'without which they would have gone under'.[62] Annoyed that Ackermann and co. had gone public without taking up the matter first with the government, the Ministry demanded that they substantiate their allegations, and the dispute dragged on over the next six months.[63]

By that time it was long since clear, as one GFP member put it, that the station 'doesn't have the same interests as commercial breeders'.[64] The station's director agreed. What the independent breeders object to, he wrote, is that the station was helping smaller breeders. But this kind of assistance was essential so long as Ackermann and his colleagues pursued price-setting policies which harmed the interests of the farming community as a whole. 'The station ...

cannot serve just the interests of a few private breeders but must instead act on behalf of the general welfare.'[65] In June 1932, presumably hoping to end the dispute, the Ministry finally replied in full to Ackermann and co., voicing the wish that a 'relationship of trust can again be established between the station and *all* Bavarian breeders.'[66]

But trust was not to be secured so easily. What went on over the next six months is not clear, but in January 1933 one of the independent breeders (who had not signed the 1931 petition) wrote to the Ministry, expressing concern at the deterioration of relations between the station and the independents and suggesting a meeting of the station's director with some of the breeders, chaired by a civil servant who enjoyed most participants' trust.[67] It was necessary, he felt, not just to calm things down but also to sort out a viable *future* relationship between the two sectors. The meeting duly took place a week later, and after a few initial skirmishes between the protagonists, the chairman managed to secure an agreement in principle over how in future public-sector breeding material would be handed over to private breeders. The process was to be overseen by a committee, consisting of the station director, a representative of *all* Bavaria's breeders (rather than just those in the small cartel), and a representative of those Bavarian small farmers who had received technical assistance from the public sector.[68] (The make-up of the committee was thus clearly stacked in favour of the station and its clientele.) This committee would propose a suitable recipient for a given variety, based on the latter's competence as a breeder as well as his or her ability to serve the general interest, and would agree appropriate compensation for the station. Final approval would lie with the Ministry. While the recipient would be entitled to name the new variety, the station would retain the right to make sure that further breeding and multiplication were being done correctly (where it was not, the station could reclaim the variety and agree a new recipient with the committee).[69]

After five years of angry exchanges, therefore, it looked as though a modus vivendi had been achieved. The state was to serve as referee in an attempt to mediate between the public interest and that of the private sector and within the latter, between the interests of large versus small breeders. Unfortunately, as we shall see in the next chapter, this arrangement didn't last long.

Conclusion

The south German stations' success in varietal development unleashed considerable controversy, for that success impinged upon others' interests. But what are the wider implications of this episode in the history of the relations between private and public sectors? One issue is the way in which we conceptualize the state's changing role in agriculture since the mid nineteenth century. It is sometimes suggested, for example, that apart from trade policy, most nineteenth century state agricultural policy was designed to assist farmers to help themselves. Only in the First World War, according to this view, did states intervene directly in the agricultural economy (as they continue to do in both the US and the EU).[70] The south German stations, however, do not fit easily into this framework. To be sure, they did foster self-help by providing information on suitable varieties and by teaching local farmers how to do plant breeding, and they were not supposed to engage in 'commercial' activity. But the aim behind the stations nevertheless had definite implications for the seed market: they were, after all, intended to free small farmers from dependence upon commercial plant breeding and to foster the growth of small breeding enterprises in their region. Accordingly, the stations encountered private-sector reservations from the start. And what began around 1900 as a relatively uncontroversial attempt to help 'little players' in the agricultural economy was proving so effective by the 1920s that the big ones began to feel the pinch. Though seemingly innocuous at first, the state's promotion of self-help eventually turned out to be market intervention in disguise. It might be worth asking, therefore, whether other forms of state policy before 1914 followed a similar historical trajectory.

The history of this conflict also helps us to think about the diversity of roles that public-sector research and development has played (and can potentially play). One of these roles is that of *complementarity* with the private sector. That is, despite the various strengths of private enterprise, not all areas of the market are profitable for private firms. If 'market failure' is deemed to be serious, the public sector may be charged with filling the gap. As we have seen, this was one of the roles assigned to peasant-friendly stations in south Germany and elsewhere. The same is true today in varietal development in the developing world (cf. Chapter 7); in Brazil and China, for example, the public sector concentrates on food staples rather than on export crops (Fukuda-Parr 2007a).

Second, even where private-sector provision for a particular market does exist, firms are still heavily dependent upon the public sector for a variety of support services. As we saw, the emergence of commercial breeding in south Germany owed its very existence to the technical assistance provided by the region's stations. Furthermore during the 1920s the survival of several large breeding firms in central Germany was only possible thanks to the work undertaken at public-sector institutes in Halle and Berlin. This relation of *dependency* is still much in evidence. Firms rely upon the public sector for research in plant genetics and breeding method, the training of breeders, the maintenance of gene banks, genetic enhancement and (in some countries) varietal testing and certification (Strosnider 1984; Frey 1992; Heisey *et al.* 2001; Franck 2007).[71]

That said, the nature of this dependency depends upon firm size. As is well known, small industrial firms with limited in-house research capacity tend to support applied research in the public sector because they are heavily dependent upon it as a source of innovation. Large corporations, by contrast, who can afford their own in-house R&D laboratories tend to argue that the public sector should steer clear of applied work – not least because this helps them to eliminate competition from 'small fry' – and concentrate instead upon fundamental research (which firms often find useful over the medium or longer term but do not wish to pay for). As we have seen, this is just what happened in Germany where the large breeders dominating the GFP called upon the state to restrict the stations to 'scientific work' relevant to breeding and leave varietal development to the private sector. This same attempt to reshape the public sector has been happening in other countries, especially since the 1980s (cf. Chapter 8). By contrast, in Germany where commercial plant breeding today is organized primarily in small- and medium-sized firms (often in family ownership), one finds strong support for the continuation of public-sector applied research and related activities which are regarded as crucial for these firms' ability to compete with giant multinational corporations. Such support was much in evidence in Baden-Württemberg during the 1970s and 1980s (Kittlitz 2005: 101ff.), and it continues up to the present (Stefanie Franck, pers. comm.).

Arguments against varietal development in the public sector have usually been justified in terms of a 'division of labour'. One well-known breeder, for example, has described public- and private-sector plant breeders as constituting a 'team' whose functions are complementary (Duvick 2002). As we have just seen, however, the relations within the 'team' have often been anything but harmonious. The third form that the relationship between sectors has taken, therefore, is *competition*. And despite Duvick's claim, there is a lot to be said for the public sector adopting this role. By competing with large commercial breeders, for example, the south German stations put pressure on these firms to keep prices low. The same thing has recently occurred in China where competition from biotechnology-based breeding in the public sector has successfully restrained the prices that multinational seed companies can charge. Furthermore, competition from the public sector does more than keep prices low; it also stimulates innovation. Evidence from the American pharmaceutical and seed industries, for example, suggests that as the market becomes dominated by a small number of firms, the amounts spent on research and development decline (Schimmelpfennig *et al.* 2004; Murphy 2007: 277–280). Apart from breeding their own varieties in-house, of course, the other way that the south German stations stimulated innovation was to offer technical support and provide breeding material to what would now be called start-up companies. This sector was deemed important since small firms were more likely to produce varieties better designed for the needs of local farmers than were the large commercial breeders of central Germany. A similar situation pertained in the US as recently as the 1980s when many small seed companies were still dependent on varieties developed in the public sector (Kloppenburg 1988: 216). And the same point is now being argued for the developing world (Morris and Ekasingh 2002). Without small seed companies, farmers will be forced to rely upon the handful of multinational corporations who currently dominate the seed market.[72]

Whenever the public sector competes in this way, of course, such activity is vigorously resisted on the grounds that the state is thereby interfering in the 'free market'. As the episode in this chapter indicates, however, this argument does not stand up to scrutiny. The GFP, for example, was highly selective in its attitude toward state 'interference'. Some forms of intervention were warmly welcomed (e.g. tariff protection, subsidies to firms in trouble) while others were condemned (public-sector varietal development in crops of commercial interest).

Another tack taken by large firms has been to say that while the blessings of the free market are generated by competition, competition from publicly funded institutions is 'unfair' since the latter do not need to make a profit in order to survive. This turns a blind eye, however, to the evidence just cited that public-sector support for small firms is important for innovation across the sector as a whole. Third, large companies sometimes object on the grounds that applied R&D (e.g. varietal development) represents 'wasteful duplication' since the private sector is allegedly better placed to carry out commercially relevant innovation efficiently (cf. Kloppenburg 1988: 109). As we have seen, however, this is by no means always the case. Public-sector breeding institutions in several countries have proved highly effective in developing commercially successful varieties. Indeed, it is for precisely that reason that they have been criticized so often. Arguments by large companies in favour of competition and the 'free market', therefore, are best understood as self-interested attempts to eliminate competitors – be they public-sector institutions or small firms – in order to dominate the market. Deciding upon the role that public-sector institutions should play is fundamentally a political matter. If one wants the public sector to stimulate the regional economy and serve the majority of farmers, however, competing with large firms is a legitimate and demonstrably effective way to do so.

So much for the 'lessons' of the interwar episode. As I noted above, the negotiated truce between private and public sectors that was agreed in Bavaria in January 1933 did not last long. What form, therefore, did this relationship take after January when a very different kind of government – not least, one that was strongly committed to the well-being of the peasantry – came to power in the Reich?

5 The fate of peasant-friendly breeding under National Socialism

At first the seizure of power by the National Socialists at the end of January 1933 made little difference to the relations between private and public sectors. In March, for example, Erwin Baur publicly called attention once again to the 'fierce competition' that private breeders faced from state institutions 'in some states' and called upon the latter to restrict themselves to the kinds of supportive R&D work which his own institute was conducting (Baur 1933). The GFP resumed lobbying for a law that would protect breeders' rights while its business manager complained to a colleague that although German plant breeding had been in continual decline since 1927, all measures, petitions and cries of alarm had achieved nothing at either Reich or state level.[1] Given the GFP's criticism of the south German stations and the new regime's declared intention to radically transform German society, it is perhaps not surprising that in April the GFP's chairman decided to send a congratulatory telegram to Hitler on his birthday, emphasizing the 'suffering' of large German breeders under the former government, and urging the new regime to restrict public-sector institutions to 'scientific studies' relevant to breeding.[2] By the autumn, however, there were signs that the handover arrangements agreed earlier that year might not be long-lived. As a Nazi farmers' union wrote to the Bavarian Agriculture Minister, the arrangements were 'no longer appropriate in the new regime'.[3] In fact, a new set of agricultural policies was then being devised which eventually opened up new opportunities while also imposing new constraints upon both public and private sectors.

On paper the prospects of the south German stations probably looked pretty good in 1933. For the Party had portrayed itself since 1930 as a strong ally of the peasantry (Franz 1963: 535ff.) and continued to do so after taking power. The general features of its agricultural policy were clear. On the one hand, price controls and other forms of state intervention in the market, along with the intensification of agricultural production, were intended to make Germany more self-sufficient in food. On the other, the biological and economic health of the peasantry – rather than of farmers more generally – were declared essential to national well-being, and various policies were designed to improve their position in that regard. Strengthened through both measures, the German economy would be better prepared for war (Kutz 1984; Tooze 2006). As we shall see, however, the Party's pro-peasant stance did not pay off for the south German stations. On the contrary; at the end of the war they emerged more subservient to the private sector than they had been in 1933.

In the first section I survey the impact of National Socialist (NS) agricultural policy in general upon peasant interests, arguing that although agricultural policy improved the peasants' economic and social status in the early years of the regime, a fundamental shift of policy from about 1936 led to a deterioration in their position. Against this backdrop the second section focuses upon plant breeding. I examine those policies designed to reform the seed market and conclude that they probably benefited farms of all sizes and that the agricultural leadership seems not to have made a particular effort to ensure that small farmers had access to high-quality seed. Turning in the third section to the regime's attempts to alter the balance of power between private- and public-sector breeding, I argue that many of the services that south German stations had been providing the region's small farmers were removed, despite opposition from public breeders. Finally, in the fourth section I outline the relations between the sectors after 1945.

1 The peasantry in National Socialist ideology and agricultural policy[4]

Although Hitler had declared in the 1920s that the maintenance of a healthy peasantry was the 'foundation of the nation' (cited in Corni and Gies 1994: 71), agrarian issues only began to play a prominent part in NS policy from about 1930 when R. Walther Darré joined the Party (see Figure 5.1). The Party programme in 1930, for example, provided for a settlement scheme aimed at the creation of new small farms, measures for debt relief, import duties, price supports, and measures to reduce the production costs of agriculture (Holt 1936: 185–187). More than simply favouring agricultural interests, however, the policy placed particular emphasis upon strengthening the peasantry.[5] Conversely, Darré was critical of large landowners. In a speech in 1934, for example, he remarked that if an East Elbian estate could survive on its own, that was OK though it would still be required to give up part of its land so that the state could establish new small farms. If an estate could not survive, the government would not aid it through subsidies (as had been done during the Weimar Republic) because that would be an irresponsible use of taxpayers' money (Corni and Gies 1994: 123–124; Farquharson 1976; Mai 2002: 68ff.). The low population density of the regions East of the Elbe, he believed, was a catastrophe because it deprived the Reich of a potential source of racial rejuvenation (Gies 2000).

One might wonder, of course, how much of this talk was mere electioneering, promises which the Party had little intention of keeping once they were in power. Given the structure of German landholding, however, if the Party was serious about increasing agricultural self-sufficiency, it would have to address the needs of peasant farmers because over 80 per cent of the acreage was in peasant hands. Moreover, support for the peasantry was predicated not just on the assumption

Figure 5.1 Richard Walther Darré (1895–1953), Reich Peasant Leader and Reich Agricultural Minister, 1933–1942 (courtesy: Bundesarchiv, Bild 119, Hauptarchiv der NSDAP).

of their racial superiority but on hard-headed economic calculation. Agricultural officials were aware, not only that yields on small farms were often greater than on estates, but also that peasant farmers were not averse to adopting new technology. As a result, officials reckoned the potential payoff from improving the performance of the weakest small farms was a good deal greater than boosting the performance of the weakest estates (Fensch 1936; cf. Lichter 1996). Accordingly, with its actual agricultural policy after 1933 the Party did attempt to strengthen the social and economic position of the peasantry, at least initially.

Over a two-year period following Darré's appointment as Agriculture Minister in July 1933, a deluge of legislation was introduced which had three basic aims (Backe 1943; Tornow 1972). First, in common with Nazi policies in other areas, centralized top-down control was imposed. All of the numerous groups and organizations concerned with food – producers, processors and distributors as well as agricultural associations, public-sector institutions for research and extension, and trade associations – were either dissolved or incorporated into a mammoth new bureaucracy headed by Darré: the corporatist 'Reich Food Estate' (*Reichsnährstand*, henceforth RNS) (Lovin 1969; Corni and Gies 1997; Farquharson 1976). The RNS was thus a vertically integrated organization which dealt with the entire range of agricultural commodities. Since it employed over 20,000 people at regional level and above, Darré boasted that it constituted the largest single economic unit in the world (Tooze 2006).

Second, in keeping with the Party's aim of achieving self-sufficiency, a 'battle for food production' (*Erzeugungsschlacht*) was declared in the autumn of 1934 (Lovin 1974; Farquharson 1976; Corni and Gies 1997). One tool by which production was to be increased was via economic inducements: the market was restructured through such mechanisms as guaranteed prices to farmers, reducing the profits to middle-men, and subsidized prices for commercial fertilizer and machinery (Corni and Gies 1997: 145ff.; Corni 1990: ch. 5). Another tool was to promote the intensification of production via the greater use of improved seed, fertilizer and mechanization (Meinberg 1937; Heim 2003: 91–102). To this end, agricultural research was generously funded but also much more tightly coordinated, namely through the creation of the *Forschungsdienst*, a system of seven committees (whose members were agricultural scientists from academe or other public-sector institutes) within the RNS under the direction of Konrad Meyer, an SS officer and professor of agricultural policy at the University of Berlin (Oberkrome 2009: ch. 4). The *Forschungsdienst*'s committees covered the major areas of the agricultural sciences, and each committee embraced several working groups, each dedicated to the solution of a particular problem deemed central to achieving autarky. The prominence of plant breeding within this campaign can be seen from the fact that of the agronomy committee's 24 working groups, about half were concerned with breeding plant varieties with increased levels of fat or protein or which were seen as prospective sources of fibrous material (Sessous 1936). More generally, agricultural scientists were constantly reminded that they did not inhabit an ivory tower and that 'science' was expected to serve 'practice' (which generally meant peasant farmers) (e.g. Backe 1936b; Meyer 1936).

Third, both the racial and economic regeneration of the German people were to be secured through legislation (the *Reichserbhofgesetz*) which protected the farms of competent and 'racially fit' peasant owners from the vagaries of the market-place in order to keep them in the family's possession. Such farms could not be sold or broken up through inheritance, nor could they be used as collateral for loans (so that they were protected from foreclosure in the event of debt) (Holt 1936: ch. 19; Corni and Gies 1994: 34–41, 103–128).

To what extent, however, did these and other policies *in fact* favour peasant farmers? Although this issue is rarely addressed in the existing secondary literature, one can nonetheless use that literature to construct a provisional answer. For a start, consider the differential impact of agricultural policy upon peasant farms versus estates. It is well established, for example, that Darré did not trust estate owners. He did not want them in the RNS, and in at least some regions it is demonstrable that the Nazi takeover of agricultural organizations led to the replacement of estate owners in leading positions by peasants (Schwartz 1990; Münkel 1996: 189). Nor, as we have seen, was he sympathetic with their problems. Accordingly, the protection provided by the *Reichserbhofgesetz* was limited to farms smaller than almost all estates (Corni and Gies 1994: 115, 121).

Second, there was the question of mechanization. Given the shortage of agricultural labour – exacerbated by rearmament and the growth of industrial employment after 1933 – Party officials were quite aware of the need to introduce machinery, but they were determined that it should also benefit small farms. Recognizing that this would not be easy, Darré urged that attention be given to how existing machines might be used more widely as well as to the possibilities of collective purchase, while Herbert Backe (see Figure 5.2) (Gerhard 2008) announced that one task of the Four Year Plan would be to provide credit facilities so that small farms could afford machinery (Darré 1938; Backe 1936a; Meyer 1934). Third, there is the fact that the costs of improved seed, fertilizer and machinery were kept low and that credit assistance was extended for the purchase of such inputs. Peasants will probably have gained something from these measures, but it is not clear whether they benefited differentially (Tornow 1972: 45; Corni 1990: ch. 11; Corni and Gies 1994: 178ff.; Farquharson 1976: 173–174). In any event, the overall consequence of such policies was that farmers as a whole were materially better off. Probably due in part to guaranteed prices, agricultural production of some crops had increased substantially by the late 1930s (Lovin 1974; Farquharson 1976: 178–179).[6] And since the cost of inputs was subsidized, the result was a sizeable increase in farmers' incomes from 1933 to the late 1930s (Petzina 1968: 95; Tooze 2006: 195; Lovin 1969; Farquharson 1976: 250).

This peasant heaven, however, was not to last. With the declaration of the Four Year Plan in 1936, a shift in agricultural policy is noticeable in which peasant farmers' favoured position deteriorated, and they came under increased pressure to boost production. As industrial growth picked up and unemployment receded, the demand for food increased, making it harder for domestic food production to stay ahead of agricultural imports. Agricultural output also suffered from the increase in industrial wages which lured workers off the land.[7] Perhaps for these reasons, in the Four Year Plan Hitler abandoned the goal of agricultural self-sufficiency, giving priority instead to preparing industry for war by 1940. Germany's conquest of territory to the East was thenceforth supposed to solve its food problem (cf. Haushofer 1958: 283–288). Moreover the appointment of Herbert Backe as the Plan's man in charge of food production reflected Party leaders' conclusion that Darré was too obsessed with maintaining the peasantry as a privileged stratum and too little concerned with the hard realities of boosting production in preparation for war (Corni and Gies 1997: 187). Henceforth the pressure upon peasants increased, partly through inducements but also through quotas and penalties. Faced with the rising cost of inputs (since industrial prices were not regulated) and food, farmers asked for commodity prices to be increased in order to maintain their incomes, but Goering (as head of the Four Year Plan) refused, wanting to keep down the cost of food to the consumer (Farquharson 1976: 173–174; Corni 1990: ch. 11; Corni and Gies 1994: 143–149). As a result, peasant incomes began to fall behind those of other sectors (Farquharson 1986; Corni 1990: ch. 11).

Figure 5.2 Herbert Backe (1896–1947), State Secretary in the Reich Agriculture Ministry from 1933, Food Commissioner for the Four Year Plan from 1936, Reich Minister of Agriculture from 1943 (courtesy: Bundesarchiv, Bild 183, Allgemeiner Deutscher Nachrichtendienst – Zentralbild).

As the pressure on peasants increased, that upon East Prussian estate owners eased. Earlier threats to break up such estates in order to create new peasant farms were rarely followed up since the Party was nervous about estate owners' open hostility to the settlement policy. Not only was Hitler much less committed to this policy than was Darré; estate owners also enjoyed the support of Goering plus numerous allies in both the military and the higher civil service. Nor did the regime promote settlement in East Prussia with the vigour they might have. Few new farms were established, partly because despite subsidies for building costs and machinery, peasants still had to pay a 10 per cent deposit on the land, and officials recognized that for many this was prohibitively expensive (Corni 1990: ch. 6; Corni and Gies 1994: 40; Farquharson 1976: 150–151; Gies 2000).[8] By 1936 there are indications that agricultural officials in the Four Year Plan organization had concluded that earlier dismissals of the productivity of large estates had been short-sighted. If the armaments industry's demand for labour was to be met while maintaining agricultural production, the only realistic way forward was by mechanization of the largest farms.

For the first three or four years of the regime, therefore, peasant farmers were clearly favoured by agricultural policy, though such privilege declined rapidly in the run-up to war. Against this general backdrop, let us turn to Nazi policy toward plant breeding, looking first at the regime's plans for the seed market.

2 Intervening in the seed market: the Seed Decree of 1934

Like other areas of Nazi legislation, the measures that were taken to reform the seed market – all of them centrally overseen and administered by branches of the RNS – grew out of movements which were already under way during the Weimar Republic (von Dietze 1934). What was new was the introduction of nationwide standards plus compulsion. As we saw in the last chapter, for example, a limited attempt was made during the 1920s to counter the proliferation of cereals and potato varieties then available on the market by publishing lists that indicated which varieties were indistinguishable from one another. Although these results allowed the farmer to make more informed choices, however, they could not of themselves reduce the chaos on the market. When the Nazis came to power, therefore, one of the measures introduced by the Seed Decree of March 1934 (Anon. 'Verordnung' 1934; Spennemann n.d.; Milatz 1937) was to set up a single compulsory varietal registering system which extended to all crops and whose published results made clear which varieties were merely duplications of well-known original varieties.

Another problem farmers had faced during the 1920s was that since there was only limited guidance available as to the agronomic properties of varieties on the market, there was usually no rational way of choosing between them. The Seed Decree set up a nationwide varietal-testing programme whose results determined which were the best varieties for particular growing areas. Only the best varieties of each crop – and in the case of largely indistinguishable ones, only the oldest of them – were allowed on the List of Approved Varieties (*Reichssortenliste*). When new improved varieties were identified, older ones of the same type were struck off the List. The List was published, and only the varieties on it could be legally sold. The result was a huge reduction in the number of varieties on the market.[9]

The third measure introduced by the Seed Decree concerned certification. During the 1920s certification had been conducted by a range of different voluntary and state organizations across the country using somewhat different criteria. Although the results had been published, a great deal of non-certified seed continued to be sold since it was cheaper. The Seed Decree established a standardized certifying system and forbade the sale of uncertified seed (at least until supplies of certified seed had run out). And in order to encourage the purchase of certified seed by as many farmers as possible, the Decree also fixed seed prices at the lowest level possible while still allowing breeders and seed merchants to make profits deemed reasonable.

Who, then, were the beneficiaries of this new system? One group which undoubtedly gained were commercial breeders. To be sure, the GFP's own power as a trade organization was curtailed. In early 1934, for example, it was forced to dissolve itself and was replaced by a new organization (*Reichsverband deutscher Pflanzenzüchter*) which was much larger than the GFP, was no longer permitted to have public-sector breeders as associate members, and was chaired by a pro-Nazi agricultural scientist from the RNS. Moreover, declaring that 'large breeders' would no longer have disproportionate influence, the RNS insisted upon equal votes for all members.[10] And in an effort to get high-quality varieties to even the smallest farmers, the RNS kept seed prices lower than some GFP members liked.[11] But there were definite compensations. The Seed Decree, for example, enlarged the market for commercial varieties and guaranteed the breeder a profit margin by allowing the sale only of certified seed, displaying the best commercial varieties prominently on the List of Approved Varieties, and fixing seed prices. But it also protected the breeder's intellectual property through two mechanisms: by creating a nationwide system which would weed out cheap imitations of original varieties and by enshrining in law the GFP's controversial practice of levying licence fees upon any farmer who sold a derivative variety under the original's name (Spennemann n.d.; Snell 1939). The considerable advantages for breeders must have been obvious since an official quickly made a point of dismissing claims that the Decree effectively granted breeders a 'monopoly' (Wick 1934).[12]

But what was the impact of these measures upon peasant farmers? As far as the varietal-registering system and varietal testing are concerned, they presumably made decisions on seed choice much easier for farms of all sizes, large and small. And since the List of Approved Varieties contained only the best varieties, farmers (of all sizes) probably gained. Unsurprisingly, the Association of Commercial Breeders declared that their goal was to sell certified seed 'even to the smallest farmer' (Anon. *Ratgeber* 1938a). And agricultural officials were keen to point out that the acreages planted in improved seed (as well as yields) had increased greatly within just a few years (Brummenbaum 1936; Tornow 1972: 14). On the other hand, when one looks closely at the acreage figures, it emerges that despite very large *relative* increases (e.g. eightfold in cereals between 1934 and 1938), the absolute acreage planted in improved varieties by 1938 was still only about 1 per cent of the total cereals acreage.[13]

If we want to know which farmers were reaping the advantages from high-quality seed, it is also important to note how the List of Approved Varieties was distributed. One year's List, for example, was not available in bookshops but had to be requested from the local and regional branches of the RNS, and the print-run for another was only 110,000 (Anon. *Sortenliste* 1940b: 4; Anon. *Ratgeber* 1938a: 8). Since this figure amounted to only 2 per cent of the number of German farmers, it is far from clear whether the information in the List reached most peasant farmers. One official conceded that 'a large proportion of the peasantry are still not using improved varieties' (Kaul 1938: 811). Although he attributed this to ignorance, others suspected that peasant farmers could not afford such seed. While emphasizing that the premium paid for improved seed had been reduced by half since 1930, for example, one spokesman acknowledged that there were still complaints that good seed was too expensive (Spennemann n.d.: 99–104). In an attempt to ameliorate this problem, a special category of farms was created in 1936 (*Vermehrungsstellen mit Eigenverkauf*) which were authorized to multiply certified seed and sell it directly to local farmers at relatively low prices (since there was no middleman and transport costs were minimal). Not surprisingly, this kind of farm was especially common in south Germany (Anon. *Ratgeber* 1938a: 9; Spennemann n.d.: 52–54).[14] Unfortunately we do not know what proportion of small farmers in fact took advantage of this opportunity, but it seems safe to assume that if the proportion had been high, agricultural officials would have broadcast it as a victory for the new order. But the fact that the regulations explicitly allowed farmers to reuse seed from the previous harvest (Buchinger 1935) – whether or not it was certified – suggests that officials may have feared that even the cheapest certified seed would still not be affordable for many.

A more affordable category of seed, of course, were local varieties. Significantly, however, the certification system placed obstacles in the way of those who wished to use these varieties. It specified that only local varieties of *clover and grasses* could be certified on the grounds that there were as yet few good improved varieties of these crops available (Spennemann n.d.: 41–42). And even in these cases such varieties were approved only for planting in their region of origin, thus turning a blind eye to the fact that local varieties had often been shown to do well outside their region of origin, so long as the soil and climate conditions were comparable. By contrast, local varieties of cereals were generally excluded from certification altogether since it was thought that in these crops there was no shortage of improved varieties available. This ruling, however, conveniently ignored the fact that some local varieties of cereals were highly valued for their qualities (baking, brewing) as well as for the fact that they guaranteed greater consistency of yield – a feature especially important to small farmers – than did commercial varieties. Overall, then, it is difficult to avoid the conclusion that while the Seed Decree paid off nicely for commercial breeders, it was not well tailored to the needs of most peasant farmers.

More particularly, in view of the well-known weaknesses of some highly centralized administrative systems, one can ask to what extent the RNS's Berlin-based decision-making was able to take into account the diverse ecological and economic conditions of south German farms. Interestingly, evidence from Bavaria suggests that when experiment stations raised limited and specific objections to the RNS's across-the-board procedures, the RNS was capable of adjusting. For example, although the Seed Decree specified that local varieties could be certified only in the case of clover and grasses, the RNS agreed in 1934 to a request from the Bavarian station and admitted a local cereal variety to the certification process.[15] And in another case when the RNS ruled that a particular Bavarian-bred rye variety should be removed from the List of Approved Varieties, an experiment station appealed successfully against the decision on the grounds that the removal would harm the economic viability of a large number of moorland farms.[16] On the other hand, where objections of a more fundamental kind were raised, the RNS proved less flexible. For example, the Bavarian station's director complained to the Bavarian Farmers' Leader (*Landesbauernführer*) that the centralized character of the certification process was out of line with basic principles of agroecology. Because the process presupposed that the value of a new variety could be adequately assessed through testing it in a small number of regions, the system was incapable of recognizing varieties that were well adapted to the soil and climate of *other* regions not included in the tests. Moreover, because new varieties were generally tested under favourable growing conditions, the tests failed to spot those varieties that did well in poor conditions. This meant that the system was biased against the needs of small farmers in regions like south Germany where both conditions applied.[17] The objection seems to have been ignored.

But apart from these examples, there are few signs of criticism of the RNS's Seed Decree from public-sector breeders. Indeed, even some breeders whose careers had suffered under National Socialism were able after the war to acknowledge the value of various Nazi measures (Kuckuck 1951: 22, 73). The reason is probably that for the most part the measures introduced had already been called for during the Weimar Republic by prominent figures in the agricultural community (e.g. Roemer 1932). Moreover, other countries were moving in the same direction. A national system of certification and a national list of approved varieties had already been established in France in 1922 (Bonneuil and Thomas 2009: 60–61), and public-sector breeders in the Netherlands and France admired features of the Seed Decree (Sirks 1936, vol. 2; Bonneuil and Thomas 2010). As we shall see below, however, not all of the RNS plans for the seed market were so well received.

3 Redefining the role of public-sector breeding

Given the stations' impact upon the interwar seed market, their support for small farmers and cooperatives who wished to take up breeding, and their commitment to increasing the use of improved varieties among smallholders, one might have expected them to be held up as a model by the NS regime. The RNS's journal of agricultural science (*Der Forschungsdienst*), for example, was full of injunctions that 'science' must serve 'practice' (e.g. Meyer 1937), and at the fourth Reich Peasants' Conference in 1936 Herbert Backe had declared that 'plant breeding must be made much more accessible to farmers in general' (Backe 1936a: 88). The following year the RNS emphasized its commitment to the well-being of 'small breeders' (Grote 1937). Moreover, the stations' expertise with high-yielding cereals varieties ought to have impressed the RNS since in planning for war, Nazi agricultural officials sought to maintain the production of German bread grains while *reducing* cereals acreage (in favour of root crops for animal feed as well as oil-bearing plants), hoping thereby to reduce the level of costly imports (Gies 1979).

Remarkably, however, despite an occasional acknowledgement of the Bavarian station's success (Schönberg 1936; Nicolaisen 1944), this is not what happened. In Baden, for example, the station's buildings and facilities were simply leased in 1938 to a commercial breeder (Bundesverb. Dt. Pflanzenzüchter 1987:

212, 87).[18] In Bavaria several of the station's previous activities were transferred to the RNS: for example, varietal testing, certifying varieties for the region, and promoting the take-up of plant breeding among the region's small farmers and cooperatives.[19] Moreover, the state-wide network of crop-improvement associations whose development the station had so energetically promoted was dissolved and its head office incorporated into the RNS (Weller and Lechner 1953: 28). As far as the stations' track record in plant breeding was concerned, the RNS's blinkered account was at times extraordinary. In an article on the role of breeding in the newly declared 'battle for production', for example, the official responsible for the reorganization of plant breeding took breeders of the pre-Nazi era to task for avoiding the difficult but all-important task of improving fodder plants. What he failed to mention was that the south German stations had taken on this very task from about 1914. Moreover, he added, it was crucial that the Association of Commercial Breeders should provide technical support to *small* breeders since it was the latter who had carried out some of the most promising work on fodder plants. But as we have seen, the stations had been providing precisely that kind of technical support since their inception. Just why this official was so economical with the truth became clear at the end of the article where he declared his 'firm conviction' that progress in breeding would be faster through harnessing private initiative than through state institutes alone (Krohn 1934). (To portray the issue as a choice between these two alternatives, of course, was a complete red herring; no one was suggesting that Germany's breeding needs should be met by state institutes *alone*.)

However shaky its premises, the aim behind the RNS's reorganization of plant breeding was to rationalize breeding work by setting up a strict division of labour between public and private sectors. Stations were no longer to 'compete' with commercial breeders, and public-sector breeding was to be subordinated to the needs of the private sector. This notion of a tidy division of labour fitted well with recurring Nazi complaints of wasteful competition, chaos, uncoordinated research effort, etc. which they believed had disfigured the economy and research policy of the pre-Nazi era. The NS 'principle of order', as one writer put it optimistically, would draw together all individual capacities in the service of the whole (Meyer 1935, 1938; cf. Spennemann n.d.: 12; Backe 1936b; Darré 1936). As we shall see, however, the new division of labour effectively undermined the stations' public-service role.

Once the Seed Decree had been declared in 1934, the RNS began to address the issue of how to reorder this relationship. Officials began to consult with public-sector breeders and to invite their views on how varieties bred in the public sector should be transferred to the private sector. Public breeders evidently accepted the need for a certain amount of central regulation of this issue since their own proposals acknowledged that such handovers ought to be overseen and approved by the RNS.[20] For the next two years little seems to have happened, but at a meeting of the Association of Commercial Breeders in 1936 Ludwig Kiessling (by then former director of the Bavarian station) presented proposals on how the two sectors might collaborate (Kiessling 1936). Accepting the need to eliminate fragmentation in research work and unproductive duplication of effort, he envisioned no problems with the idea that handovers might be centrally directed by a Reich agency, but he insisted that public-sector institutes must have some say over whom they collaborated with and how. That the relationship between the sectors was still tense, however, is evident from his remark that since plant-breeding stations were being accused of 'unfair competition' and promoting an 'unnecessary proliferation of breeders', he fully expected his proposals to be criticized.

In the event, it was not until early 1937 that Darré finally distributed his first tentative proposals – intended, as he put it, 'to avoid a lack of coordination between state and private breeding' – to the directors of the main public-sector breeding institutions for comment.[21] His proposals, however, took remarkably little account of what public-sector breeders had been saying over the previous few years and essentially granted all key powers to the Association of Commercial Breeders (*Reichsverband deutscher Pflanzenzüchter*, RdP). Public-sector varieties of various kinds were to be made available to members of the RdP at a price to be agreed with the RdP. The RdP (NB: not the public breeder) would designate which firm should receive the variety and would ensure that it carried out further breeding correctly. And the RdP would agree the licence fee that the recipient would pay to the public-sector institute. Darré concluded the letter, declaring that his proposals would prevent individual public-sector institutes from selling their varieties to the highest bidder, a practice which was 'not in the interests of German plant-breeding as a whole'.[22]

Predictably, public-sector breeders were angry, but perhaps more surprisingly, they did not hesitate to voice their objections. Kiessling declared that the proposals were rather less far reaching than his own previous suggestions and was wary of granting the RdP so many powers since it was 'after all an association of interested private bodies'.[23] Institutes, he suggested, should be free to decline to hand over one of their varieties to the nominated private breeder where they had reservations about the latter's scientific competence. And it was preferable that a state agency – not the RdP – should oversee the handover as well as the way in which the recipient firm subsequently dealt with the variety. Theodor Roemer was similarly suspicious. The proposals were unworkable, he argued, since they did not make clear how the RdP was to distribute public-sector varieties to one or other of its members. 'Can the state hand over the results of its work to private interests and allow the latter to divide these up among themselves? That is simply unacceptable.' The public-sector breeder had to be granted some influence upon the way in which the variety was subsequently handled by the recipient firm.[24] Furthermore, he objected, the proposals were plainly demeaning to the public sector, not least Darré's claim that public breeders had tended to sell their varieties to the highest bidder. This ignored the fact that public breeders in Bavaria, Baden and Württemberg had worked 'for the common good rather than for the interests of capitalism'.[25]

Interestingly, the RNS seems to have taken such criticism on board, for the official Guidelines finally declared nearly a year later granted fewer powers to the RdP and were noticeably more accommodating toward the public sector (Anon. 'Richtlinien' 1938b).[26] To be sure, the RdP was misleadingly presented in the new Guidelines as the 'mediator and honest broker' (*Treuhand*) between public and private sectors. But the final decision on how a variety was to be transferred now lay with a four-member 'Handover Committee' (*Abgabe-Ausschuss*) for each station or academic institute, consisting of representatives of the station in question, the RdP, the RNS and the coordinating body for the agricultural sciences (*Forschungsdienst*). Following consultation with the relevant public-sector institute, the RdP was to make proposals to the Handover Committee as to how a public-sector variety was to be dealt with. The Committee was then obliged to take the wishes of both the institute and the RdP into account before making its final decision.[27] The Guidelines, however, were also quite general, leaving a great deal of discretion to the Committee to decide precisely what the terms of the handover would be. And given the history of mutual recrimination between public and private sectors, it was likely that the two sides would often disagree on those terms. In such cases the deadlock was supposed to be broken by the RNS since its representatives held half of the seats on the Committee.[28] In either event, the Committee's decision had to be approved by the Ministry of Agriculture, guaranteeing that one way or the other, Darré's people held ultimate control.

Although the RNS had made some attempt to meet public-sector breeders' objections in reformulating the Guidelines, the fact that the Guidelines left so much discretion to the Handover Committee means that we need to examine the deliberations of the Committee before we can be sure how the system actually worked. Few records of this kind appear to have survived,[29] but the handful concerning the Bavarian station's dealings with the Committee are illuminating. They indicate that the director, Theodor Scharnagel (see Figure 5.3), continued to voice his doubts to the Bavarian authorities concerning the adequacy of the system. Part of the problem, as he saw it, was that the centralization of decision-making in the Handover Committee did not serve Bavarian farmers well because it was ill suited to meeting regionally diverse needs. A key bone of contention was that for each variety developed in a public-sector institute, the Committee alone had the power to determine which private breeder would receive it.[30] From time to time, for example, Scharnagel – representing the Bavarian station on the relevant Handover Committee – found himself at odds with other members of the Committee over the issue of who constituted a suitable recipient. On one occasion he was able to overcome his fellow members' doubts,[31] but in a number of instances when representatives of the RNS and the *Forschungsdienst* sided with the RdP's representative, he was simply outvoted.[32] Scharnagel found this irritating because in his opinion such decisions were not always 'scientifically' justified, a problem aggravated by what he regarded as the weak qualifications of some RNS and *Forschungsdienst* representatives as well as by a certain antipathy toward Bavarian breeders which he perceived in some of the RdP's representatives. In one case, for example, the Committee had decided – evidently on the basis of drawing straws – to hand over to a breeder in Mecklenburg (north

Figure 5.3 Theodor Scharnagel (1880–1953), director of the Bavarian Plant Breeding Station, 1928–1950 (source: Weller 1951).

Germany) a potato variety which the Bavarian station had developed for very different growing conditions in Austria. Scharnagel stressed on several occasions that he had no objection in principle to non-Bavarian recipients as long as the decision made agro-ecological sense.[33]

Although Scharnagel's complaints found support at the Bavarian Ministry which forwarded them to the Reich Ministry of Agriculture and to the RNS, he seems to have got nowhere with the latter. A Reich Ministry official insisted that all representatives of the RNS and the *Forschungsdienst* on the Handover Committee were technically expert as well as neutral, and he denied the possibility that the Committee advantaged north German breeders at the expense of south German ones. A RNS official similarly defended the Committee's procedures, claiming that there had been no complaints so far. He then went onto the offensive, taking the Bavarian station to task for failing to abide by the Guidelines in recent years.[34]

So, returning to the central question, was NS policy toward public-sector plant breeding peasant-friendly? As we have seen, many of the provisions of the Seed Decree either failed to prioritize the needs of smallholders or even made things more difficult for them. Moreover, while one might have expected the south German stations' track record in serving their regions' farmers to have resulted in praise and imitation after 1933, that did not occur. Instead their powers were cut back; the effect of some RNS policies was that the stations' intimate familiarity with local varieties, growing conditions, farmers and breeders was pushed aside in favour of a highly centralized system – the RNS – whose experts, even when technically competent, were unlikely to know the territory. If the Bavarian experience was representative, peasant-friendly breeding did not fare well under National Socialism.[35]

4 After 1945

In the chaos following the war, it is hardly surprising that both East and West Germany initially retained various agricultural policies from 'the Third Reich'. The List of Approved Varieties published in the western zone, for example, was explicitly based on the Seed Decree of 1934, retaining the varietal categories which had been used in the 1942 list (Anon. *Sortenliste* 1949). And with the formal establishment of the two German states in 1949, both countries set up national authorities which regulated the seed market in much the same way as had been done before 1945. Each agency oversaw a nationwide programme of varietal testing; the results were used to place the best varieties on a list of approved varieties, and only listed varieties could be sold (Anon. 'Der Sortenschutz' 1987; Kratzsch 2008). In formal legal terms the Seed Decree of 1934 actually remained in effect in West Germany until the passage of a new seed law in 1953 (Büttner 1954: 124), and some of the Decree's provisions – for example, a national system of certification and a national list of approved varieties – still continue today (Albrecht Meinel, pers. comm.).

In respect of continuity and change, of course, the main question of interest from my point of view concerns the south German stations. Given the radical shift in the political environment after the fall of National Socialism, one wonders how the movement for peasant-friendly breeding fared in Germany after 1945. Since this is too large a task to take on here, I have asked a more limited question. In view of the Nazi regime's success in altering the division of labour between public and private sectors – so as to restrict the south German stations' ability to serve their regions – I focus in this section upon the nature of the stations' relationship to the private sector after 1945.

As we have seen, Baden no longer had a public-sector station after the late 1930s. The only existing history of Württemberg's station (Kittlitz 2005) indicates that its post-war history was one of gradual detachment from the concerns of the region's breeders, of whom only a handful of relatively small firms had survived the 1920s depression as well as the early post-war period. Due to extensive wartime damage, the station's activities during the first decade were relatively limited, and breeders expressed dissatisfaction. Ties to the latter improved during the 1960s, but Germany's integration in the European Community brought with it major challenges for the region's breeders as well as new pressures upon the station. Presented with a wider range of markets, Württemberg's breeders no longer concentrated upon developing varieties mainly for the region's farmers. At the same time, however, facing more intense competition both from large north German breeders as well as multinational seed companies, the region's breeders were now more heavily dependent upon the station than ever. Accordingly, during the 1970s and 1980s the regional breeders' association proposed much stronger ties to the station in an attempt to monopolize control over access to its services and breeding material. That they failed in this attempt owed much to the fact that the station increasingly found itself operating within a larger universe, spanning Germany as a whole as well as Europe. This meant a wider range of potential publics to be served, funding bodies on which to draw, and potential collaborators from the private sector. From the station's point of view, however, the close collaboration which regional breeders had requested looked as though it would curtail the station's freedom to take advantage of the widened range of opportunities. The breeders' proposal was thus declined, and by doing so the station was in effect abandoning its strong pre-war commitment to regional service.

In Bavaria the number of breeders who had survived by the mid 1950s was much larger – about 50 (Weller 1953: 259) – and a large majority of these had been in operation since *c.*1920 (Weller and Lechner 1953: 13). By the summer of 1958 for reasons that are not clear, the Bavarian government called upon the station to alter its handover arrangements. After consulting public-sector institutions in other states about their handover rules, the station's director drafted new guidelines which specified that the station would not release its own varieties in order to maintain relations of trust between the station and breeders. (This represented a retreat from the station's position in the 1920s when it had come under attack for doing just that.) Nor would it develop near-finished varieties; instead it would concentrate upon supplying breeding material and unfinished varieties to the private sector.[36] (This represented a shift from the 1920s in the kind of material released, reflecting the fact that by the 1950s virtually all of the remaining Bavarian breeders were 'independent'. Though keen to obtain raw materials from the station, none depended upon it for finished or near-finished varieties.) To qualify for receipt of such material a breeder had to be based in Bavaria (thus the station was continuing its regional service role), but material was no longer to be handed over free (thus the station was no longer subsidizing small breeders who lacked in-house breeding capacity as it had done in the 1920s). The charge levied, however, would be modest since the point of the station's work was to serve Bavarian agriculture, not to extract maximum income.[37]

The question of how to deal with handovers arose again in 1966 when the head of one of the station's departments asked the director what he should do with promising grass varieties that were nearly finished but in which no Bavarian breeder was interested. It seemed to him that it was no longer realistic for the station to cooperate solely with breeders in Bavaria, especially where varieties of importance for Bavarian agriculture were concerned.[38] The director then drafted new handover guidelines which stated once again that, *as far as possible*, the station would not develop finished varieties and that where it did so, its varieties would not compete with those of regional breeders. (This meant that the station would not restrict itself to developing materials of interest only to regional breeders; where the breeders had failed to develop varieties of a type reckoned to be important for the state's agriculture, the station would be entitled to take on this task.) Significantly, the draft also advanced an argument then being voiced in Württemberg: that if Bavaria's breeders were going to compete effectively with more powerful firms from north Germany and abroad, they would have to organize themselves into larger economic formations which would need to collaborate closely with the station.[39] The draft was circulated among the station's department heads for comment, several of whom agreed with the draft that the station's primary responsibility was to Bavaria's farmers and consumers (i.e. rather than merely to its breeders). If the station developed a promising variety, they maintained, it should go to the best recipient, even if the latter was a non-Bavarian breeder. Conversely, if a variety from a non-Bavarian breeder looked promising for Bavarian agriculture, it should be promoted by the station, even if this harmed the interests of one or other Bavarian breeder.[40]

A version of these proposals then went to the breeders' association (*Verband bayerischer Pflanzenzüchter*) who took the station's advice and decided, as had their counterparts in Baden-Württemberg, to organize themselves as a commercial cooperative (established in 1970 as the *Bayer. Pflanzenzuchtgenossenschaft GmbH*). The Bavarian Agriculture Ministry then consulted with its counterparts in other states to find out how they dealt with the handover of material from public-sector institutes, and a 'framework-contract' was drafted, specifying what the station's relationship with the new cooperative would look like.[41] The station would hand over promising breeding material to the cooperative, with the exception of parent lines suitable for hybridization or inbred lines suitable for heterosis breeding. (This meant that the station did not grant a monopoly to Bavarian breeders over access to valuable breeding material; it retained the right to offer such material to *any* breeder who promised to develop it in a way the station saw fit. This represented, therefore, a dilution of the station's original commitment to regional breeders, though not of its commitment to serve Bavarian farmers and/or consumers.) The station would also have a say on which breeder should receive it and could intervene should it feel that the further breeding was being done badly. (The station was thus able to exercise quality control over the breeding process comparable to its role at the start of the century.) For its part, the cooperative undertook to continue breeding the variety in the direction(s) initiated by the station until it qualified for inclusion on the List of Approved Varieties. (This meant that if a breeder accepted material which the station deemed potentially important for Bavarian agriculture, s/he was not free to let it remain unused, perhaps because it threatened to compete with the breeder's existing varieties.)

Overall, therefore, the surviving south German stations certainly regained some of their autonomy after 1945. They were no longer obliged, for example, to hand over breeding material exclusively to a single national organization of commercial breeders. Nor were decisions over which individual breeder was suitable to receive a particular unfinished variety any longer largely controlled by a national agency such as the RNS. And stations reverted to being solely accountable to the agriculture ministry of the state in which they resided. But they never quite regained their interwar role as energetic promoters of smallholder agriculture and breeding. To be sure, in Bavaria the station retained a strong commitment to the needs of the state's farmers and consumers (which it was obligated to do as a state institution), but it was no longer permitted to fulfil this aim by competing with the larger independent breeders as it had done during the 1920s. By the post-war era the vast majority of breeders who had survived were independent, and the station's original role in helping small farmers and cooperatives to take up breeding was no longer required. Moreover by the 1970s the nature of the stations' relationship to breeders in Bavaria as in Baden-Württemberg had changed. Although they continued to provide technical support and breeding material, the stations were no longer prepared to restrict themselves to collaboration solely with regional breeders. In both respects the station's role as an agent of peasant-friendly breeding had weakened.

Conclusion

During the Nazi era, as we have seen, the Party's relentless championing of the peasantry was not of much use to the south German stations. In terms of general policy, peasant farmers appear to have gained economically and socially over the first three or four years of the regime, though those gains were sharply eroded thereafter. And as far as reforms of the seed market are concerned, the Seed Decree's promotion of high-quality seed probably benefited farms of all sizes (insofar as they could afford it). But some provisions of the Decree made it harder for small farmers to obtain good quality local varieties at low prices, and there are signs that the RNS did not try very hard to ensure that small farmers used certified seed. Turning to the reform of public-sector breeding, the RNS managed to take away several of the services that south German stations had been providing to the region's small farmers and breeders since early in the century, though its attempt to harness the stations more tightly to the needs of the private sector also ran into opposition. On balance, peasant-friendly institutions did not find the support from the regime which one might have expected. And while some of the stations' powers were restored after 1945, they never regained the regional role which they had enjoyed during the 1920s as champions of smallholder agriculture and the dominant force in south German breeding.

Elsewhere, however, the need for peasant-friendly breeding did not go away. By the 1970s, as we shall see in the next chapter, development experts were concluding that something rather similar would be essential if the gains of the Green Revolution were to be extended to small farmers in the developing world. In the event, they took no notice of the pre-war movement in Central Europe.

6 The Green Revolution and its critics

After 1945 the Central European movement for peasant-friendly plant breeding, in Germany at least, went into gradual decline. And by the 1950s in both North America and Western Europe it was becoming clear that in agriculture small was *not* beautiful. 'Get bigger or get out' was the slogan of the day. While state efforts to foster agricultural change in the developed world from the 1950s were particularly intense, intervention as such was nothing new. As we saw in Chapter 1, it had been going on in Europe since the late nineteenth century. Furthermore, over the same period 'development' was under way in various colonial empires as European powers sought to boost production, usually focusing upon cash crops designed for markets in the 'mother country' but sometimes including staple crops with an eye to improving the standard of living of the indigenous population (Rist 2008: 56–58; Cooper 2010; Unger 2010b). After the war 'development' resumed in European colonial empires, taking the form of large-scale state-led projects (Bonneuil 2001) which proved popular with some post-colonial elites (Speich 2009a). Emerging from the war both economically and politically stronger, the United States greatly expanded its development activity. Central to this was a series of agricultural programmes which eventually became known as the 'Green Revolution'. Devised initially by US foundations and complemented by US government initiatives, these programmes sought to stimulate economic development in Latin America and Asia under the banner of 'modernization'.

The term 'Green Revolution' is generally thought to have been coined in the late 1960s by an official in the U.S. State Department who suggested that green revolutions could be a non-violent way to avert 'red revolutions' among Third World peasants (Gaud 1968). Since then a surprising number of commentators have misleadingly claimed that the Green Revolution as an *undertaking* originated during the 1960s. It is quite clear, however, that the GR programmes in Asia of the 1960s were modelled in large part upon similar programmes in Latin America during the 1940s and 1950s. Since then, in its basic features the Green Revolution has remained largely unchanged right into the present. All GR programmes old and new, for example, have sought to improve agricultural productivity not through fundamental socio-political change (e.g. land reform) but via institutional reform and 'getting the technology right'. Nevertheless, the Green Revolution has also undergone significant changes since the beginning which are important in what follows. In this chapter and the next, therefore, I will distinguish between a first, second and third generation of green revolutions. By the 'first generation' I mean the programmes of the 1940s through 1960s whose principal aim was to boost production in suitable regions with little concern for the social consequences of the intervention. The transition to the next phase was marked by a wave of criticism of the Green Revolution around 1970 which left its impact upon the 'second generation' of programmes during the 1970s, 1980s and 1990s. These programmes were much more concerned with poverty alleviation, focusing especially upon the needs of small farmers and the rural population. The transition to the 'third generation' (dating roughly from 2000) has been marked by the emergence of the new biotechnology which proponents of the Green Revolution now claim provides a necessary foundation for all future programmes. And as I will argue in chapter 7, this most recent phase shows signs of a return to the productionist emphasis which marked the first generation.

After outlining the nature of the Green Revolution in the first section and the criticism which it attracted in the second, I turn in the third section to the green revolutionaries' response to that criticism during the 1970s and 1980s. Reflecting upon the first generation of programmes, experts sought to pin down why so many programmes had failed to reach small farmers. Ironically, the conclusions that they reached were not new. As I show in section four, many of the undesirable social consequences of the first generation of programmes could have been anticipated by anyone familiar with the transformation of US agriculture in previous years. And most of the features of development programmes which green revolutionaries concluded were necessary for helping smallholders were already to be found in peasant-friendly institutions at the start of the twentieth century, in Central Europe and elsewhere. Evidently unaware of their predecessors, green revolutionaries had rediscovered the wheel.

1 The first generation, 1940–1970

The declared aim of the GR programmes was to reduce world hunger (which in 1960 was generally reckoned to affect about one-half of the world's population). This goal was endorsed during the 1960s, for example, by Rockefeller Foundation officials and their advisors (Stakman *et al.* 1967; cf. Oasa 1987: 17) as well as by scientists at the international agricultural research centres (IARCs) who implemented the Green Revolution (Anderson *et al.* 1991). In the 'war on hunger', agricultural science seemed to provide the necessary weapons, and optimists reckoned by the late 1960s that victory was in sight (Wharton 1969: 464).[1] While acknowledging the various social and economic conditions that limited food production, for example, J. George Harrar – an agricultural scientist and president of the Rockefeller Foundation – nonetheless concluded that 'today the greatest cause of malnutrition is underproduction'.[2]

The goal of poverty alleviation gave the Green Revolution a humanitarian justification, but proponents also argued that it conferred political benefits. As we saw in Chapter 1, the prospect of widespread peasant unrest in late nineteenth century Prussia prompted a series of agricultural reforms. Post-1945 agricultural development programmes were little different. As the largest social class throughout the Third World, the peasantry was of strategic political importance for both sides in the Cold War. In order to counteract the Soviet bloc's offer of land reform, it was argued, the West needed to take measures that could substantially improve peasants' well-being. As long as Third World peasants remained hungry, they would be vulnerable to the allure of communism.[3]

Beginning in the 1940s, therefore, the US government, along with the Ford Foundation and the Rockefeller Foundation, funded programmes which allowed American agricultural scientists to spend periods of time in Latin America and subsequently in Asia, training native-born agricultural scientists and advising governments on the reform of research and extension institutions as well as the establishment of agricultural universities modelled upon the US land-grant universities (Goldsmith 1997). But the better-known – and more controversial – feature of the GR programmes was its 'package' of cultivation technologies and practices which were designed to boost cereals production. (As will become evident, almost all parts of this package had been central to 'Europe's Green Revolution' two generations earlier.) Key to the package were high-yielding plant varieties (sometimes called 'modern varieties') which responded well to mineral fertilizer. In addition, because they were relatively insensitive to day length and temperature, they could be planted at various seasons, enabling farmers to harvest several crops in a single year.

The Green Revolution's impact was complex. On the positive side there is widespread agreement that it substantially increased cereals production in many countries. Across the world average grain yields more than doubled between 1950 and 1985 (Wolf 1986: 9). In Colombia between 1966 and 1974 the acreage planted in the new rice varieties increased to 99 per cent; yields increased by 60 per cent; and total production increased two-and-one-half fold (Ruttan 1977: 19). In India, Pakistan and Nepal between 1965 and 1975 farmers in some regions increased their wheat yields by two to threefold. In the Punjab, for example high-yielding varieties spread over more than 70 per cent of its farmland; wheat yields more than doubled between the mid 1960s and 1980 while rice yields increased by 174 per cent (Lipton with Longhurst 1989: 4). As a result India, which had relied upon imports of US wheat throughout the 1950s, became self-reliant in cereals (Farmer 1986; Parayil 1992; Perkins 1997). In the case of rice, yield increases were less impressive. In Bangladesh by 1970, for example, results were not living up to the International Rice Research Institute's (IRRI) expectations (Anderson *et al.* 1991). Nevertheless, in 1969 the Philippines claimed it had become self-sufficient in rice; Malaysia predicted that it would do so in 1971; and Indonesia achieved that goal in 1984 (Wharton 1969: 472; Staatz and Eicher 1998: 22). The new varieties proved popular with farmers, at least in some regions. By the early 1980s approximately 80 per cent of the wheat acreage in Asia and Latin America and over 40 per cent of the rice acreage in Asia had been planted in the new varieties (Dalrymple 1985: tab. 1; Pinstrup-Andersen and Hazell 1985: tab. 1.).

Some of the Green Revolution's social consequences were also positive, at least initially. On the one hand, the new varieties required more labour – for example, for weeding, water control and fertilizer applications during growth as well as for harvesting larger crops – which created more rural employment. On the other, increased production also brought down food prices, benefiting the urban poor as well as the landless rural population (Dalrymple and Jones 1973; Sisler and Colman 1979; Staub and Blasé 1971; Falcon and Naylor 2005). As a result, the Green Revolution was probably one of the factors contributing to a global reduction in the number of poor from 1.1 billion in 1975 to 800 million in 1995 (Hazell 2003; Manning 2000: 4). Over the last 20 years, for example, the proportion of the South Asian population that lives in poverty has declined from 45 to 30 per cent (McIntyre *et al.* 2009b). Lastly the proportion of the population suffering from chronic hunger has declined from 37 per cent in 1960 to about 17 per cent today (Conway and Waage 2010: 127).

2 The revolution attracts criticism

Despite these favourable impacts, from the late 1960s the Green Revolution became the subject of an 'intense controversy' (Ruttan 1977: 16; Byres 1972) which continued over the following decade. One of the criticisms was that GR technology was proving to be environmentally damaging. The heavy use of mineral fertilizers was contaminating water supplies and reducing soil fertility; pesticides were damaging farmers' health; irrigation was lowering water tables and leading to salinization; and the reliance upon oil-based inputs put pressure upon fossil fuels (Dahlberg 1979; Conway 1997; Pretty *et al.* 2006). But the issue that attracted most critics' attention was that despite large increases in cereals yield and production, in many areas the Green Revolution seemed to make relatively little impact upon rural poverty and malnutrition. In the Punjab, for example, despite the large-scale uptake of new wheat varieties, the proportion of the population who could not afford a minimum diet was only slightly smaller (Lipton with Longhurst 1989: 4). In India as a whole, hunger persisted despite a threefold increase in wheat yields (Nair 1979: 92, 135). In the Philippines between the early 1960s and 1976 both rice yields and overall production had gone up by about 50 per cent, yet 60–70 per cent of young people were still judged to be undernourished, and agricultural wages fell (Pearse 1980: 213–214; Griffin 1974: 148ff.). The story in other regions was similar. By the early 1970s, according to one economist, it was not yet possible to claim that GR varieties had reduced malnutrition; in some regions poverty had instead increased (Griffin 1974: xii–xiii).

One reason for the Green Revolution's failure to alleviate poverty is almost certainly that this had never been a principal aim of its donor agencies.[4] For some individuals involved in GR programmes, to be sure, solving the problem of world hunger was undoubtedly a genuine ethical commitment which they felt lent meaning to their work. But if poverty alleviation had in fact been an aim of those funding most GR programmes, it is difficult to understand why they were designed with so little regard for their impact upon the rural poor (as we shall see below). Despite all of the rhetoric about alleviating hunger, the major aim of the programmes' donors was instead to 'modernize' Third World economies. In the 1950s development economists – and many Third World governments – thought agriculture to be an unlikely engine of growth because 'traditional' farmers were believed to be unresponsive to economic incentives. It would be better, it was thought, to concentrate upon promoting industrialization; agriculture's role in such a scheme would be merely to provide capital and labour (Schultz 1983). To that end, the sensible strategy seemed to be to provide advanced technology to the minority of 'modern' or commercial farmers in the developing world who were keen on increasing yields and who produced for the market. Increasing their productivity would aid capital formation and liberate labour from agriculture. By lowering the price of food, it would keep industrial wages down and might also alleviate poverty.[5] That this was the underlying aim is evident from various programmes funded by the World Bank and USAID (Griffin 1974: 235, ch. 3; Pearse 1980: 208–210). At the Rockefeller Foundation, too, several figures were taking the view by about 1950 that the Green Revolution would necessarily take peasants off the land and put them to work in factories (Oasa and Jennings 1982; Cullather 2003). A decade later the Foundation's vice-president was taking this process for granted:

> Presumably farming as a way of life will give way to agricultural production as a strictly business enterprise with significant increases in land holdings and comparable decreases in the number of individual land owners and the size of the farm labour force. Only when the business of farming assumes a financial and corporate structure similar to that practiced in industry can it become truly competitive. ...These developments will first come to the more advanced areas of the world but will later have to be applied generally.[6]

By the late 1960s this view of how to bring about 'modernization' was widely endorsed. Ford Foundation officials were saying that peasant agriculture had to disappear if agriculture was to be fully modernized (Perkins 1997: 114), and the European Commission's 'Mansholt Plan' was dismissing family farms as unproductive and calling for much larger 'production units' of 1000–2000 hectares (Kluge 1989; Ingersent and Rayner 1999: 155ff.).

The point behind modernizing agriculture in the developing world, however, was not merely to make it 'competitive'. As western government officials often pointed out, agricultural development aid offered other advantages. Growth of the US economy after the war, they argued, would depend on economic development in the Third World (Millikan and Rostow 1998). As an executive director at the World Bank noted, for every dollar lent to the Bank, ten dollars came back to firms in industrialized countries.[7] The largest part of the United States' Aid to International Development funding by 1968, for example, took the form of loans to allow the purchase of fertilizer (Gaud 1968). American firms were quick to take advantage of the new markets for agricultural inputs and very supportive of GR programmes (Wharton 1969; Perkins 1997: 257–258; Staples 2006: 113; Unger 2010b).

Why, then, did the Green Revolution's increased production fail to alleviate rural poverty? The answer emerges when we look at who adopted the new technology and why. By the 1970s it was clear to most observers that throughout the developing world the Green Revolution's package was taken up earlier and more often either by large landowners – who gained the most from the technology – or by those smallholders with access to good land and irrigation.[8] Smallholders on poor or rainfed land gained much less, and the landless poor were often worse off (Dalrymple and Jones 1973; Ruttan 1977; Lipton 1978; Perrin and Winkelmann 1976; Hazell and Ramasamy 1991; Byres 1972). The overall effect of introducing the technology was thus to widen the gap between rich and poor in the countryside. In some places this resulted in violent clashes between estate owners and the rural poor (Wharton 1969; Frankel 1971; Anderson *et al.* 1982), thus undermining the Green Revolution's claim to produce political stability in the countryside. To be sure, consumers – including the urban poor and the landless rural population – gained since increasing yields boosted production, resulting in lower prices on the cereals markets.[9] But lower prices hurt the large number of smallholders who produced for the market but were unable to adopt the technology; losing income, many were forced to sell their land.

By choosing to focus upon food crops, the Green Revolution was a great improvement over the colonial period when breeders had devoted most attention to export crops. Thus the Ford and Rockefeller Foundations' decision in the late 1950s to fund work on rice was a recognition of the fact that rice was then the most important food crop in the world (Weaver 1970: 101). Similarly, the Rockefeller Foundation's initial cereals improvement programme in India from 1956 was intended to include sorghum and millet which are major food staples in semi-arid regions (Goldsmith 1997). But the particular food crops that breeders chose to improve were not always those most important for the diet of the poor (e.g. Chambers 1983: 78–79; Harwood 2009). Some have argued, for example, that the Indian programme's emphasis upon wheat and rice had undesirable nutritional consequences since their acreage grew at the expense of more high-protein pulses, millet and sorghum.[10] More generally, across the developing world very few improved varieties of sorghum, millet, the major pulses or root crops (also eaten predominantly by the poor) were bred until the 1980s (Evenson and Gollin 2003; Lipton 1978; Falcon 1970).

But for those who did grow GR crops, in order to understand why the Green Revolution's package was taken up and exploited more readily by large farmers, we need to look at the properties of the high-yielding plant varieties at its core. Just as in Germany 50 years earlier, a major obstacle for smaller farmers wishing to adopt them was cost. Part of the increased cost was for seed. Most estimates suggest that the cost of high-yielding varieties was about twice that of traditional varieties (Hayami 1971; Olea-Franco 2001: 663) though sometimes quite a lot more (Gaud 1968; Staub and Blasé 1971). In some cases (maize), the new varieties (unlike local ones) were generally unstable hybrids whose yields fell off substantially the following generation, obliging the farmer to purchase fresh seed every year or so. But the GR package consisted of more than just seed; the improved varieties only produced high yields where adequate water was provided and large amounts of mineral fertilizer were applied. And the cost of fertilizer is generally recognized as one reason why GR varieties have not made as much impact as they could.[11] Moreover since some of the high-yielding varieties were vulnerable to disease, pesticide had to be applied. When the cost of seed, fertilizer, pesticide and water are added up, estimates of increased cost varied greatly from one crop and region to another: from 50 to 75 per cent (Dalrymple 1969: 45) to as much as tenfold more (Wharton 1969; Fowler 1994: 243).[12] Since this was more than many peasants could afford, host country governments often offered subsidies to ease the cost, but this proved to be a serious financial burden and had to be abandoned (Hazell *et al.* 2007: 20).[13] In Bangladesh during the 1970s, for example, government subsidies for fertilizer, seed and machinery had to be paid for out of development grants from foreign donors whose own corporations were normally the suppliers (Anderson *et al.* 1991: 322).[14] Where government subsidies were not available, smallholders nonetheless keen to try the new technology had to take out loans. But in many countries the limited amounts of credit at favourable rates of interest available from lending organizations in rural areas tended to go to low-risk borrowers, namely large farmers. Smallholders, therefore, had to turn to moneylenders who invariably charged much higher rates (Conway 1997: 59). Finally, as in Europe *c.*1900, large farmers generally enjoyed lower input prices because they could buy in bulk (Lipton 1978). Thus right across the board the question of cost was systematically biased against small farmers.

A second obstacle was the new varieties' requirement for abundant amounts of water. As a result, the first generation's programmes were targeted at irrigated regions like the Punjab in India or Central Luzon in the Philippines (Griffin 1974: 207–208). But these constituted only a fraction of the cultivated acreage; only one-quarter of the cultivated land in developing countries is irrigated on average, and the proportion in Latin America and Africa is much smaller (FAO 2004: table A5). Thus the vast majority of farmers were left out of these schemes (Ladejinsky 1973; Sisler and Colman 1979: 13–14). Since larger farms were more likely to be irrigated (or to be able to pay for wells to be sunk), this meant that those farmers who already enjoyed the possession of one scarce resource also gained differential access to another (Chambers and Farmer 1977; Byres 1972; Alauddin and Tisdell 1991: 115).

A third feature of the package was that GR varieties, like commercial varieties in Europe *c.*1900, were 'fussier' than local varieties. With high-yielding rice, for example, the seedbed had to be prepared with care; transplanting seedlings had to be done on time; fertilizer had to be applied precisely; water levels had to be controlled; and weeds and pests had to be eliminated (Hayami 1971; Nair 1969). Initially this was good for the rural economy: more work to grow the new varieties (as well as to harvest several crops per year) meant more jobs. And the demand for labour led to increased wage rates. But the response by landowners was that those who could afford to mechanize – the larger farmers – did so in order to keep labour costs down. By the early 1970s, for example, the use of machinery to grow wheat in Mexico had reduced labour requirements by over 30-fold (Dalrymple and Jones 1973; de Alcántara 1976: 318–319), and the same trend was evident in India (Pearse 1980: 150–151; Byres 1972). Where small farmers were unable to mobilize sufficient family labour, they either had to sacrifice income in order to pay the costs of outside labour or avoid the new varieties altogether.[15] The effect of displacing labour, in turn, was to drive down the wages of farm workers, such that rural poverty actually increased in some places.[16]

A fourth problem was that GR breeders' preoccupation with increasing yield meant that they neglected other traits which were just as important for smallholders, if not more so (Chambers 1977). The general problem had been recognized in Europe much earlier:

> The more we demand that a variety performs extremely well on one criterion, the greater the dangers which threaten it in other respects. Various kinds of vulnerability, not evident at first, can suddenly lead to enormous damage.
>
> (Weck 1927: 671)

GR varieties, for example, were much more sensitive to weather conditions, pests and disease than were local varieties (Pinstrup-Andersen and Hazell 1985; Olea-Franco 2001: 664; Oasa 1987). Similarly, having been bred for monoculture, they did not do well in mixtures with other crops (inter-cropping), a form of cultivation much favoured by small farmers because of its lower risks and higher yields (Donald 1968; Manning 2000: 156). And in focusing upon the quantity of a crop's output, breeders forgot about its quality (again reminiscent of European commercial varieties half a century earlier). Thus the IRRI's first high-yielding rice variety turned out to have inferior taste and to spoil quickly after cooking while Norman Borlaug's famous semi-dwarf wheat varieties were criticized for their poor baking and milling properties (Frossard 2002; Ishikawa 1981: 154; Danbom 1990: 149–50).[17]

Finally, GR varieties were often inappropriate for most small farms because breeders had pursued the cosmopolitan breeding strategy which we encountered in Chapter 2. They assumed that it would be possible to develop new 'miracle' varieties (Byres 1972). IRRI's first big variety, IR-8, for example, was much heralded on its release as a 'rugged' variety which would yield well almost anywhere in the tropical zone (Farmer 1979; Cullather 2004: 244).[18] Accordingly, such varieties were transferred to other regions where they were expected to do well with little need for further adaptive breeding (Hayami and Ruttan 1985: 255–256; cf. Chambers 1983: 206). The trouble with this assumption was that, as in south Germany, tropical climates and soils vary a great deal over short distances, for example compared to the US (Johnson and Ruttan 1994).[19] In the

Philippines local agronomists tried in vain to warn IRRI that its new varieties would be inappropriate for most cultivation conditions (Anderson *et al.* 1991: 66; Cullather 2004), but the release went ahead. In the event IR-8 proved to be poorly adapted to actual growing conditions in many places. That it and the other varieties bred at IRRI did badly in rainfed areas, for example, was not really surprising; they had been designed to be grown in irrigated paddies, and as dwarf varieties, they were vulnerable to periodic flooding in monsoons (Farmer 1986; cf. Dalrymple 1977). By the late 1960s breeders at IRRI were acknowledging that their varieties were mainly suited for the 10–20 per cent of the Asian rice acreage that was irrigated.[20] Compounding the embarrassment was the fact that in some regions local varieties were *preferred* to IRRI's since although yielding less, they required no fertilizer, were pest-resistant, needed little irrigation and were better tasting (Anderson *et al.* 1991: 347). If some breeders were surprised by this preference, south German peasants would have understood it very well.

All in all, therefore, the Green Revolution's limited impact upon rural poverty derived from its differential uptake. Its package suited large commercial farms nicely, but from smallholders' standpoint the technology which it offered was inappropriate.

3 Revolutionaries reflect: the 1970s and 1980s

Criticism prompted a reconsideration of the Green Revolution's aims and a fundamental shift of orientation from the 1970s onward. One agricultural economist at the time nicely captured this change of perspective which was to characterize the second generation of the Green Revolution:

> Technological change in agriculture has traditionally, and perhaps too often, been viewed solely in terms of its immediate impact on increasing food output. Yet technology must be studied in broader terms if second generation problems and issues are to be avoided or mitigated.... [E]mphasis on economic growth to the exclusion of other factors can lead to serious social costs.
>
> (Dalrymple 1969: 1–2)

But just how were these social costs to be mitigated? Given the sheer scale of the criticism, major GR institutions were under pressure to respond.[21] Interestingly, the responses diverged sharply and often along disciplinary lines (e.g. Ruttan 1977: 16; Lipton 1978: 335; Freebairn 1995: 267). This was probably to be expected since some members of the agricultural science community (e.g. agricultural economists) had been quite closely involved in GR programmes while social scientists of other kinds had played a much smaller role and were prominent among the critics.[22] And as we shall see, this disciplinary conflict has persisted, arguably into the present.

On the one hand, quite a few agricultural economists reacted with irritation, exempting the Green Revolution's technological package from responsibility for the failure to alleviate rural poverty (Barker 1970; Dalrymple 1979; Ruttan 1977; Pinstrup-Andersen and Hazell 1985; Hayami 1981; Falcon 1970; Eicher 1995). The technology itself, they insisted, was 'scale-neutral' such that farmers large and small could benefit equally from it. The problem, they maintained, lay instead with host countries' inadequate institutions, infrastructure, and policies (e.g. land taxation, input subsidies, price supports). Thus environmental damage from the Green Revolution was attributed not to the fact that high-yielding varieties depended upon mineral fertilizer and pesticide but to farmers' ignorance, inadequate extension systems, and excessively cheap inputs (Hazell 2003). Similarly, the fact that farmers in many areas found the new varieties inappropriate for their growing conditions was not perceived as a deficiency of the varieties; instead the solution recommended was to develop improved irrigation and pest-control systems (Falcon 1970).[23] Thus the Green Revolution's defenders assumed that there was simply no technological alternative to the particular package that was offered. Since it was the most powerful way to boost productivity, it was up to 'society' to find ways to use it to best effect. As three champions of the Green Revolution put it, 'Science can show society what to do technologically, but society must enable science and technology to function and must contrive to distribute their contributions fairly among its members' (Stakman *et al.* 1967: 301; cf. Parayil 1992).[24] From this perspective, therefore, it was politicians, not agricultural scientists, who had to sort out the problems.

Here was a classic example of what has been called the 'use-abuse model' of science. Scientific knowledge itself is seen as neutral; whether it is applied for good or bad, therefore, is not the responsibility of the scientific community but of the society that deploys such knowledge. The trouble with this model is not that its claims about neutrality are necessarily wrong; sometimes a technology can indeed be used for good or ill. The problem is that the model is politically naive. It treats scientists and technologists as though they worked in a social vacuum in which no one could anticipate what use would be made of his or her work.[25] In the case of the Green Revolution, for example, various critics agreed with the agricultural economists that *in principle* the technology was indeed scale-neutral so long as access to all other relevant resources – irrigation, credit, extension services, transport systems, etc. – was equal. *In practice*, however, access was rarely equal since institutions and policies were almost always biased against smallholders. For the Green Revolution's defenders to insist upon the technology's scale-neutrality, therefore, was to turn a blind eye to the actual economic and political context within which it had to operate.[26]

As should be clear by now, however, the Green Revolution's package was by no means the only way to boost productivity or to extend the advantages of plant breeding to small farmers.[27] That there might be alternative approaches to development soon found support in other quarters of the development community. This group – consisting mainly of rural sociologists, anthropologists and geographers along with a few agricultural economists – did not react by defending the Green Revolution's package but rather took the criticisms *c.*1970 as grounds to reflect upon the first generation of programmes in order to see what could be learned.[28] They agreed with the defenders that host countries' institutions, infrastructure and agricultural policies often aggravated the inequitable outcome of the Green Revolution, though there seemed to be little that most development agencies could do about this. But they parted company with the defenders on two counts. First, they rejected the latter's implicit assumption that the GR package was the only rational way to boost production (an issue we will consider in Chapter 7). And second, taking time during the 1970s and 1980s to reflect upon the first generation of the Green Revolution, they displayed a greater concern to rethink the ways in which development programmes were designed so that they might in future better serve the smallholder. In this section I survey this 'diagnostic' literature, noting what its authors took to be the strengths and weaknesses of previous programmes.

The first conclusion, unsurprisingly, was that **the way in which programmes were organized** was crucial. It was important, for example, that they should be decentralized. Because both the ecological and economic conditions of farms in many areas of the developing world are very diverse, research and development activity needed accordingly to be decentralized. It was quite unlikely, for example, that any single plant variety, bred at some central experiment station, would do well everywhere. As we have seen, this was one of the weaknesses of the early work at IRRI where 'The planners ... were interested in ... universal, not local, solutions' (Anderson 1991: 85).[29] But the point was a more general one. Chambers, for example, argued that decentralization of decision-making was essential in order that scientists can respond flexibly to the diversity of local conditions (Chambers 1989). A case in point was the organization of agricultural research and education. One of the reasons cited for the success of the Japanese and US public-sector systems – as well as the weakness of some of their counterparts in the developing world – was that the former were far more decentralized (Hayami and Ruttan 1985: 423–424; Farrington 1988).

Organizing the research effort properly, however, was not enough; its intended beneficiaries also needed to be organized. With large commercial farmers this was not generally a problem since they tend to be relatively well organized (and thus able to cream off the benefits of new technology quite readily) (Anderson *et al.* 1982: part 1; Hayami and Ruttan 1985: 362). The issue was crucial, however, for peasant farmers who are rarely organized and thus less able to voice their needs and lobby for resources as well as to demand better terms from commodity distributors and processors (Ruttan 1983; Tendler 1993). In addition, as the south German stations were aware, an organized peasantry makes extension easier. Thus organizing peasants into cooperatives, for example, was often cited as a key to the success of development programmes (Flora and Flora 1989; Chambers 1983: 214–215; Johnston 1983; cf. Antholt 1998).

The importance of extension is another recurring organizational theme in this literature. Although it was occasionally argued that applying *known* techniques could have a greater impact than developing sophisticated new ones, extension was still the stumbling block since as the poor relation of research during the 1970s and 1980s, it received lower funding and enjoyed less status (Francis

1990).[30] Moreover extension services were often badly organized. Incentive systems did not encourage fieldwork; large amounts of paperwork kept officers at their desks; linkages between research and extension were weak; little attention was paid to feedback from farmers; and where officers were responsible for dispensing inputs, corruption was common (Chambers and Wickremanayake 1977; Chambers 1989; Compton 1989b). As some experts had learned the hard way, there wasn't much that donors or programmes could do to put this right since extension services were provided by the host country (Harwood 2009), but it was advisable in designing a programme to take the existing extension arrangements into account.[31]

Even when programmes were well designed, however, they still required sponsors who managed them sensibly and could take the long view. But as various practitioners saw it, donor agencies don't like risk and so favoured projects that were likely to 'succeed' quickly (Porter *et al.* 1991: 60; Flora and Flora 1989; Farrington and Bebbington 1994).[32] As Chambers put it,

> There is a recurrent tendency to home in on a single, preferably technical and physical, objective [such as increased food production] .... Narrow professionalism here combines with practical imperatives, the need to do something and to be seen to have done it, regardless of who gains.
>
> (Chambers 1983: 35; cf. Leonard 1986 and Johnston 1983)

When speed was of the essence, experts tended to concentrate on tasks that were relatively straightforward (such as varietal development) rather than more difficult and time-consuming activities (such as extension) though the latter was arguably more important for smallholders.[33] Furthermore, various authors warned, even where programmes had chosen to take on the difficult task of extension, they still often ended up neglecting the needs of peasant farmers since dealing with a relatively small number of enthusiastic commercial farmers was so much easier (Pearse 1980: 16; Chambers 1983: 24).

The second theme that recurs in the diagnostic literature is that the scale of a programme's impact hinged on **the knowledge and attitudes of experts** who dealt with farmers day to day. Arrogance – for example, a boundless confidence in 'science' and an inclination to dismiss local knowledge, whether from scientists or farmers – was not uncommon and had hampered more than a few programmes.[34] Conversely, the successes of several GR programmes in producing effective cultivation practices for smallholders were attributed in part to experts' willingness to listen and observe, to consider the rationale for long-standing indigenous practices, and sometimes to draw upon both indigenous and 'scientific' methods in order to devise hybrid approaches.[35]

Arrogance was not always the problem; sometimes it looked as though experts were simply ignorant of the problems faced by small farmers. Where experts from abroad intended eventually to return home, their decisions upon the appropriate technology to develop were sometimes affected more by the need to impress prospective employers rather than the farmers they currently served

(Goodell 1984). But the clearest evidence on this score concerns experts from the developing world. Unlike the US where throughout most of the twentieth century agricultural scientists came from farm backgrounds and thus identified with farmers there, in the developing world few peasants have been able to send their children for higher education. As a result agricultural scientists tend to come from relatively well-off urban backgrounds and thus have no experience of either manual labour or agriculture, nor much sympathy with the needs of peasants (Cotter 2003; McCook 2002: 114; Gupta 1989; Warren 1989; Stross 1986: 192–95; Shepherd 2006; Flora and Flora 1989; Millikan and Hapgood 1967). And the problem was compounded when higher agricultural curricula in the developing world convey primarily theoretical knowledge rather than practical skills (Goldsmith 1997). In sum, even a well-designed programme is apt to make little impact if its staff are poorly trained and/or insensitive.

Calling for better designed programmes and more informed and sympathetic experts implied that flawed GR programmes could be fixed through more careful organization and recruitment. While these factors probably helped to explain why some programmes had failed to reach peasant farmers, a few commentators drew attention to a third and more fundamental weakness; namely, that many **programmes failed to take into account the political implications of development**. In Chambers' view, for example, during the 1970s it became evident that the most serious limitation of most development programmes was their failure to consider what was politically feasible. Though there were well-established procedures for assessing technical and financial feasibility, political feasibility was not part of standard programme appraisal, even though it was known, for example, that projects aimed at resource-poor farmers were typically captured by local elites (Chambers 1983: 160–163; Barkin 1987; Johnston 1983). Kenneth Dahlberg agreed:

> those development aid programs that seek to improve the condition of the small peasant farmer must necessarily address the question of how existing administrative and political power patterns need to be changed if such programs are to have any chance of success. [That means they] need to become much more sophisticated regarding the real redistributive effects of various kinds of policies and technologies.
>
> (Dahlberg 1979: 131, 180; cf. Ladejinsky 1977; Farmer 1986; Leonard 1986)

Others pointed out that projects were often undermined because supportive measures from the host government were opposed by large landowners. Commercial farmers, for example, tended to block measures in aid of the landless because they wanted to preserve a cheap labour force (Pearse 1980). Policies that could limit the inequitable effects of new technology (e.g. land reform, land taxes) were seldom passed because it was not in the interests of the groups that control government (Griffin and Khan 1978; Johnston 1983; Porter *et al.* 1991; Eastman and Grieshop 1989).[36]

But why had programmes managed to overlook the political dimension? Several observers pointed to the fact that most programmes were dominated by natural scientists who tended to be overly optimistic about the prospects for change and only dimly aware of the economic, social or environmental implications of the Green Revolution (Chambers 1984; Pearse 1980: 216–217; Farmer 1986; Lipton 1978; Lipton with Longhurst 1989: ch. 6). Others noted, however, that social science perspectives, while useful, were no panacea. A technocratic mentality was not unknown even among social scientists, especially agricultural economists (Dahlberg 1979: 182; Chambers 1977). In any event, the conclusion was that until designers had learned to anticipate how their programmes were likely to be opposed or co-opted – and to restructure them accordingly – peasant-friendly interventions were unlikely to have much impact.

The diagnostic literature reviewed above is rich in insights for those in development studies, whether academics or practitioners. It offers, on the one hand, a set of hypotheses about the factors that can divert a development programme away from the needs of smallholders. And by drawing upon a wide range of localities, programmes and sponsors, it also constructs a composite picture – in effect, an ideal type – of how *not* to organize a programme. Though useful, however, types are both abstract and ahistorical. They give us little sense of how concrete programmes actually functioned or the changes that they underwent. If we 'test' the diagnostic literature's ideal type against a particular case study, however, it stands up remarkably well. Elsewhere I have reconstructed the early history of the first of the post-war GR programmes: the Rockefeller Foundation's Mexican Agricultural Program (MAP) (Harwood 2009). What I found was that those who planned the MAP in the early 1940s – academic advisors as well as Foundation officers – were concerned from the outset to develop technologies that would be useful to Mexico's peasant farmers in order to alleviate large-scale rural poverty there. Their concern is evident, for example, in the initial ways in which the Program went about developing improved maize varieties. By the late 1940s, however, there are various signs that MAP staff were finding it very difficult to improve the condition of peasant agriculture. Under pressure from the Foundation nonetheless to produce results, they had begun to concentrate upon easier tasks (such as developing wheat varieties suited to large commercial farms). There appear to have been several reasons for this change of direction. For one thing, Mexico's extension service was weak; its staff were not only ignorant of peasant farmers' needs but unsympathetic as well. Moreover the Ministry of Agriculture showed little interest in strengthening the service, evidently out of deference to large landowners who wanted public expenditure channelled toward other things. In addition, MAP staff did not give as much attention to the social impact of their work as they might have because the Foundation was very slow to appoint an agricultural economist to the Program (and even slower to appoint a rural sociologist or anthropologist). Finally, although MAP staff felt that peasant farmers would be in a stronger position if the government's land reform programme distributed somewhat larger tracts, they hesitated to raise the issue for fear of being accused of interfering in Mexico's political affairs. Thus the MAP embodied many of the weaknesses later identified in the diagnostic literature.

4 Reinventing the wheel

The period of reflection during the 1970s and 1980s was certainly not a waste of time. Analysing one's past mistakes is no bad idea. But one can only wonder whether some of the shortcomings of the first generation of GR programmes might have been avoided, had green revolutionaries of the 1940s and 1950s looked further afield before setting out to transform the developing world's agriculture. Because the odd thing about the diagnostic literature is that for the most part it did not uncover anything new. Virtually all of the problems which experts identified had already been encountered (and solved) in other places previously.

One of those places, of course, was south Germany whose peasant-friendly stations embodied nearly all of the 'hallmarks of success' which the diagnostic literature later identified. The stations' breeding and varietal-testing work, for example, was decentralized. Staff did not restrict themselves to plant breeding or other forms of research but also devoted a great deal of time to extension as well as to organizing farmers into crop-improvement associations. (Organizational work of this kind was necessarily slow and labour-intensive, thus not something the stations were likely to have indulged in, had they been under pressure to generate quick results.)[37] Moreover, the evidence suggests that staff were neither ignorant of peasant agriculture nor arrogant in their dealings with smallholders. Finally, the history of the south German stations clearly illustrates the importance of political support for the success of a development project. The stations owed their existence to state governments' willingness to set up institutions that would invigorate the agricultural economy (while keeping peasants quiet). And because large farms were relatively rare in south Germany, the movement for peasant-friendly institutions could not be subverted by estate owners. As a result, experts did not have to steer clear of agricultural issues which had political implications; the stations were 'politics-driven' right from the start.

But south Germany was not the only historical episode that seems to have gone unnoticed. A number of recent studies of development policy in the British, French and Dutch empires have drawn attention to the experience acquired there between the wars by colonial agricultural officers. To be sure, one could hardly claim that these colonial agricultural services were 'peasant-friendly'. They tended to focus on export crops rather than domestic food production, and there was no shortage of disasters caused by the introduction of inappropriate technology (Henry 1995; Hodge 2007: ch. 7; McCann 2005: ch. 7). Nonetheless, as a result of the failure of numerous development schemes, quite a few individual officers reversed their original assumption that indigenous cultivation practices (e.g. inter-cropping, shifting cultivation, minimum tillage) were backward or irrational. Some concluded that they were actually superior and called for more research on them (Richards 1985; Warren 1989; Bonneuil 1999; Tilley 2010; Uphoff 2007). Mechanization, mineral fertilizers and improved varieties central to intensive cultivation schemes, they concluded, were too expensive for most peasant farmers and often led to serious environmental damage (thus anticipating later criticisms of the Green Revolution). If they were to succeed, development schemes needed not only to take much more account of local variations in soil and climate but also to be based upon consultation and cooperation with peasant farmers. Commenting in 1909 on a proposal to build a model rice farm in Java, one Dutch expert remarked:

> The question is not what maximum possible amount of rice can grow on a certain area, but how it will be possible, once acquainted with rice cultivation as conducted by the Javanese, to increase together with the Javanese farmer his rice yields economically, taking into account his development, workforce and his capital.[38]

That colonial agricultural officers were able and willing to learn from indigenous farmers is perhaps not surprising. Unlike many development practitioners today, they were obliged to learn the local language as well as the history and geography of the region to which they were posted and often spent a decade or more in one place (Kothari 2005). Albert and Gabrielle Howard, for example, felt this kind of grounding was crucial for their work in India (Gieryn 1999: 291). Unfortunately, for reasons as yet unclear, these colonial peasant-friendly insights seem to have been eclipsed after 1945. Few of them played much of a role, for example, in the huge post-war development schemes in British and French colonial Africa.[39]

Yet more striking, however, is another pre-war context from which lessons might have been learned: the remarkable agricultural development of Japan from about 1880 to 1920. Since Japanese farms were very small (1–2 hectares), there was little point in investing large sums in machinery; draft power was instead supplied by animals (Johnston and Kilby 1975: 422). In any event, neither the capital nor the foreign currency required to purchase equipment were available. Nevertheless the introduction of intensive cultivation was possible thanks to the availability of an extensive irrigation system, state subsidies for fertilizer purchase, a well-developed extension service, and a network of local cooperatives with which extension agents could work (Sawada 1965).[40]

At the start of the Meiji Era (1868–1911) Japan had sent agricultural experts abroad who brought back the technology used on large American and British farms, but most of this – including imported high-yielding plant varieties – failed. From the 1880s, therefore, the government decided to develop new agricultural technologies which would be better adapted to Japanese growing conditions. This was done by starting with the best practices already in use by farmers and improving them with the aid of western agricultural science. These 'hybrid' practices proved much more successful.[41] Varietal testing, for example, was conducted by local farmers' organizations who examined a range of native varieties, discovered several excellent ones which had been bred by skilled farmers during the 1870s, and distributed the best of these throughout the relevant region.[42] From about 1900, however, both the breeding and distribution of rice and wheat were taken over by public-sector experiment stations and seed-growing cooperatives. As in south Germany, the stations began by analysing the best of the improved native varieties (Ishikawa and Ohkawa 1972; Ishikawa 1981: 163–164). Moreover, the breeding process was organized in a decentralized manner. Although the initial crosses and selection in the first few progeny generations took place at the national experiment station, subsequent selection was carried out at the regional experiment stations, and the resulting lines were then distributed to provincial stations for testing (Ogura 1967: 325, 647; Hayami *et al.* 1975). Breeders also took into account the circumstances in which farming took place, aiming not just to increase yield but also to develop rice varieties suited to marginal growing conditions, including regions previously thought to be quite unsuitable for rice cultivation (Ogura 1967: ch. 1 and part 3; Francks 1984: 77–80, ch. 5; Ushiyama 2008). Similarly, when Japan began to improve rice cultivation in colonial Taiwan after 1918, breeders noticed that high-yielding Japanese varieties did not do well in the semi-tropical conditions there. Adopting much the same strategy that had been used in late nineteenth century Europe, therefore, they chose to cross the imported varieties with well-adapted local ones, producing the very successful 'Ponlai' variety (Hayami 1971). The successes of Japanese breeders were not unknown abroad. Japanese dwarf wheat varieties were chosen as parent lines by a major Italian breeder in the 1920s and by the Mexican Agricultural Program in the 1940s (Saraiva 2010; Dalrymple 1985; Perkins 1997).

Japan's agricultural productivity grew rapidly from the 1880s, and the agricultural economy generated the capital necessary for industrialization. Moreover the 'Japanese model' proved very successful when transplanted to Taiwan over the period *c.*1900–1940 (Johnston and Kilby 1975).[43] To be sure, the model was not so readily applicable to other Asian countries after 1945. Japan's population growth rate between 1880 and 1930 was lower; its research and extension system as well as its irrigation network were unusually well developed; and levels of literacy were high (Francks 1984). Nevertheless several features of the model were still relevant for conditions in the developing world. It demonstrated, for example, that even small farms' productivity could be greatly increased, that improvements to cultivation did not require importing expensive technology from abroad but could be based on promising indigenous practices, and that such practices could absorb large amounts of rural labour.[44] As a result, agriculture was able to earn foreign exchange, provide investment for the emerging industrial sector, supply raw materials for industry, and meet the growing urban demand for food (FAO 1966). Despite this relevance, however, the first generation of Green Revolutions seems to have borrowed very selectively from the 'Japanese model', adopting its plant varieties while turning a blind eye to the other features which had made it so successful: its breeding practices, institutions and policies.

5 Why?

The key question is obviously 'why?' How could a development programme which had declared war on hunger have fallen so short, given what might have been learned from its historical predecessors? There are several possibilities to consider.

The first GR programmes, for example, were planned and implemented largely by American agricultural scientists. Could it be that these experts were simply unfamiliar with the problems of peasant agriculture?[45] The average farm in the US, after all, was 20 times the size of most smallholdings in the developing world and used intensive cultivation methods which were still relatively rare in the global South during the 1950s and 1960s. This is unlikely for several reasons. For one thing, it is not the case that the US had no tradition of peasant agriculture; such farms were common in the South and in the Appalachian mountains into the 1950s (Gregg 2004; Magdoff *et al.* 2000: 13), though there are hints in the literature that the problems faced by such farmers were largely ignored by the US Dept. of Agriculture as well as by most land-grant universities (Flora and Flora 1989; Daniel 1985).[46] For another, at least some American experts will have encountered peasant agriculture through working in US territories in the Pacific or Caribbean, though the evidence also suggests that those employed in Cuba or Puerto Rico devoted very little attention to the needs of smallholders (McCook 2002). Was such experience with peasant farming not widely known among US agricultural scientists, or was it ignored?

Similarly, if any American expert in the 1950s had wanted to anticipate the possible social consequences of the Green Revolution, they need only have looked at the history of agriculture in the US. The picture they might have found was sobering. Between 1890 and 1920 agricultural scientists had tried to talk up the merits of boosting productivity through mechanization, higher yielding plant varieties, and the use of commercial fertilizers, but farmers were not interested. Adopting these methods, they objected, would increase their costs, and unless demand continued to rise, boosting production would only lower prices, so why bother? Indeed, during this period the champions of high-tech agriculture were to be found, not in the farming community, but among urban interest groups keen to stimulate industrial growth (Danbom 1979). As the advocates of high-tech agriculture saw it, an abundance of cheap food would enable factory owners to keep wages under control. And if the growth of high-tech methods led to a reduction in the number of farms, that was OK since those farmers forced out of business might come to the cities in search of work. Eventually the onset of the Second World War, however, gave a huge stimulus to the economy, persuading farmers to switch to the new technology on a large scale. Capital investment in farming rose quickly: the number of tractors on farms doubled during the 1940s; the use of nitrogen fertilizers increased 18-fold over the next 30 years; and monoculture necessitated the widespread use of pesticide (Cochrane 1979: 126–127; Kloppenburg 2004: 116ff.).[47] Average farm size steadily increased as large commercial farms – which were better able to invest in machines and inputs and better served by the extension service (Buttel and Busch 1988; Hightower 1973) – bought up the smaller ones which had failed. By reducing labour costs, however, mechanization also brought unemployment; the farm labour force fell by 70 per cent between 1940 and 1970, and more than half of American farms closed. During the 1950s alone 1.5 million farm families were forced out of business (Cochrane 1979). The result, as even those economists who endorsed the high-tech revolution agreed, was appalling levels of poverty. By 1965 there were 14 million people in rural areas living in poor housing with high rates of unemployment and illiteracy and 'shockingly high' rates of infant mortality (Cochrane 1979: 148). Were green revolutionaries unaware of what was going on in US agriculture? Or did they regard rural poverty as collateral damage?

Another apparently plausible explanation is cultural and geographic isolation. On the face of it, the world of most green revolutionaries in the 1950s and 1960s would seem to have been linguistically remote from both Central Europe and Japan. And since many of these scientists were from the US, one can imagine that few of them would have been familiar with the work of European colonial agricultural services. This explanation, however, also fails. For one thing, the work of colonial agricultural scientists was not unknown in the major development agencies after 1945. Many of the experts who had acquired experience of peasant farming while working in British colonies during the 1930s to 1950s, for example, were subsequently employed by the World Bank, FAO and similar organizations. And several of them evidently looked on in disbelief as post-independence programmes simply ignored the colonial experience and proceeded to make the same mistakes all over again (Kothari 2005, 2006; Hodge 2007, 2010; cf. Frey and Kunkel 2011).[48]

Nor were US agricultural scientists ignorant of what was going on in early twentieth century Germany. Numerous institutions facilitated transatlantic contact. The first international conferences for genetics and plant breeding (held in London and New York in 1899, 1902 and 1906), for example, were attended by a large contingent from agricultural colleges in all countries, including plant breeders from Germany and the US (Wilks 1906). Another such channel may have been the International Congresses of Agriculture which took place at regular intervals from 1889 until 1939 (Aldenhoff-Hübinger 2002). Of particular importance was the Rockefeller-financed International Education Board (IEB) whose fellowships enabled European agricultural scientists in a range of fields to visit US agricultural colleges and experiment stations.[49] Moreover the IEB was particularly important as a conduit for information about plant breeding. The key figure was undoubtedly Albert R. Mann, dean of Cornell University's School of Agriculture and the IEB's director of agriculture from 1924 to 1926. Already in 1908 he had visited Europe in order to learn about work in barley breeding, in the process meeting Carl Kraus, director of the newly established Bavarian station.[50] During the 1920s Mann again visited the major European experiment stations on behalf of the Board, alert to the possibility that American agriculture might have something to learn from them. Among the institutions he visited were the south German stations, and although his diaries do not single out their peasant-friendly orientation for comment, he seems to have been aware that European peasants had different needs from most American farmers. Mann also became acquainted with the problems faced by peasant farmers through the IEB's support in 1924 for a 'Cooperative Plant Improvement Program' in China which sent breeders from Cornell University to the University of Nanking. Since average farm size there was only one or two hectares, the breeding programme had to be tailored to the needs of small farmers (Stross 1986; Love and Reisner 1964). After the IEB was wound up in the late 1920s, Mann continued to play a major role in Rockefeller agricultural programmes, eventually becoming a vice-president of the Foundation. As the Foundation began to explore the possibility of a Mexican development programme in 1941, therefore, it was natural that he was consulted on its design, and from 1946 he served as part-time officer in the division of Natural Sciences with responsibility for the Mexican Program (Harwood 2009). Through Mann, therefore, all of the Foundation's pre-war knowledge of peasant agriculture, in south Germany as well as in China, was available to those who designed the earliest of the post-war GR programmes, even if they made little use of it.

Finally, consider the case of Japan. Despite the apparent linguistic barrier, the country's rapid agricultural growth by the start of the twentieth century soon attracted interest in the US. Already in 1911, for example, a division chief at the U.S. Department of Agriculture was urging American readers to take note of the very high yields that farmers in Japan and elsewhere in East Asia were getting, using neither mechanization nor mineral fertilizer. Their apparently crude and simple tools were cheap, efficient, and ideally suited to the conditions, he argued, and their cultivation methods made good use of the rural areas' abundant labour (King 1911). Although US agriculture chose not to follow that particular path, by the 1950s and 1960s Anglophone development experts were again calling attention to Japan as a potential model for other Asian countries. In 1951, for example, Bruce Johnston suggested that western planners might do well to draw upon Japan's experience because

> The Western world has only begun to develop an alternative formula [to that of the USSR] for fostering economic progress and raising levels of living in [Asian] countries, and experience in the West is in large part inapplicable to Asiatic conditions.
>
> (Johnson 1951: 498)

From the 1950s through the 1970s articles and books on what was called the 'Japanese model' were common in the Anglophone development literature. Nor was this view confined to individual authors; in 1963 the FAO published a report on the Japanese model – in English, French and Spanish – for its World Food Congress (FAO 1966). No development expert then interested in Asia, therefore, can have been unaware of the Japanese experience. As far as I can see, however, from about 1980 the Japanese model largely disappeared from the development literature. The most likely explanation would seem to be that state-led development programmes – especially a successful one like Japan's – would have seemed anomalous against the backdrop of the emerging Washington Consensus. The 1980s was a period in which experts in development studies were losing confidence in the state as an agent of development (Cameron 2005; Lewis 2005). And at the World Bank officials were finding it difficult to accept the continuing emphasis in Japanese aid policy upon state-led development (Wade 1996). If this interpretation is correct, the development industry's neglect of the Japanese experience since the 1980s was due not to ignorance but to ideological predisposition.

If the GR programmes of the 1950s and 1960s took little note of early twentieth century work, therefore, it was certainly not because no information was available. Alternatively, it might be argued that green revolutionaries did have access to this information but did not make use of it because the Central European context (or that of Japan) were too different from those of the developing world after 1945. It is certainly true that the early twentieth century cases were different from the global South in key respects. Japan had a much more extensive irrigation system than most developing countries, and both Japan and south Germany possessed well-developed systems of extension and primary education. Moreover neither country had to reckon with resistance to reforms from a powerful class of estate owners. But these differences do not mean that the European or Japanese experience was *irrelevant* for GR planners. As a general principle, even cases which are quite distinct can still be useful for comparative analysis as long as similarity exists at some level. Besides, it is clear from the development literature of the 1950s to 1970s that experts at that time believed that the 'Japanese model' *was* relevant for devising strategies for other Asian countries. And as we shall see in the next chapter, many of the approaches to improving smallholder agriculture which are being proposed at the moment are little different from those successfully pursued in south Germany and Japan a century ago. If planners of the 1950s and 1960s failed to learn from past programmes, therefore, it is not because they were irrelevant.

Finally, could it be that development experts were indeed aware of early twentieth century work and did regard it as relevant, at least in part, but that peasant-friendly practices are just very difficult to implement in most countries? If this were the case, however, what we would see is not that the wheel would keep being reinvented, but that experts would struggle to improve and adapt methods which they borrowed from the past. Thus there ought to be some signs of a learning process going on in which experience is collected and reflected upon. The development literature would refer to long-standing attempts to use particular approaches, would note the difficulties in achieving success with them, and would suggest how such approaches might be modified in order to work better. I find nothing of this kind. Recommended peasant-friendly approaches are presented instead as though they were new. It is thus difficult to avoid the conclusion that development experts have not so much been ignorant of past experience as indifferent to it.

Conclusion

All in all, it appears that green revolutionaries chose to learn the hard way. Some might want to argue that the first generation of GR planners took no notice of successful earlier approaches to development because those GR programmes were primarily concerned to boost production rather than alleviating poverty. Even if that could be demonstrated, however, one still has to account for the fact that the European and Japanese episodes found only occasional mention in the diagnostic literature of the 1970s and 1980s whose authors *were* concerned with assisting the rural poor as well as committed to learning from the mistakes of the past.

Nevertheless, criticism of the Green Revolution did more than simply trigger a period of reflection and self-criticism. It also had tangible consequences, prompting a reorientation of breeding programmes and the development of new peasant-oriented approaches to agricultural research. In the next chapter, therefore, we will look at the emergence of these approaches and assess their impact upon the development community.

7 Reforming the revolution

Peasant-friendly innovation, 1970–2010

As we have seen, criticism of the Green Revolution prompted green revolutionaries to think more carefully about the design of development projects. Did criticism, however, have any impact upon actual development practices? Were new cultivation technologies devised or different strategies of development pursued in order to make the Green Revolution more peasant-friendly? In this chapter I trace the emergence of several such new approaches during the 1980s and 1990s, noting how they have fared since then. In the second section we shift to the present day, looking at a series of recent assessments of development problems in order to gauge what importance is now being attached to peasant-friendly approaches by experts and major development agencies. And in the final section I turn to the latest innovation in breeding method – biotechnology – in order to test the claim that it can generate the improved varieties for an equitable and sustainable agriculture which will at last alleviate rural poverty.

1 New technologies and approaches since the 1970s

In response to the critiques around 1970, GR experts' attention began to shift toward the needs and circumstances of small farmers. Part of the reason for this was that the relationship between food supply and poverty was being reconsidered. Economists had begun to rethink the nature of rural poverty. Hunger, they concluded, could not simply be due to a shortage of food because in numerous cases hunger persisted despite food surpluses (not least in the United States where during the 1980s 12 per cent of poor children were said to be suffering from malnutrition (Lappé and Collins 1986: 14)).[1] Surplus grain, for example, might be exported or could end up as animal feed; it was rarely handed over to the starving (cf. Stone 2002). The problems, the U.S. Department of Agriculture concluded, 'are not those of increasing the production of food but of distributing it' (cited in Perelman 1977: xii). The poor could not either grow enough food for themselves or afford to buy it (Wittwer 1983; Staatz and Eicher 1998; Dalrymple 1977: 203). Moreover studies were showing that whereas increasing productivity among subsistence farmers improved the average nutritional status of the population, boosting the productivity of commercial farms did not (Pearse 1980: 212).

By about 1970 it had also become apparent that the Green Revolution's 'package' of technologies was not being adopted as readily by small farmers as by large ones. And even those small farmers who adopted it did so only partially. In the past such behaviour would usually have been attributed to ignorance or conservatism, but by this time evidence had accumulated, demonstrating that smallholders were not irrational or incompetent in their choice of technology. Small farms were actually quite efficient, once one took into account the constraints under which they had to operate. Moreover, they did not passively adopt technology but rather experimented in order to improve it (Schultz 1964; Biggs 1980; Richards 1985, 1994). From this point of view, it looked as though peasants were responding selectively to the Green Revolution, choosing those parts of the package that looked suitable for local conditions (or adapting them to that end) and discarding those that produced poor results (e.g. Dalrymple and Jones 1973; Farmer 1986). The problem, therefore, was that experiment station research seemed to be out of touch with farmers' needs.[2]

Some economists went further, arguing that the common distinction between aiming for 'efficiency' versus 'equity' (often formulated as 'growth' vs. 'poverty alleviation') was misleading and unnecessary. Because the output per hectare on small farms was almost always greater than on large ones, *both* of these aims could be achieved by an agricultural strategy which made it easier for small farmers to adopt improved (and appropriate) technology. This was the most efficient way to extract the maximum social and economic benefit from the Green Revolution (Johnston and Kilby 1975). Nor were these ideas persuasive only among experts. By the early 1970s the president of the World Bank was writing that 'In my view ... there is no viable alternative to increasing the productivity of small-scale agriculture if any significant advance is to be made in solving the problems of absolute poverty in the rural areas.' (McNamara 1987: 426; cf. Finnemore 1997; Staples 2006: 40ff.). Some UN agencies were arriving at the same conclusion (Esteva 2010). Eventually the Consultative Group on International Agricultural Research (CGIAR) fell into line (Oasa and Jennings 1982; Buttel *et al.* 1985; Redclift 1983), and the Rockefeller Foundation and the U.S. Department of Agriculture conceded that their earlier enthusiasm for the Green Revolution had been excessive (Perelman 1977: 161).

But how to do this? As we saw in the last chapter, some thought the solution was to stick with the existing GR technology while redirecting policy so that smallholders had easier access to the technology as well as improved infrastructure and supportive institutions (Pinstrup-Andersen and Hazell 1985; Barker 1970). Another approach was to broaden the scope of the existing technology to include crops and traits which were of particular importance to smallholders but had so far been neglected. From the mid 1970s this kind of breeding work was taken up at various international agricultural research centres (IARCs) on crops such as sorghum, millet, pulses, cassava and yams (Lipton 1978; Evenson and Gollin 2003; Wolf 1986: 15). In addition breeders at these institutions began to place less emphasis upon improving yield and paid more attention to traits important to smallholders – such as yield stability (Lipton 2001) – as well as those important to rural consumers such as taste or cooking properties (Dalrymple 2006).[3] The IARCs also became more concerned with breeding varieties that were less dependent upon commercial inputs. Already in 1969, for example, agricultural economists at IRRI were concluding that in the long run pest-resistant varieties would be better than those dependent upon insecticide because of the latter's cost and uncertain benefits (Oasa and Jennings 1982). Breeders at various institutions also sought to develop varieties better adapted to sub-optimal environments. At IRRI, for example, the main emphasis from the 1970s was upon rainfed rice and drought tolerance, (Lipton 1978; Lipton with Longhurst 1989; Anderson *et al.* 1991). From the late 1960s, similarly, the International Centre for Maize and Wheat Improvement (CIMMYT) established a programme to develop maize varieties for small farms in non-irrigated regions of Mexico (Lewontin 1983). In the 1970s the body that funded the IARCs – the CGIAR – also established new research centres to address the problems of dryland agriculture: one at Hyderabad for the semi-arid tropics and another at Aleppo for temperate dry areas (Mann 1997; Evenson and Gollin 2003; Sanghi 1989). By the 1970s, too, the IARCs were coming to the conclusion that it was impossible to breed centrally for a huge range of diverse growing conditions. They decided, therefore, that henceforth they would not hand over finished varieties but rather unfinished germplasm to national breeding stations which would then select the material further under local conditions before release (Dalrymple 1977; Anderson *et al.* 1991; Farmer 1979). (Although this represented an important shift in the IARCs' breeding policy, one wonders why it took so long to get there; a decentralized breeding strategy of this kind had been standard practice at the south German stations 50 years earlier.)

By the mid 1970s, however, some experts were conceding that the idea that technologies developed in the global North could be adapted for conditions in the South – heretofore 'an article of faith' – was looking increasingly doubtful (Johnston and Kilby 1975: 441). Not everyone thought it sufficient merely to extend the scope of the existing technology. Some wanted more radical change: an 'appropriate' technology designed specifically for resource-poor farmers.[4] By the late 1970s, accordingly, new approaches to technical innovation were emerging, characterized by a reappraisal of indigenous cultivation practices (once thought to be 'backward') used on small farms in difficult environments. Robert Chambers' argument – that agricultural experts should put 'farmers first'– was particularly influential (Chambers 1983; Chambers *et al.* 1989; cf. Arndt *et al.* 1977: preface) He called for a wide-ranging change of perspective among agricultural scientists which entailed not just taking poor farmers' practices seriously but also enlisting their participation in both problem definition and research and devising solutions that would be sustainable in unfavourable growing conditions.

Over time, work in this general area became more specialized, being pursued by distinct groups of experts. In this section I will focus on three such strands now known as farming systems research (FSR), participatory plant breeding, and agroecology. To call these approaches 'new', however, is slightly misleading.

Within the context of the post-war Green Revolution, to be sure, they represented a marked shift of interest away from the large commercial farms and toward small resource-poor ones. But viewed in the larger context of the twentieth century, they were not entirely novel since several of their features were already part of early twentieth century peasant-friendly work. There are obvious similarities, for example, between FSR and the interest in peasant cultivation practices displayed by some colonial agricultural officers in Africa and Asia before the Second World War (Tripp *et al.* 1990). And one reason for this may be that a good deal of the early work in FSR was done by scientists who had become acquainted with peasant farming while working in the colonies (Collinson 2000; Joseph Hodge, pers. comm.). Similarly the decentralized breeding strategy pursued both in south Germany and in Japan *c.*1900 – applying modern methods to improve the best of the available local varieties – represented a hybrid of western expertise with indigenous knowledge of the kind also sought by farming systems researchers.[5]

One thing that was new about these three approaches, however, was the greater prominence given to the social sciences. During the 1960s it was becoming clear to some experts that social, cultural, economic and political constraints substantially influenced whether a technology was adopted and if so, how it was modified by farmers and with what consequences. Since natural scientists based at experiment stations often seemed unaware of these factors (Norman 2000; Collinson 2000), the new approaches often called upon social scientists to understand existing cultivation systems and their rationale and to suggest ways in which they could be improved so as to better meet farmers' needs. Accordingly major donors such as the U.S. Agency for International Development (USAID) began to employ more social scientists, as did the CGIAR system. In 1971, for example, IRRI convened for the first time a meeting with rural sociologists and other social scientists to consider research which would 'improve our understanding of the changes occurring and the problems associated with the adoption of the new rice technology....' (Oasa and Jennings 1982: 40–41). They also launched joint projects with social scientists and in subsequent years held several conferences on the economic consequences of the new technology. At CIMMYT in the late 1960s, similarly, various staff began to think that socio-political obstacles were preventing peasants from benefiting from the centre's crop improvement work and called for more research aimed at small farmers. Prompted by the Ford Foundation (one of its major donors), CIMMYT's response in 1971 was to establish an economics section to study the social consequences of new technologies (though some suspected its actual role was to protect the Centre against criticism) (Jennings 1988; Oasa and Jennings 1982). By 1974 the CGIAR had concluded that the social sciences needed to play a larger role within the IARC system as a whole. The representation of these disciplines was subsequently strengthened at several of its institutes, and in 1979 the International Food Policy Research Institute (IFPRI) was incorporated into the system. (Ruttan 1989; Warren 1989; Norman 1978).

During the 1980s these new approaches attracted substantial support since donors were prepared to fund an 'almost bewildering array' (Farrington 1988:

271) of projects aimed at the problems of resource-poor smallholders (Krishnamoorthy 1982; Allen 1987; Chambers *et al.* 1989; Hazell and Ramasamy 1991). And although public-sector agricultural institutions in the developing world were at that time becoming weakened by the policies of the World Bank and other major donor agencies (see below), donors instead directed funding toward non-governmental organizations (NGOs) whose concern with the environment and poverty alleviation made them particularly receptive to peasant-friendly approaches (Tripp *et al.* 2006). Let us now look in more detail at each of these three new approaches.

Farming Systems Research (FSR)

The basic problem that FSR tackled was the frequent mismatch between the available technology and the huge diversity of conditions in which smallholders worked (Ashby 1990). FSR practitioners began by looking closely at small farming systems, focusing not upon single crops but upon the whole system of production used on such farms. But the 'system' was seen to consist not just of the biophysical factors in production but also the socio-economic context of the farm. This meant that FSR was necessarily multidisciplinary. Social scientists were needed to make sure that attention was paid to a system's economic return and risk as well as its social acceptability and impact. It also meant that FSR's recommendations sometimes included institutional reforms targeted at small farmers. Once the system had been defined – including farmers' perceptions of the problems they faced – FSR identified those of its elements that limited production. By doing this, it became clearer why some GR technologies introduced in 'top-down' fashion had not been adopted: they were incompatible with constraints – be they economic, ecological or cultural – which had shaped the existing system. Armed with this knowledge, the researcher went back to the experiment station in order to improve some of the existing cultivation practices. Finally the eventual improvements were brought back to the farm to be tested (Norman 1978).[6]

One common practice that FSR analysed was inter-cropping (or 'mixed-cropping') in which smallholders plant particular combinations of crops in a single field. Often this is a cereal with a legume (such as the well-known combination of maize and beans in Central and South America). In search of a rationale for this practice, FSR scientists calculated that although the yield of a given crop in such a field was lower than when grown in monoculture, the total yield of *all* crops in the field was substantially higher (Chambers 1983: 85–92; Krishnamoorthy 1982). In addition inter-cropping reduced the fluctuation in yields from one year to the next. The efficacy of the practice was found to have a biological explanation: given the right combination, the two crops do not compete for resources such as light, water or soil nutrients but instead make more efficient use of them so as to reduce or eliminate the need for expensive inputs (Francis 1990).

Some of the earliest work in FSR took place at IRRI during the 1960s when the soil scientist Richard Bradfield was arguing that more attention should be paid to the diversity of growing conditions worked by small rice growers so that new cropping systems could be developed to meet their needs. Initially he seems to have been ignored, but by the end of the decade, with increasing recognition that the Institute's new rice varieties were not proving universally successful, an IRRI committee recommended that more work be conducted on the (sub-optimal) growing conditions – ecological, social and economic – in which rice was actually grown, leading in 1972 to the establishment of a Cropping Systems Program (Anderson *et al.* 1991). During the 1970s work in FSR also took off elsewhere, and the approach was established at several national and regional institutions as well as at other CGIAR institutions (Harwood, R. 2000; Baker and Norman 1990; Norman 1978; Waugh *et al.* 1989). The 1980s was thus a period of considerable expansion when some regarded FSR as having become a 'dominant concept' (Tripp *et al.* 1990: 384).

Participatory Plant breeding (PPB)

The growing interest in working closely with farmers extended to plant breeding as well. As we have seen, the breeding strategy of the Green Revolution had been a cosmopolitan one (as discussed in Chapter 2) which had worked well for commercial farms in good growing conditions, but was less successful with resource-poor farmers operating in highly diverse environments. GR breeders attempted to solve this problem by testing their centrally bred lines in marginal conditions in order to identify some that were well adapted. But this strategy was still plagued by low adoption rates of such varieties. Part of the problem was that breeders tended to select for yield while ignoring other things that farmers wanted (Maurya 1989; Biggs 2008; Witcombe 2001). Farmers are sometimes unhappy, for example, with the dwarf cereals varieties released by experiment stations because these provide less straw which can be fed to their animals. But at the other extreme, breeders were often discarding lines that farmers actually preferred.

Critics of this breeding strategy (like their counterparts in south Germany *c.*1900), therefore, argued that a more effective way to obtain well-adapted varieties was to breed locally in the relevant marginal conditions (Ceccarelli 1989; Simmonds 1991; Ceccarelli *et al.* 1994). If one took already adapted local varieties as the starting material and established through consultation with farmers the kinds of varieties they value, there was much less risk of developing varieties whose traits were of little interest to the locals. The extent of farmer participation in PPB varies widely. Often the breeding is conducted by experts with farmers merely providing breeding material or being consulted on choice of variety (Sperling *et al.* 2001). But where varieties for a wide range of growing conditions are required, this arrangement can be prohibitively expensive. It then makes both ecological and economic sense for such breeding to be carried out by the farmers themselves working in those environments (Ceccarelli n.d.).[7] The justification for proceeding in this way is the evidence that farmers are effective and thoughtful experimenters (Richards 1985) who typically possess more extensive knowledge of their local growing conditions, crops and farming systems than do experts (though it has often taken the latter a while to realize this: Ashby 1990; Marglin 1996).

The results have been promising. Studies of PPB from various regions indicate that farmers who were taught how to do mass selection were able to develop varieties which proved popular with farmers, in part because their yields were equal to (and occasionally greater than) those of high-yielding varieties while their input requirements were lower (Cleveland and Soleri 2007; Frossard 2002). The time required for the development and dissemination of improved varieties is also substantially shorter than with conventional breeding (World Bank 2007: 160).

Precisely when the movement for PPB emerged is not clear. Some see it beginning at some IARCs in the late 1970s (Biggs 2008) while others regard it as taking off a decade later (Ceccarelli 2009; Chambers 1990). During the 1990s there is general agreement that participatory approaches to technology development were flourishing (Harriss 2005). For Gordon Conway this approach had 'finally challenged the traditional top-down process that has characterized so much development work.' (Conway 1997: 199). Practitioners of PPB were also optimistic, regarding it as having become 'a popular component' of international agricultural research (Cleveland and Soleri 2002: 11; cf. Chambers 2005: 78; Rhoades 2006: 417). PPB was finding support from donor agencies who were keen to encourage social scientists to collaborate with natural scientists on this and similar problems (Biggs 2008). And by 2000 there were about 50 programmes around the world, most them located in NGOs, a few IARCs, or universities (Ceccarelli n.d.; Ashby 1990).

Agroecology

The central problem informing agroecology is the vulnerability of industrial agriculture and the damage it tends to inflict upon the environment. Sustainable cultivation practices are seen as particularly important to poor farmers because, having few options, they stand to lose most from environmental degradation. Like FSR (and unlike conventional agronomy), agroecology does not focus upon single crops but rather upon the complex *systems* of crops and other species (including trees, animals, and soil flora/fauna) which interact on a given field. And as with FSR, the scientist's task is to look carefully at existing small farming systems, select those that are ecologically sound and promising, and use them as a basis for developing improved ones whose disruption to the agricultural ecosystem will be minimal (Altieri and Hecht 1990). 'Improvement' generally means 'sustainable intensification' (Pretty 1997). Yields are increased by developing cultivation methods which rely less upon inputs and thus are both cheaper and less damaging to health and the environment than are conventional GR methods. The process is supposed to produce an array of options from which farmers can select what is most appropriate for their circumstances. Some proponents argued that although marginal land had often been written off by green revolutionaries, agro-ecological technologies were able to increase the productivity of such land by as much as three- or fourfold, resulting in marketable surpluses (Conway and Barbier 1990).

One example is 'integrated pest management'. In an attempt to reduce farmers' dependence upon pesticides, conventional chemical and genetic approaches to pest control are combined with changes in cultivation practice as well as the introduction of the pest's natural enemies. A closely related approach is 'low external-input agriculture' in which cultivation methods are developed which make use of local renewable resources rather than externally supplied ones, as far as practicable, on grounds of both sustainability and cost (Tripp *et al.* 2006). A more recent (and controversial) example is the 'system of rice intensification'. SRI is a cultivation method based on farmer practices in Madagascar which have been improved by systematic experimentation. By deploying specific ways of planting, watering, fertilizing and weeding, SRI has in some areas produced rice yields twice as high as those achieved with the usual GR packages while reducing water requirements and lowering production costs (since mineral fertilizer and herbicides are avoided). One drawback of the method is that it is more labour intensive than conventional production methods, though this is not an obstacle in those rural areas where surplus labour is abundant (Uphoff 2007; Stoop *et al.* 2002; Lines and Uphoff 2005).

Although integrated pest management began to be used in the developing world already in the 1960s (Conway 1997), agroecology more generally seems to have taken off in the 1970s (Hogg 2000: 18), prompted presumably by critics of the Green Revolution who had drawn attention to its ecological impact, the oil crisis of the early 1970s, and the emergence during that decade of environmental movements in many countries.[8] During the 1980s and 1990s sustainability became a prominent theme in development research and funding (Chambers *et al.* 1989; Staatz and Eicher 1998; Woodhouse 2009), among other places at the World Bank (Goldman 2005) and the Rockefeller Foundation under the leadership of the agricultural ecologist, Gordon Conway (Paarlberg 2008).

Ten to 20 years ago, therefore, FSR, PPB and agroecology were evidently flourishing. But what is the extent of their impact upon the development industry today? One measure of that impact might be the CGIAR's new strategy (2000) which emphasized food security, poverty alleviation and sustainability (Kassam 2006). Certainly 'sustainability' and 'participation' are by now to be found in the reports and strategy documents of virtually all donor agencies (Rist 2008: 178ff.). Furthermore in the mid 1990s some were claiming that the 'farmer-first' approach had led to a 'virtual revolution in the agricultural sciences' (Scoones and Thompson, 1994b: 2). On the other hand, in the same year Chambers himself reckoned that the changes required were 'still nowhere near' being achieved (foreword to Scoones and Thompson 1994a: xiii). And some observers would say that the meanings of 'sustainability' and 'participation' have been greatly diluted since their original formulation. How, then, has each of these three approaches fared since the 1990s?

FSR seems not to have fared particularly well. Despite its high profile during the 1980s, FSR was weakly institutionalized from the outset. Often perceived as a 'social scientists' invention', it has had to deal with scepticism from both natural scientists and economists, rarely securing the kind of interdisciplinary collaboration which it sought (Tripp *et al.* 1990; Rhoades 2006; Biggs and Farrington 1991: 79). At IRRI, for example, the new Cropping Systems Division encountered suspicion from other divisions. According to one account, breeders – who had learned nothing about ecosystems in their training, were little interested in existing rice cultivation systems, and saw their own scientific status hinging on the production of new varieties – feared that the growth of the new division would diminish the importance of their own work. Although there was awareness at IRRI throughout the 1970s that work on the station was remote from the actual growing conditions of most farmers, defenders of the conventional style of centralized research insisted that their kind of work was the only way to attract money and good people (Anderson *et al.* 1991).[9]

Getting plant scientists to cooperate was not FSR's only problem. Although funding from various large donors (e.g. the World Bank) was for a time substantial, donor interest in FSR seems to have peaked around the mid 1980s. Thereafter a number of projects were phased out by major donors. USAID decided to shift its funding back toward traditional agronomic research as well as biotechnology, and Canada's International Development Research Centre also sharply reduced funding (Baker and Norman 1990; Waugh *et al.* 1989). In the developing world the widespread perception that on-farm research is relatively expensive (Norman 1978), combined with the destructive effect of Structural Adjustment programmes (see below) upon public-sector agricultural research, has meant that many national agricultural research systems (NARS) could no longer afford FSR (Ian Scoones, pers. comm.). Just why donors lost interest is not clear, but proponents of FSR have argued that while the kinds of improvement that FSR recommends necessarily take a decade or two to make a significant impact upon the region targeted, both donor agencies and national agricultural ministries have grown impatient, returning instead to conventional research because of its short-term payoffs (Baker and Norman 1990; Collinson and Lightfoot 2000).

Finally, by the 1990s even proponents of FSR were acknowledging that its payoff had fallen short of expectations because practitioners often failed to abide by FSR's principles. Either they focused narrowly upon crops while leaving relations to livestock out of account, or they failed to take sufficient account of farmers' needs, or they lost sight of the institutional and policy environment in which an innovation had to operate, such that relatively few successful technologies were developed (Baker and Norman 1990; Tripp *et al.* 1990; Chambers 1990; Ashby 1990). Quite how much impact FSR has made upon development practice by now is hard to say. Some argue that despite opposition in some quarters and major donors' lack of enthusiasm, the basic principles of FSR are now widely diffused (Bellon *et al.* 2006; Rhoades 2006; McIntyre *et al.* 2009b: 65), for example in southern African development projects (Bowman 2011), thanks to continuing support from NGOs.

Agroecology's status today appears ambiguous. To be sure, there is evidence that sustainable cultivation practices can substantially increase yields in unfavourable growing conditions (Pretty *et al.* 2006). The field's achievements are recognized in a recent report from the Royal Society (Royal Society 2009), and the World Bank cites conservation tillage among promising new technologies (World Bank 2007: 163–164). But it is not clear how many people in development circles are actually listening. In the mid 1990s, for example, a 'very small proportion' of the IARCs' budgets was devoted to developing low external-input technologies (Pretty and Chambers 1994: 196), and the situation seems not to have changed much since then (Tripp *et al.* 2006: 207). This institutional weakness is also reflected in agroecology's public visibility. One search of the *New York Times* archive for the period 1981–2008 found only 10 citations to 'agroecology' or 'agroforestry' versus several thousand to 'genetic engineering' (Vanloqueren and Baret 2009: 976). As one advocate put it, although 'sustainability' is by now a commonplace in development talk, it is much less clear that it has made a substantial impact upon development practice (Hogg 2000: 70).

The fate of participatory technology development is rather similar. Although 'participation' is now commonly endorsed by virtually all development agencies, its impact upon mainstream development work is not so obvious (Biggs 2008; White 1996; Scoones and Thompson 1994b). By the early 1990s, for example, one advocate of participation felt that it had been oversold and had had little effect on the R&D establishment 'which goes about business as usual' (Bentley 1994: 142). Developments at the CGIAR research centres would seem to confirm these doubts. At IRRI, for example, the extent of farmers' participation grew gradually during the 1980s until they were involved in experiments, their knowledge of local varieties was sought, and their criteria were increasingly being adopted in the breeding process. By the mid 1990s, however, pressure from some donors prompted a change of policy within the CGIAR which meant that centres were expected to concentrate upon 'upstream' and 'strategic' work while leaving 'applied' work (including participation) to NARS (Fujisaka 1994). And although participatory work continued at a few centres, the scientists undertaking it 'have tended to be isolated and marginalized within their institutes' (Pretty and Chambers 1994: 196).[10]

As far as PPB is concerned, debate on this subject over the last decade has been sharply polarized, and PPB has been strongly resisted in various places, by conventional breeders as well as by some ministries of agriculture (Biggs 2008; Ceccarelli 2009; Biggs and Gauchan 2001; Conroy 2009).[11] Although breeders in some NARS are receptive, most public-sector breeders both within and outside the IARCs are said to consider PPB as an unnecessary alternative to conventional breeding (rather than as an additional option). Worse, some senior research managers evidently doubt PPB's scientific credentials. From my correspondence with half a dozen proponents of PPB, it is clear that while opinions differ somewhat over the extent to which this approach has made headway in mainstream institutions, no one is any longer optimistic about the future, and almost all believe that this approach now has a low priority on the development agenda. The CGIAR's budget for biotechnology, for example, is now far higher than that for PPB, and some see a similar trend at foundations and UN agencies (Cleveland 2006).

Perhaps the starkest evidence that all is not well with the post-1970 innovations comes from the status of the social sciences within the IARCs. In 1974 things were looking up. For it was in that year that the Rockefeller Foundation established a fellowship programme to place post-doctoral fellows from the social sciences at various of the IARCs where they were to work on the design of technology (rather than merely its social impact). Over the next 20 years about 100 such fellows were awarded two-year placements (Conway *et al.* 2006). By the 1990s, however, there were few signs that the social sciences were finding acceptance within the agricultural science community. One survey of the social sciences in agriculture noted that the social and natural sciences remained largely isolated from each other and possessed conflicting perspectives on technical change (Biggs and Farrington 1991). The idea of the Rockefeller programme, for example, had been to 'prime the pump', but by the time the programme closed in 1996 only a small number of institutes chose to offer positions to the fellows at the end of their placements, so the initiative was discontinued (Guggenheim 2006). Recent evidence from some centres points in the same direction. At IRRI social scientists believe that their qualitative methodologies are not being taken seriously by biologists or by agricultural economists, and when social sciences are drawn into projects it is still mainly to assess the social impact of technology rather than to play a role in shaping it. Thus they see their role as little more than window-dressing (Paris *et al.* 2006). The situation at CIMMYT seems to be much the same. There the institution has been dominated by breeders who favour 'hard' over 'soft' methods, such that they are more receptive to economists than to other kinds of social scientist (Bellon *et al.* 2006; Guggenheim 2006). At the International Centre for Potato Research, on the other hand, social scientists have enjoyed a relatively prominent position, but even they feel they have to contend with condescension and prejudice from biologists (and especially from economists who felt particularly threatened by the 'non-economist' newcomers) (Rhoades 2006). In 2000 it looked as though the situation would improve since the CGIAR's new strategy emphasized increased investment in the social sciences at the research centres and called upon the latter to take action accordingly. The strategy was endorsed by various external review panels, and two years later the CGIAR convened a conference on the role of the social sciences in agricultural research, but data presented there indicated that since 1995 the number of social science staff at IARCs had declined by nearly one-quarter. CIMMYT had previously had four or five anthropologists and sociologists but by 2002 had none; other centres were replacing social scientists with economists; and no centre had appointed an additional social scientist in the previous two years (Kassam 2006). It looks, therefore, as though the social sciences are once again becoming marginalized within the IARCS (Cernea and Assam 2006).

That peasant-oriented innovations have encountered resistance within the agricultural science community is not really surprising. For the challenge (or threat) posed by these approaches is not confined to their unorthodox practical recommendations (for example, that SRI calls for early transplantation of seedlings and widely spaced planting) nor their success in achieving high yields without the use of standard GR technology. As Jefferson (1993) has pointed out, these new approaches also call into question ways of thinking and working which are conventional within most fields of the natural sciences. With the emergence of these innovations, for example, scientists were being asked to spend more time in farmers' fields, to take seriously farmers' cultivation practices, to give farmers some say over the research agenda, and to collaborate with those from other disciplines (including the lowly social sciences). Moreover they were being called upon to approach problems from a systems perspective rather than a reductive one and to abandon the aim of finding one universally applicable solution in favour of generating many locally appropriate ones. None of this is easy.[12]

To conclude, the second generation of the Green Revolution was clearly a period of innovation as far as peasant-friendly development strategies are concerned. And by the 1990s it looked as though this kind of work was on track to become part of mainstream agricultural research. Since then, however, the momentum seems to have been lost, and the picture is a good deal less encouraging. At least this is the message one gets from a particular kind of source, namely accounts 'from below' written by practitioners of peasant-friendly persuasion. Could it be that they are unduly pessimistic? In the next section I turn to a different kind of source: assessments of the current state of development policy and practice, prepared by major development organizations as well as by experts in development studies.

2 Recent assessments of the problems of development

Over the last decade several major development organizations have issued high-profile assessments of the state of agricultural development, suggesting measures that are thought likely to alleviate rural poverty (World Bank 2007; IFAD 2001; FAO 2009b; McIntyre *et al.* 2009a, 2009b). Significantly, all of them identify smallholders as the key social group whose livelihoods must be improved if development is to be achieved. In this section I ask: what approaches to development do these organizations think are most promising, and to what extent are they the product of reflection upon past practice?

All of the reports cover much the same ground. For example, they all recommend the following solutions:

- Decentralizing development efforts in production, regulation and provision of services so as to increase accountability to farmers.
- Empowering farmers' organizations (especially those of smallholders) so that their market position is stronger and they can exert greater impact on policy formation, research and extension.
- Increased public expenditure on agriculture in developing countries, including greater investment in R&D in particular.

- The development of technologies more appropriate for the needs of smallholders.

What is striking here, of course, is that none of these solutions is new. Not only are they to be found in the green revolutionaries' diagnostic writings of the 1970s and 1980s; they were also central to the work of the south German stations by the First World War. Has nothing changed?

Equally remarkable are some of the specific recommendations aimed at helping smallholders. Discussing the provision of financial services, for example, the World Bank states that

> the root of the problem is that ... wealthier farmers can obtain larger loans at lower cost from formal lenders because they can credibly pledge assets or future cash flows. Asset-poor households, by contrast, are limited to considerably smaller loans at much higher rates....
>
> (World Bank 2007: 144)

One can only wonder who needs persuading on this point. In making it the Bank is merely restating a fact well recognized by Central European agricultural experts a century ago. Exactly the same could be said about the championing of 'financial cooperatives' (World Bank 2007: 146, 154–155), praise for the merits of marketing cooperatives (IFAD 2001: 9), the value of smallholders organizing themselves in growers' associations (World Bank 2007: 153–155), or the emphasis upon the importance of yield stability for subsistence farmers (World Bank 2007: 161). In short, the overall strategies offered by these reports, as well as some of their specific recommendations, are all too familiar.

One set of reports, however – those produced by the International Assessment of Agricultural Science and Technology for Development (IAASTD) – departs from the others in significant respects. First, in contrast with the others, the IAASTD reports were based upon expert opinion from an unusually wide range of development organizations (including farmers, NGOs, civil society organizations and the private sector). Second, they assessed not only formal science and technology but also local and indigenous knowledge. Third, they drew more heavily upon the social sciences relevant to agriculture (i.e. not just agricultural economics). Finally, despite being commissioned by a number of major development organizations (e.g. the World Bank, FAO), drawing upon an international panel of 400 experts, being subjected to two peer reviews, and being approved by the governments of 58 countries (Australia, Canada and the US declined their endorsement), the IAASTD proved highly controversial. It is not difficult to see why. Unlike the other reports, the IAASTD confronted head-on a major division within the agricultural science community over development strategy, arguing that 'the current model [of agricultural science and technology for development] requires revision. Business as usual is no longer an option.' Achieving development in an equitable and sustainable way, it argued, would require a 'fundamental shift' in science and technology as well as in policy and institutions (McIntyre *et al.* 2009a: 3 and 4).

The radical challenge posed by the IAASTD is especially visible in its account of the history of agricultural development since 1945. In it the limitations of industrial agriculture are identified, some of the major international organizations that underpinned the Green Revolution are criticized, and a case is made for promising alternatives (McIntyre *et al.* 2009b: ch. 2). The report distinguishes between two 'paradigms' of agricultural science and technology. A specialized 'productivist' paradigm whose concerns are exclusively economic is contrasted with an 'integrative' one which treats agriculture as a multifunctional activity whose social and ecological dimensions are also important. Significantly, the IAASTD perceives the productivist approach to have been institutionally dominant since 1945 while the integrative approach, emerging since the 1970s, is said to have so far enjoyed less visibility and influence (McIntyre *et al.* 2009b: 58–59, 116).[13] In saying this, the IAASTD thus captures the contrast between what I have called the first and second generations of the Green Revolution. Cautiously trying to be even-handed in its discussion of these paradigms, the IAASTD authors emphasize that these paradigms offer 'diverse and conflicting interpretations of past and present events which need to be acknowledged and respected', that the productivist paradigm has been highly effective in some times and places, and that the IAASTD presents 'options rather than recommendations' for action (McIntyre *et al.* 2009a: viii).

Nevertheless, on balance it is quite clear that the reports' sympathies lie with the integrative paradigm. The social and ecological problems deriving from industrial agriculture are said to have seriously weakened the productivist paradigm, and the authors' aim is to present the integrative approach as a corrective (McIntyre *et al.* 2009a: 25 and *passim*; Feldman and Biggs 2012; Feldman *et al.* 2010). Thus the IAASTD strongly endorsed participatory plant breeding, agroecology and integrated pest management, and local and traditional agricultural knowledge, but in each case these are presented as supplementing – not replacing – the tools of the productivist paradigm. What seems to have provoked the most criticism, however, was the IAASTD's position on agricultural biotechnology. Unlike the other reports, which stressed biotechnology's potential while urging increased public investment in it,[14] IAASTD was more restrained in its endorsement. Although the authors accepted, for example, that genetic modification could be used to breed stress-tolerant varieties for marginal growing conditions, they worried that biotechnology would not address smallholders' needs (for reasons that I discuss in the next section). Once again attempting to tread the tightrope, the report concluded that the use of biotechnology in plant breeding may make sense in some contexts but not in others (McIntyre *et al.* 2009a: 40–45).

The IAASTD's attempts to produce a 'balanced' report were no doubt an expression of the fact that the experts consulted came from both paradigms of agricultural science as well as farmers' organizations. This made for 'lively debate', to say the least (cf. McIntyre *et al.* 2009a: 81). During the discussions that formed the basis of the Assessment, it proved exceedingly difficult to arrive at a consensus within the various committees charged with drafting sections of the several reports (Jiggins 2008; Scoones 2009). Given the IAASTD's reservations about the adequacy of the productivist paradigm in general and biotechnology in particular, it comes as no surprise that particular 'stakeholders' protested vigorously. Before the consultations were concluded, for example, representatives from Monsanto and Syngenta withdrew from the process, complaining that the benefits of genetically modified organisms were not being objectively assessed (Stokstad 2008; Keith 2008). Similarly, a year before the Assessment was published the director general of CropLife International (a trade association for the agrochemical sector) wrote to the World Bank, objecting that a draft report inadequately 'recognized the past contributions [and] potential of plant science technologies' (Feldman and Biggs 2012). Then a few months before publication, the director of the CGIAR's 'Challenge Fund' wrote to the IAASTD secretariat, complaining that a draft report

> understates the past benefits and future potential of modern agricultural science and technology and overstates the potential ... of indigenous and local knowledge and participatory approaches. A more balanced and integrated treatment of these worldviews would have been highly beneficial.
>
> (Emile Frisson, cited in Feldman and Biggs 2012)

Once the IAASTD was released (2008), it again came under fire. Among the criticisms was the charge that the Report gave too much credence to 'social science' perspectives and too little to those of agricultural economics or plant sciences (Keith 2008; cf. Scoones 2009). This objection, along with the generally hostile reception of the IAASTD reports by mainstream organizations,[15] bears out the analysis presented in section 1. Namely, that the peasant-friendly (or 'integrative') innovations which emerged from the 1970s and thrived in the 1980s and 1990s have nonetheless failed to overcome significant resistance from the mainstream agricultural science community and are now tolerated as a minority strand within development thinking.

In addition to reports from various development agencies, the future of smallholder farming has also been addressed over the last decade by many development academics and practitioners. While I cannot review this literature exhaustively here, it is worth examining what a few prominent figures have had to say on this question, for it echoes in many respects the reports discussed above. Several authors, for example, identify smallholders as a key target for intervention,[16] drawing attention to the ever-increasing proportion of small farms in poor countries, their efficiency compared to large farms, and the lack of employment opportunities outside agriculture in rural areas. Accordingly, many call for agricultural research to better address the needs of small farmers (for example, more attention should be given to the breeding of orphan crops and to improved services for this group) and that greater emphasis be placed on organizing small farmers so that they can put pressure on governments to deliver (e.g. Lipton 2010; Hazell *et al.* 2010; Wiggins 2009; Falcon and Naylor 2005). As with the reports discussed above, many of the specific recommendations in this literature are reminiscent of much earlier development programmes. Calls for the development of labour-intensive technology, for example, merely reiterate arguments dating from the 1970s (e.g. Chambers and Farmer 1977). Similarly, in his conclusion to a thoughtful and measured assessment of the value of low external-input technologies, Robert Tripp argues that the current assortment of projects in this area is too small, short-termist and uncoordinated to make much impact upon smallholder agriculture. The way forward, he suggests, lies in much more general strategies which build institutional capacity for the longer term. That means strengthening rural education and extension as well as promoting the formation of broadly based and viable farmers' organizations which can put pressure on governments, donor agencies and NGOs to provide the desired services. 'The major responsibility', he concludes, 'lies with national governments who must foster strong institutions that support the interchange of information and experience' (Tripp *et al.* 2006: 217; cf. 213ff.). His analysis could serve as a summary of why the Central European (and Japanese) approaches to small farm development were so effective a century ago.

Reading this literature, therefore, one has a strong sense of déjà vu. Strategies for promoting development among smallholders seem to have changed relatively little since the early twentieth century. Many of the conclusions reached by green revolutionaries in the 1970s and 1980s as to how best to serve small farmers are now being voiced once again as though nothing much had been achieved in the meantime. This reiteration again suggests that the once promising development of peasant-friendly innovations since the 1970s seems to have lost momentum. Over the last generation, however, one new approach to plant breeding has been much more successful in attracting attention from agricultural scientists as well as support from donor agencies. Moreover its advocates argue that this powerful approach will enable breeders to develop varieties well suited to the needs of resource-poor smallholders. Is this perhaps the long-awaited breakthrough?

3 Biotechnology: a peasant-friendly innovation?

The advent of agricultural biotechnology marks the beginning of what I have called the 'third generation' of the Green Revolution. (The technology's proponents often call it 'the gene revolution' or the 'second Green Revolution'.) Whether the new agricultural biotechnology should be introduced in the developing world is a controversial and multifaceted issue. Most of the discussion has centred upon whether or not genetically engineered organisms pose a threat to the environment, to biological diversity, and to human health. These are all legitimate issues for debate and investigation, but they are not my concern here. Instead I want to consider the large claims which have been made over the last decade or so for the importance of this new approach to breeding in alleviating poverty in the developing world. Biotechnology's advocates have argued that because the gains using conventional GR methods have slowed substantially – for example, yield improvements through breeding are levelling off, productivity has fallen through soil degradation – the only way to boost food production sufficiently in the global South over the next few decades will be through biotech-based breeding.[17]

The enthusiasm for this new technology is by no means confined to representatives of the biotech industry.[18] It has been endorsed, for example, by recent chief scientists of the British government (Randerson 2008; McKie 2011; Sample 2008) as well as by politicians in the US, Britain and Germany (Jowitt and Vidal 2010; Anon. 'Gentechnik' 2010; Murphy 2007: 214).[19] Prominent agricultural scientists have also been among the product champions. The Nobel Prize-winning plant breeder, Norman Borlaug, has been a tireless defender of the new technology's importance for Third World farmers, as have several agricultural economists (Borlaug 2000; Paarlberg 2008; Collier 2009; cf. Conway and Waage 2010; Serageldin and Persley 2000; Victor and Runge 2002).[20] Furthermore, as we have just seen, several major development organizations regard it as promising.

In examining these claims I want to answer two questions. Is biotech-based breeding likely to alleviate rural poverty among smallholders? And to what extent have the lessons from earlier phases of the Green Revolution been taken on board? These questions are quite important for two reasons. First, biotechnology[21] opens up new and powerful technical possibilities for breeding. For example, genetic modification, unlike conventional hybridization, allows the breeder to draw upon a much wider range of genes since they can be taken from quite unrelated organisms. It also permits the breeder to work more quickly. And although genetic modification has often been criticized on the grounds that it allows for the transfer of single genes – whereas most agriculturally important traits are polygenic – some important traits which were previously assumed to be affected by many genes have turned out to be surprisingly simple. Single genes, for example, have been identified which have quite a large effect upon drought resistance, such that engineering new varieties could prove easier than once thought (Reynolds and Borlaug 2006).

Second, as moderate advocates have pointed out,[22] some claims for biotechnology's peasant-friendly potential are entirely plausible. Several of the new varieties, for example, are being designed to require fewer chemical inputs. This would be desirable on environmental grounds but could also be valuable in countries where inputs are expensive or where distribution systems are weak (Santaniello 2002). Insect-resistant cotton varieties already on the market are said to be helping small farmers by reducing the scale of harvest losses as well as the cost of pesticide (FAO 2004: 6; Coffman 1983), though the benefits for human and animal health would also be sizeable (Herring 2007). Similarly, attempts are under way to engineer some crops so that they can fix their own nitrogen (i.e. take it out of the air and convert it into a chemical form which can be absorbed by the plant), thus reducing the need for nitrogen fertilizers (Raven 2003). Varieties that tolerate herbicide would require little or no tillage, thus reducing soil erosion aggravated by ploughing (Royal Society 2009: 30). Crop varieties are also being bred which can better handle the stresses imposed by difficult growing conditions. These include salt tolerance, the ability to survive occasional flooding, and drought resistance. The latter two are especially important since the majority of Third World farmers do not have access to irrigation (and climate change is apt to worsen the problem). Water shortage has been identified by the World Bank as the biggest environmental problem faced by the poor (IFAD 2001).[23] Finally, biotechnological methods have been applied to developing nutritionally improved varieties. For example, the prospect of iron-enriched varieties may be significant since one-third of the world's population is anaemic, as are two-thirds of the children in Southeast Asia (Pinstrup-Andersen and Schiöler 2001: 61–62). The best known enriched variety developed so far is the so-called 'golden rice' which is designed to reduce the incidence of blindness due to vitamin A deficiency (which leads to blindness in 500,000 Third World children annually).

Because of this promise, some experts have argued that 'biotech has given us a tool to accelerate the development of improved crops which ... is easily directed towards the crops and diseases of the poor' (Conway and Waage 2010: 363–364). The key question, however, is what is the *likelihood* that this potential payoff will be realized? There is certainly no doubt that the giant multinational seed companies are interested in the developing world; by 2000 all of these corporations had expanded into the global South. Their focus so far, however, has been on the big markets in Brazil, Argentina, China, India and South Africa where firms aim to sell to large commercial farmers rather than to small and poor ones. And even that investment is modest; the proportion of the global private sector's R&D budget which is spent in the developing world is only about 7 per cent.[24]

Furthermore the private sector's lack of interest in the needs of small farmers becomes clear as soon as one looks at the breeding work which is going on in firms. As far as crops are concerned, the overwhelming majority of work has been devoted to cash crops: maize, cotton, soybeans and rapeseed (canola). Of these only maize is also a dietary staple (e.g. in sub-Saharan Africa), but the varieties of it that are being improved are not those which are grown by African smallholders (Pinstrup-Andersen and Cohen 2001; Chrispeels 2000; Paarlberg 2008: 160–164; Ruttan 1999). And since maize not only requires a lot of mineral fertilizer but is particularly vulnerable to insects and disease, the farmer is faced with relatively high costs for inputs (Hubbell and Welsh 1998). Very little breeding effort, on the other hand, has so far been conducted on staple crops – those which are central to the diets of large parts of the developing world and which are primarily grown for subsistence rather than sale – such as millet, sorghum, or cassava.[25] Similarly neglected are sweet potatoes, a major food crop in East Africa, as well as chickpeas, lentils and other pulses which are high in protein as well as drought-tolerant. (Falcon and Naylor 2005; Fowler and Lower 2005). That these crops should be regarded by firms as 'minor crops', therefore, is not because they are unimportant economically or nutritionally. It is simply that there is no money to be made from them.[26]

A case in point is rice, the main staple food in large parts of Asia and the one that provides three billion people worldwide with 80 per cent of their calories.

Nevertheless, as of 2004 no genetically modified varieties of rice had been planted anywhere in the world (Hucho *et al.* 2005: 306; FAO, 2004: 38). Agribusiness corporations routinely cite 'golden rice' as an example of how bioengineered crops could reduce malnutrition and improve health. As it happens, however, this variety was developed, not in a company laboratory, but at the Technical University in Zurich in collaboration with the University of Freiburg. Indeed, when the inventors of golden rice originally sought industrial support for their work, they found that none of the large agribusiness firms was interested (Pringle 2003: 20–21, 29; Chrispeels 2000). The reason for this lack of interest is partly because rice – unlike wheat, cotton or maize – is largely a subsistence crop: less than 10 per cent of world production is traded on international markets (Bello 2009: 54). But it is also because rice is a self-fertilizing crop which maintains its properties from generation to generation. Thus farmers only occasionally need to buy fresh seed; most of the time they can save a part of each harvest as seed for the next planting. Low seed sales hold few attractions for the private sector (Morris and Ekasingh 2002). (And this is also why, even before the advent of biotechnology, the breeding of rice in the US was concentrated in the public sector while firms focused upon more profitable crops: Frey 2000; Heisey *et al.* 2001: tab. 1; Knight 2003).[27]

But it is also worth considering the traits that are being bred in the private sector. Among transgenic plants under cultivation worldwide, the traits that have most often been bioengineered are almost exclusively herbicide tolerance or insect resistance (Ruttan 1999; Pingali and Traxler 2002). Among the genetically engineered traits tested in the developing world during the 1990s, these two traits constituted two-thirds of the total (FAO 2004: fig. 3). That agribusiness firms should be devoting so much attention to herbicide tolerance makes good sense in large-farm contexts where farmers want to reduce the labour costs of weeding. But traits that reduce labour costs are actually harmful in many parts of the developing world where high levels of rural unemployment make labour cheap and plentiful, and people are seeking work (Lipton 2010). Furthermore, breeding varieties that require herbicide use is not helpful for small farmers who often rely upon inter-cropping since herbicides are designed to eliminate all plants other than the crop that is tolerant (Hogg 2000: 134–135). Finally, agricultural ecologists have expressed concern that the planting of herbicide-tolerant varieties may encourage heavier use of what is already an environmentally problematic agrochemical, not least because such use is expected to induce the formation of herbicide-resistant weeds whose elimination will require the use of an additional kind of herbicide (Pimentel *et al.* 1989; Hubbell and Welsh 1998). Unsurprisingly, the firms that are breeding for this trait also tend to be manufacturers of the herbicide in question.[28] Varieties that require the purchase of commercial inputs, however, may well be out of reach for resource-poor farmers. Accordingly, among the traits being bioengineered by breeders in *developing* countries, herbicide tolerance ranks very low (Cohen 2005).

Similarly, insect-resistant varieties (e.g. those producing a bacterially derived substance called 'Bt toxin') could in principle offer certain advantages to smallholders, as noted above, but they also embody some of the same drawbacks as herbicide-tolerant varieties. For example, it is not clear whether the introduction of Bt varieties will in fact reduce pesticide use since within a few years their widespread use is expected to promote the emergence of Bt toxin-resistant insect populations which will necessitate the reintroduction of insecticide (Soleri *et al.* 2008; McAfee 2008: 71). In some places use has already increased (McIntyre *et al.* 2009b: 106). For this reason the technology has been described as little more than a 'delaying tactic' (Hubbell and Welsh 1998: 48). In addition, the increased yields reported so far for Bt cotton in several countries have usually been modest or even negative (Qaim and Zilberman 2003; cf. McAfee 2008). In any event, the insect-resistant varieties developed so far have been designed for use in North America (Morris and Ekasingh 2002; Pinstrup-Andersen and Cohen 2001). Far more important for smallholders than insect resistance or herbicide tolerance would be crops that can cope in stressful environments. (As we shall see, these are being developed but not in the private sector.)

And finally, there is the little matter of cost. Even if genetically engineered seed were to require neither mineral fertilizers nor pesticide, it would still need to be purchased. As we have seen, until well into the twentieth century it was standard practice for most European and North American farmers to save seed rather than relying upon commercial seed, and the same is true today for most smallholders in the developing world, 80–90 per cent of whom save seed (Paul and Steinbrecher 2003: 184).[29] The problem for smallholders is that bioengineered seed is even more expensive than conventional commercial seed.[30] As a result the main purchasers of GM seeds in Brazil, India and South Africa have been relatively well-off, commercially oriented farmers (Scoones 2008). The additional cost of GM seed, aggravated by poor credit facilities in many rural areas, makes it doubtful whether most small farmers will be able to afford the new technology.[31] Thus one study of three developing countries found that when presented with a choice between GM and local maize varieties, most peasant farmers favoured the latter (Soleri *et al.* 2008).

The conclusion that most observers – including the UN's Food and Agricultural Organization as well as the World Bank – have drawn from such evidence is that biotechnology is unlikely to produce much of interest to Third World smallholders in future because, unlike the work of the European peasant-friendly stations *c.*1900 or the first and second generations of the Green Revolution, biotechnology-based breeding is now concentrated in the private sector (e.g. FAO 2004; World Bank 2007; IFAD 2001; Paarlberg 2000: 33–34; Goodman 1985). If, therefore, biotechnology is to alleviate rural poverty, greater investment in public-sector research will be essential, for it is there that the great majority of peasant-friendly work is being done.[32]

This is evident as soon as one looks at attempts to improve the world's major food staples. In the 1980s, for example, the Rockefeller Foundation chose to allocate about one-half of its funding for agriculture to rice biotechnology because, as one of its agricultural directors remarked, lack of market incentives meant that 'very little was being done and was likely to be done on rice' (cited in Smith, E. 2009: 467; cf. Pringle 2003). GM rice varieties have also been developed in the NARS of at least half a dozen Asian countries (Cohen 2005). In Brazil the public sector is focusing upon staples such as cassava and beans (Fukuda-Parr 2007b) while in Africa several countries have developed GM varieties of millet and sweet potatoes (Cohen 2005). The Rockefeller Foundation and several of the IARCs are also working on cassava, millet and sorghum (Wolf 1986: 37; Stone 2002).

The breeding of traits useful for resource-poor farmers has also been concentrated in the public sector. At the John Innes Centre in Britain, for example, breeders are working on drought tolerance, salt resistance, and reducing the nitrogen requirements of high-yielding varieties. At Cornell University breeders are studying a gene which should increase tolerance to drought, cold and salinity (Uphoff 2007) while at other American institutions attempts are under way to engineer a nutritionally enriched variety of cassava and a perennial variety of wheat which will require much less fertilizer. Again, the Rockefeller Foundation has supported the development of rice varieties that can tolerate drought, soil toxicity and prolonged flooding.[33] In the developing world biotechnology research in India is placing a high priority on resistance to pests and disease as well as on drought tolerance (Ramaswami and Pray 2007), and in 1999 China was conducting field trials on a salt-tolerant rice (Huang *et al.* 2002). And in recent years some international research centres have done important work. IRRI, for example, released a variety that can tolerate flooding, and varieties are also being developed which are salt- and drought-tolerant (Baier 2009). At CIMMYT maize varieties are being bred which are drought-resistant and can cope with poor soil fertility (Reeves and Cassaday 2002).

Nevertheless, while the nature of the work going on in these institutions is encouraging, public-sector agricultural research worldwide is facing serious problems. During the 1960s and 1970s global expenditure on agricultural research increased nearly threefold (Arndt *et al.* 1977), but from the 1980s funding fell sharply. Between the mid 1980s and 2004, for example, the USAID cut its funding for agricultural R&D in the developing world by *three-quarters* (Pardey *et al.* 2006: 21). A major cause of this decline in research funding has been an unwillingness among western countries in recent decades to provide aid for agricultural development of *any* kind. Although global foreign aid to agriculture in developing countries had grown rapidly during the 1970s, it began to drop from the 1980s, falling by at least 50 per cent by 2000 (Pinstrup-Andersen and Cohen 2001; FAO 2004: fig. 31; Pardey *et al.* 2006: tab. 8; Ashley and Maxwell 2001) and much further by 2007 (FAO 2009b: 18). The collapse of US funding was particularly stark.[34]

At the World Bank agriculture made up a smaller percentage of lending, and support for agriculture and rural development reached an all-time low in 2000 (Morris *et al.* 2006; Pardey *et al.* 2006; Falcon and Naylor 2005). Between 1982 and 2007 none of the Bank's annual *World Development Reports* had agriculture as its subject (World Bank 2007: 1). In addition a major shift of policy at the

World Bank and the International Monetary Fund entailed the imposition of 'Structural Adjustment' programmes which required a reduction of public expenditure by recipient country governments as a condition of loans (Staatz and Eicher 1998; Rist 2008).

Drastically reduced funding has had a serious impact upon public-sector agricultural research of all kinds. The network of international research institutes funded by the CGIAR is particularly important because of their disproportionate impact upon agricultural innovation. But some who point hopefully to those institutes' smallholder-oriented work on biotechnology (Wield *et al.* 2010) neglect to mention that the CGIAR's budget is relatively small. It funds only about 3 per cent of total global agricultural R&D and has been in decline for some time.[35] Although its funding rose rapidly during the 1970s, it began to level off about 1990, aggravated by sharp reductions in funding from both the US and Japan (Pardey *et al.* 2006: 19–20; Morris *et al.* 2006). During the 1990s, for example, the budgets of its institutes for rice research (IRRI) and for wheat and maize improvement (CIMMYT) declined by nearly one-third (Mann 1999; Fan *et al.* 2005: 378), and further cuts have been imposed in the last few years (Knight 2003). As far as work in biotechnology is concerned, the CGIAR's total budget for such work is a tiny fraction (1 or 2 per cent) of the private sector's (FAO 2004, 33; Pinstrup-Andersen and Cohen 2001; Spielman *et al.* 2006).[36]

The situation is even worse in the developing world. If peasant farmers are to reap the benefits of biotechnology, NARS in those countries are of crucial importance. But these institutions have been hit very hard (Kousari 2004; Mellor 1998), especially in sub-Saharan Africa where government cutbacks driven by Structural Adjustment programmes have left NARS heavily dependent upon donor aid (Byerlee 1998). This is particularly clear in the case of biotechnology research. Although a few countries in the developing world have strong research programmes in biotechnology – China, India and Brazil – such work is generally more expensive than conventional breeding (Ndiritu 2000). In sub-Saharan Africa, therefore, despite donor aid, most countries can still only afford work of an elementary kind rather than genetic modification (Santaniello 2002; Nwalozie *et al.* 2007), and even capacity in conventional plant breeding – a necessary prerequisite to make use of biotech – is limited (Byerlee and Fischer 2002; Pray and Naseem 2007; Morris *et al.* 2006).[37] Since agriculture is generally the largest sector of most Third World economies, funding for agricultural research in such countries ought to be proportionately high. Yet the sums that are in fact being spent there on research are a much smaller fraction of GDP than is spent in rich countries.[38] And while the latter are infinitely better off, even their spending on public-sector agricultural research (and on plant breeding in particular) has been in decline for a generation (cf. Chapter 8). One consequence has been that various institutions in the global North have stopped breeding work on crops of importance in the global South (Just and Rausser 1993; Heisey *et al.* 2001: 6–7; Frey 1992).[39]

Funding, however, is not the public sector's only problem. If public-sector breeders are to apply biotechnology to Third World problems, they will have to find some way through the dense thicket of intellectual property rights. For as is well known, in order to protect their investments in plant breeding, the multinationals have been allowed to patent those innovations that look commercially promising.[40] As a result, in the US about 75 per cent of patents in agricultural biotechnology are now owned by private firms (Graff *et al.* 2003: 990) and of those patents three-quarters belong to just five companies (Kloppenburg 2004: 328–329). Although a long-standing justification for the patent system has been that it stimulates innovation, many observers are now concerned that it is instead restricting public-sector research. The invention of 'golden rice', for example, used techniques covered by 70 patents owned by over 30 different organizations. Asked how he managed to negotiate this minefield, one of the academics who invented it is reported to have said 'So many fields of research are blocked by corporate patents. I had to ignore them or I couldn't move at all' (cited in Falcon and Fowler 2002: 207). Not everyone evidently feels free to do this; a director of the biotech unit at CIMMYT has complained that the cost of royalties which must be paid in order to do research is becoming prohibitive (Poitras 2008).[41] Others have warned that the patents granted to multinationals have in some cases been so wide-ranging that their effect may be to shut out public-sector breeders from working on some crops at all (Hogg 2000: 229). In the case of golden rice several of the giant corporations, no doubt sensitive to public opinion, agreed to waive their rights to licence fees wherever golden rice is given to Third World farmers earning less than $10,000 per year (while retaining them on sales to rice farmers in the developed world). For public-sector institutions to have to rely on largesse from the private sector, of course, hardly provides a secure basis on which to plan future research.[42]

Furthermore, it is not just the private sector which is protective about intellectual property. With the decline in funding for public-sector agricultural research in rich countries, accompanied by the rise in private-sector spending, universities have been forced to look for alternative sources of money. One of these has been to patent their own work and license it, generally to the private sector.[43] But the patents on basic techniques are now scattered across a wide range of institutions, private and public. Since no single institution possesses the entire toolbox needed to develop a new variety, each project requires access to processes or materials patented by others. With an increasing amount of time being spent on negotiating rights of access, plant breeders and others are worried about the consequences for research (Barton and Berger 2001; Frey 1992; Graff *et al.* 2003).[44]

Faced with the collapse of public funding for agricultural research worldwide plus the private sector's dominance in biotechnology, numerous experts and some major donor agencies have concluded that the only way to assure that biotechnology will be applied to the problems of Third World agriculture is through 'public–private partnerships' (Cantrell and Reeves 2002; Tripp 2002; Pingali and Traxler 2002; IFAD 2001; World Bank 2007). In principle such arrangements might be easier to set up than they were in interwar Germany where private- and public-sector breeders were competing. For today, as we have seen, there is a growing division of research labour between the sectors which might make collaboration easier. In the developing world collaboration has so far taken a number of forms. In some cases a public-sector institute has acquired a patented technique or material from a corporation – either free or for a modest royalty – on the condition that the resulting variety will be made available only to farmers in the developing world.[45] In other cases where the corporation sees commercial potential in a public-sector institute's work, it may subsidize that work in exchange for exclusive rights to a licence (Barton and Berger 2001).

As of a few years ago, however, few such partnerships – especially compared with those in the international health sector – had yet been set up in developing countries, even in countries like India or Brazil which have very large public-sector research programmes in biotechnology (Cohen 2005; Chataway 2005; Lipton 2001; Byerlee and Fischer 2002; Ramaswami and Pray 2007; Fukuda-Parr 2007b). And even those optimistic about the potential of such partnerships concede that only a few of them have produced the claimed benefits for smallholders (Spielman and von Grebmer 2004; cf. Poitras 2008). According to a 2004 survey the reasons for this dearth were usually reservations among firms (Spielman *et al.* 2006). Many of the respondents said that their firms' accountability to shareholders meant that they could not get involved unless there was the prospect of a commercial payoff. This is understandable, for apart from providing access to germplasm, the public sector often does not have a great deal to offer its partners. Although there have been a few high-profile cases where firms have secured favourable publicity through handing over intellectual property at no charge, the terms on which they enter collaborations are normally self-interested. Royalty-free licences are generally granted to public-sector researchers either in order to demonstrate the value of a new product and to create a demand for it or on the condition that the technology will only be used for subsistence crops in particular regions (Spillane 2002; Spielman and von Grebmer 2004: 33).[46]

Given private-sector motivations, the terms of such collaboration will have to be negotiated very carefully. In the international health sector, for example, there has been criticism of pharmaceutical companies' influence upon World Health Organization policy. In agriculture, similarly, NGOs have already expressed concern that some of the CGIAR centres' existing partnerships threaten to divert the centres' work away from the needs of small farmers.[47] The public sector's vulnerability on this score is aggravated by the fact that its bargaining power is limited through lack of experience in handling patented material and in dealing with the private sector. This may change over time, but some observers have suggested that public-sector institutions in the developing world may need to patent much more of their own work and/or form consortia among themselves in order to strengthen their hand in dealing with corporations (Byerlee and Fischer 2002; Barton and Berger 2001).

Finally, even when firms enter such a partnership, it is not clear that they will be prepared to remain in it for the duration of the project. The history of collaborative arrangements at universities in the North – in which multinational firms fund academic research in exchange for exclusive access to the results (Kenney

1986) – is instructive in this regard. At the University of California, Berkeley, for example, the Department of Microbial and Plant Biology signed a contract in 1998 with the Swiss pharmaceutical/plant-breeding corporation, Novartis (later to become Syngenta), which provided one-third of the department's funding for basic research (Press and Washburn 2000). An evaluation of this arrangement by external reviewers, however, was quite critical, and Syngenta terminated the deal after five years. In dozens of cases by now collaborations between public-sector research institutions and biotech firms in the North have been terminated at short notice by firms whose financial status had changed (Murphy 2007: 146–149).

Given firms' evident reluctance to enter such partnerships, various mechanisms have been suggested by which they could be induced to collaborate.[48] One common proposal has been that public funds be used – via prizes, research grants, contracts or tax incentives – to attract private-sector participation in projects aimed at resource-poor farmers (World Bank 2007: 169; Spillane 2002; Tripp and Byerlee 2000; Lipton 2010; Pray and Naseem 2007; Pinstrup-Andersen and Schiöler 2001: 143). Although the idea sounds plausible in the abstract, it is remarkable that none of its proponents have yet suggested where the money for such schemes is supposed to come from, especially at a time when global funding for agricultural research is severely reduced. All in all, the prospects of public–private partnerships in order to develop new varieties designed for Third World smallholders do not look good.

So what are the prospects for biotech-based breeding helping resource-poor farmers in future? I have argued that the concentration of biotechnology in the private sector, the weakness of public-sector agricultural research, and the very few public–private partnerships so far all suggest that these prospects are poor. Indeed, some argue that the diffusion of biotechnology in the developing world is likely to *widen* the gap between commercial and subsistence farmers (Swanson and Goeschl 2002), and the evidence from Latin America seems to bear this out. In several countries there agbiotech has so far benefited primarily large commercial farmers who can afford to invest in the new technology while gaining from reduced production costs (Otero 2008). Large farmers in Argentina, for example, have abandoned other crops in order to plant herbicide-tolerant soybeans which now make up nearly half of Argentina's grain exports. But the technology offers little or nothing to smallholders, and many have been forced to abandon farming in search of work elsewhere. As a result, although in 2001 the country had the capacity to produce enough food for several times its population, one quarter of its population was actually living in extreme poverty, unable to afford a basic diet (Teubal 2008).

Faced with such evidence, many advocates of biotech insist that the technology will be capable of serving smallholders once a series of 'obstacles' have been removed. It is essential, they argue, that (a) the private sector be encouraged to invest in this area (e.g. through protection of intellectual property rights, reasonable biosafety regulations, and state-funded incentives which encourage pro-poor research) and (b) governments in the South fund public-sector biotech research and provide input subsidies to small farmers (Janssen and Gupta

2009).[49] This bright-eyed view, however, turns a blind eye to the serious limitations of both private and public sectors which I have outlined above. By arguing that governments in the global South merely need to get their policies on biotechnology right, advocates have missed the wood for the trees. They have simply ignored the fundamental political fact that the public sector has been deliberately run down in all countries since the 1980s and that there are as yet no signs that this is changing (cf. Scoones 2002).

If one wants some indication of what a strong public-sector effort could do for peasant farmers, one need only look at China. Public funding there for agricultural biotechnology is comparable to that of a giant multinational, and public-sector staff numbers in biotechnology are comparable to those of the US public sector (Hucho *et al.* 2005; Fukuda-Parr 2007b). In the mid 1980s China began to invest heavily in rice genomics; almost all of the work on sequencing the (*indica*) rice genome, for example, was conducted in China (Hucho *et al.* 2005). The Chinese public sector also began to develop transgenic varieties, prompted by the vulnerability of previous hybrid varieties to pests as well as by growing health problems from heavy pesticide use (Shen 2010).[50] Although its transgenic rice varieties have not yet been released, in 1997 the public sector did release transgenic Bt cotton varieties. And though the evidence that Bt cotton may be of advantage to smallholders is contested, Bt cotton's impact in China is regarded as one of the few big payoffs so far from GM in the developing world (Pray and Naseem 2007). Increasing yields by 10 per cent while reducing pesticide use by 50 to 80 per cent, these varieties have proved popular with China's cotton growers (whose average farm size is less than one hectare).[51] Within seven years of release, about two-thirds of the cotton acreage was planted in Bt cotton (Huang *et al.* 2002, 2007). Thanks to seed subsidies, farmers' profit from Bt cotton is five times higher in China than in India and ten times higher than in Argentina (Fukuda-Parr 2007b; cf. FAO 2004: 49–51).[52]

But if the evidence is so unconvincing that the developed world's biotechnology will alleviate rural poverty, why do the proponents of biotechnology never seem to tire of advancing this claim? For the scientists among them, the explanation is probably to be found in the apparently irresistible lure of a technical fix for all manner of socioeconomic problems. Biotechnology is yet another of the flawed but 'technically sweet' technologies which in the past have proved so hard for engineers to resist. As far as industrial enthusiasm is concerned, some argue that the notion that biotechnology might offer a solution for Third World poverty was originally concocted (at Monsanto at least) in the 1980s – after little or no serious reflection – in an effort to make a business case to investors that the firm's huge investment in biotechnology was a sound commercial move (Glover 2010). Since about 2000, however, the poverty-alleviation claim has come to occupy a very prominent place in the industry's advertisements, websites and publications, almost certainly in an attempt to change tack after the public outcry in Europe against GM in the late 1990s (Stone 2002). In an attempt to put moral pressure upon biotechnology's critics, industry spokesmen – echoed by prominent politicians – argue that by making ill-founded criticisms of biotechnology's safety, well-fed critics in the North are undermining the possibility that GM seeds could alleviate hunger in the South (Janssen and Gupta 2009; McMichael 2004; Buttel 2003). Finally there are the major donor agencies. One might have hoped that these organizations' familiarity with the complexities of development might have immunized them against overly simple solutions. If 'miracle seeds' could not deliver the poverty alleviation that was promised in the first generation of the Green Revolution, why should one expect a 'miracle technology' to do so in the third generation? The answer here, I think, requires us to reflect upon the nature of the development industry, a task I defer to the next chapter.

Conclusion

The criticism to which the Green Revolution was subjected around 1970 had a substantial impact. In part it triggered a period of reflection upon the first generation of programmes during which experts tried to pin down why their work had so often failed to help smallholders. But criticism also prompted several decades of innovation, generating a number of new approaches to development which paid far more attention than heretofore to the circumstances in which small farmers worked, be they economic, ecological, social or cultural. The new approaches also took seriously the cultivation practices that farmers had devised and sought to collaborate with them in developing improved ones. This second generation of the Green Revolution was thus characterized by a wholesale shift away from the largely productivist orientation of the first generation toward a concern with the welfare of resource-poor peasant farmers.

As we have seen, however, the earlier hope and expectation that these peasant-friendly innovations would become central to development practice have not been realized. While both farming systems research and participatory plant breeding continue to find a place at some IARCs and in various development projects, they have also met with considerable resistance from mainstream experts, and donor support is noticeably weaker than during the heydays of the 1980s and 1990s. Agroecology looks to be in somewhat better shape, but there are no indications that it has become central to 'the new normal'. And over the last decade advocates of all three approaches seem to have been drowned out by the chorus of enthusiasm for biotechnology coming from donors, politicians, plant scientists and agricultural economists. To be sure, this addition to the breeder's toolkit could in principle ease the task of developing varieties suited to resource-poor smallholders. But as we have seen, the weakness of public-sector research as well as current donor perspectives on development mean that this potential is very unlikely to be realized. For this is a classic case, as one nineteenth century political economist would have put it, where the relations of production are constraining the forces of production. Virtually everyone recognizes that public-sector breeding is essential if the rural poor are to be helped. But strengthening the developing world's public sector will require a major shift of policy at the World Bank and other donor agencies. Should they fail to make this shift, the most likely outcome is that the gains from biotechnology will go to large commercial farmers. It looks, therefore, as though the third generation of the Green Revolution is turning away from a concern with equity and welfare in order once again to embrace productivism.[53]

8 Three conclusions

Having reached the end of the story, it is time to spell out its meaning. Looking back upon this material from three different perspectives, I draw conclusions which are respectively analytical, hopeful and bleak. The first of these is an attempt to place the material I have surveyed within a wider theoretical framework. The second is a normative conclusion in which I outline what I believe the implications of my historical analysis are for development policy. And the third is a rather pessimistic series of reflections upon the nature of the development industry today whose implication is that, given the way in which it currently operates, 'the lessons of history' are unlikely to make much impact upon development strategy.

1 Analytical

One of my aims in this book has been to deepen our understanding of green revolutions by placing them against a longer historical backdrop than is usual. This made it possible to use comparative analysis in order to identify some of the key features of the post-1945 Green Revolutions. As a result, it became clearer that the Green Revolutions of this period went through three phases or generations. The first was characterized by a heavy 'productivist' emphasis, the main aim being to boost agricultural production in order to better feed the population and reduce the need for imports. Few GR planners were then concerned with the social consequences of the new technology. A concern with 'equity', however, came to dominate the second generation of the Green Revolution, in response to criticism *c.*1970. Accordingly, the peasant-friendly innovations of the 1970s through 1990s all sought to devise more appropriate technology and institutions so as to better assist the resource-poor smallholders who had gained least from the first generation of programmes. Characteristic of the third generation, as I argued in Chapter 7, is that the peasant-friendly approaches of the second generation are being sidelined at precisely the time when a new 'miracle technology' is attracting much enthusiasm – unrealistic as I believe that to be – from both donors and plant scientists. It looks, therefore, as though the third generation of the Green Revolution will once again display a productivist orientation.

That the development process has passed through phases of this kind is reminiscent of the ideas of Michael Cowan and Robert Shenton (1996). In their model the growth of a capitalist economy in many countries – whether in Europe in the nineteenth century or the global South in the twentieth – is profoundly destructive, dismantling traditional social relations while imposing capitalist ones. Faced with the resulting disorder and political instability, the state typically introduces a series of palliative measures in an attempt to alleviate social misery and re-establish social order. Although in Cowan and Shenton's usage the term 'development' applies only to this ameliorative process, one could just as well stretch the term so that it also embraces the promotion of capitalist growth in the first place (Bernstein 2006). Indeed, the enthusiasm of development agencies in the 1950s to foster 'modernization' as well as the declared aim of the first generation of green revolutionaries to 'bring peasant farmers into the market' invite precisely this wider use of the term. Interestingly, among the palliative visions of development Cowan and Shenton discuss are some that place value upon indigenous knowledge, small-scale solutions, ecological concerns and participation to compensate for the disruptive force of capitalist transformation (ibid.: 452ff.). It looks, therefore, as though what I have called the productivist first generation and the peasant-friendly second generation of the GR map quite closely onto Cowan and Shenton's model.[1]

Others have remarked upon similar phases in the history of development. Reflecting a generation ago upon the Green Revolution in India, for example, Deepak Lal observed that over the course of the twentieth century the aims of Indian agricultural policy had shifted back and forth. During the first two five-year plans (1949–1959), for example, the primary emphasis was 'distributional' (i.e. upon equity) while from the early 1960s the emphasis shifted toward 'growth'. By the early 1970s, having successfully increased production, 'the pendulum swung once again to distributional aspects' (Lal 1975: 390). In his study of the agricultural service in British colonial Africa, similarly, Joseph Hodge (drawing upon the framework of Cowan and Shenton) identifies a growing concern among colonial officials from the 1920s for native welfare (Hodge 2007). Migration to the towns, combined with unemployment and food shortages, were aggravating urban unrest, prompting one official to warn of the possibility of 'social revolution' (ibid.: 180). In an attempt to restore order, 'rural stabilization programmes' were recommended in order to create more stable and self-sufficient peasant communities.[2] Greater emphasis was placed upon indigenous cultivation practices and food staples in order to strengthen local food production and buffer smallholders from the hazards of international markets. After the Second World War colonial policy reverted to a series of large-scale development schemes relying upon mineral fertilizers and mechanization. More recently, others have begun to draw attention to similar historical shifts in development strategy (Frey and Kunkel 2011; Bowman 2011).

One of the interesting features of the developmental phases in Lal's and Hodge's accounts is the fact that they *alternate*, just as in the three generations of the Green Revolution which I have identified. Several years ago Michael

Watts suggested that development might best be seen as a modernizing movement of a dialectical kind whose history represents a shifting balance between competing strategies which wax and wane as the system is periodically beset with crisis (Watts 1995). This conception of development as dialectic, in turn, bears a definite resemblance to Karl Polanyi's classic analysis of the economic and political history of Britain and the industrialized world from the enclosures to the rise of fascism. A recurring phenomenon in his analysis was what he called the 'double movement':

> It can be personified as the action of two organizing principles in society, each of them setting itself specific institutional aims, having the support of definite social forces and using its own distinctive methods. The one was the principle of economic liberalism, aiming at the establishment of a self-regulating market, relying on the support of the trading classes, and using largely laissez-faire and free trade as its methods; the other was the principle of social protection aiming at the conservation of man and nature as well as productive organization, relying on the varying support of those most immediately affected by the deleterious action of the market....
>
> (Polanyi 2001: 138)

If we think about the Green Revolution in this way, the key problem becomes understanding why competing strategies of development ebb and flow, each predominating for a time before giving way to the other.[3] Though this is not something which I can pursue here, it suggests that comparative histories of development in different contexts over longer time spans could be illuminating.

2 Hopeful

As we saw in the previous chapter, any prospect that genetic modification might contribute to developing varieties of use to resource-poor smallholders is being undermined by the weakness of public-sector biotech research. This weakness, however, is not peculiar to biotechnology but rather part of a longer term run-down of public-sector plant breeding over the last 30 years.[4] And this latter process has serious implications since genetic modification is only useful as an *adjunct* to conventional breeding, not as an alternative to it (Duvick 1996).

The changes that public-sector breeding have undergone, however, have not just been quantitative; the nature of the work conducted in that sector has also changed. In the UK during the 1980s, for example, large agribusiness corporations, keen to eliminate competition from the public sector, found a natural ally in the Thatcher government which decided that state-funded research institutions such as the Plant Breeding Institute (PBI) at Cambridge should be sold off (Webster 1989).[5] Soon privatization embraced most British public-sector agricultural research, and similar attempts to 'restructure' the public-sector spread to other countries. In the US there were calls for the land-grant (agricultural) universities to abandon varietal development and concentrate instead on basic plant science (Kloppenburg 1988; Busch *et al.* 1991).

Twenty years ago a prominent British plant breeder warned that this process would eventually be seen as a 'disastrous error' (Simmonds 1990: 340). He may well have been right. Some observers are now arguing that the result of driving the public sector out of varietal development in the UK has been a growing focus in that sector upon basic plant genomics (rather than applied work) while the private sector has substantially cut back its expenditure upon varietal development. The result is that there is very little attempt in *either* sector to bring advances in basic plant biology to bear upon crop improvement (Murphy 2007). Furthermore, since the small amount of work in varietal development that still survives is concentrated in the private sector, varieties that might be of social benefit (e.g. for sustainable agriculture) but are unprofitable are no longer being bred (AEBC 2005).[6] Indeed, in recent years there have been suggestions that the private sector is instead breeding for 'planned obsolescence', based on the facts that the average lifespan of varieties in the UK has declined sharply as has their disease resistance, both of which oblige the farmer to purchase new seed at shorter intervals. In addition there has been a tendency in the private sector to select for input traits (e.g. responsiveness to fertilizer, ease of machine harvesting) rather than traits such as crop quality (Murphy 2007: 118–123). Finally, the privatization of the PBI, at least, seems not to have brought the gains that were expected. According to a preliminary assessment of the impact of privatization, the sell-off of the PBI did not save the government very much (since the Institute had been making a profit), nor was it clear whether privatization had increased the efficiency of breeding research (Pray 1996). Furthermore none of the corporations that bought the PBI seem to have made much money from it, as reflected in the fact that by 2004 it had been resold twice (Murphy 2007: 133). By 2006 it was beginning to look as though Simmonds was right. Several British research councils and advisory bodies were voicing concern at the undesirable impact privatization was having upon British plant breeding.

Those who decided to run down public-sector breeding had evidently chosen to ignore its remarkable success. We have already seen what was achieved in south Germany by the 1920s (as well as in Japan about the same time). During the 1930s in some regions of Switzerland, similarly, wheat varieties bred by the experiment station at Mont Calme were being planted on 70 per cent of the relevant acreage (Moser 2006). In Austria the improved local cereal varieties bred at Vienna's seed-control station between the wars were still being planted on a substantial scale in the 1950s and in some cases as recently as 1980 (Pammer, F. 1952; Hron 1981). In the UK, too, 90 per cent of the wheat varieties grown after 1945 were derived from varieties bred at the PBI (Pray 1996). In the US as recently as 1980 nearly three-quarters of commercial hybrid maize varieties grown there were based on inbred lines developed in the public sector, and about 90 per cent of soybean, wheat, oat and rice varieties were also of public origin (Kloppenburg 1988: 301–302, 145). China provides a final example. Since 1949 strong state support for rice research and varietal development has brought results. The technical difficulties of constructing hybrid rice varieties (i.e. F1 hybrids) meant that for many years this was commercially impractical. But in 1964 Chinese breeders set out to do this, and in 1976 China became the first country to release such a hybrid rice variety (Dalrymple 1986; Fan *et al.* 2005). By the late 1990s hybrid rice, yielding 30 per cent more than conventional rice varieties, covered one-half of the Chinese rice acreage (Normile 1999).

The effectiveness of public-sector plant breeding, of course, is just one illustration of the more general historical phenomenon that public-sector agricultural research has had a major impact upon the agricultural economy. It is well established, for example, that expenditure on agricultural research has proved an extraordinarily good investment, yielding annual returns through increased production of at least 20–30 per cent and often much more.[7] Furthermore there is substantial agreement that (publicly funded) national agricultural research systems (NARS) have been and are likely to be important in addressing the needs of smallholders. This is partly because they can, at least in principle, be more responsive to small farmers' needs than is the private sector. Some NARS, for example, have developed higher yielding varieties of orphan crops (Goldsmith 1997). Similarly, improved cultivation *practices* designed for such farmers are only likely to be developed in the public sector because, unlike new varieties or other 'products', they are not easily patented and thus have little appeal for the private sector. But NARS are also necessary for adapting international agricultural research to local conditions, both ecological and economic (Hayami and Ruttan 1985: 271–272; Staatz and Eicher 1990; Pearse 1980: 233–234).[8]

Again, China provides a good example, partly because of its small average farm size but also because of its nearly thousand-year history of state (or imperial) support for technical change (Bray 1979, 1986; Stavis 1975). Given population pressure and land shortage, the cultivation technologies developed there have aimed for increasing yield. Adoption of improved varieties by small farmers was possible thanks to a dense network of local technical advisory services, state-regulated prices for rice and cotton, and subsidies for seed, fertilizer and pesticide. State coordination also meant that irrigation water was distributed more rationally and less wastefully than elsewhere. Since the revolution state support for smallholder agriculture as well as a policy of food self-sufficiency has made for rapid increases in the growth of production (from 2.6 per cent per year in 1978 to 7 per cent in 1989 (Conway and Waage 2010: 96), enabling the country since the late 1970s to become a net food exporter. More significantly, it also led to dramatic reductions in rural poverty; in just six years (1978 to 1984) the proportion of the rural population living in poverty declined from 33 to 11 per cent (Fan *et al.* 2005: 370).

Of course, the fact that NARS are potentially of crucial importance for smallholders doesn't mean that in practice they will all manage to do the job. Many have been criticized, for example, for neglecting farmer needs, above all those of smallholders (Poitras 2008; Jain 1982; Goeschl and Swanson 2002; Zerbe 2001; Chambers 1984; Carney 1995; Tripp 2002; Scoones 2006: 357) or of marginal growing conditions often occupied by resource-poor farmers (Raina 2006;

Conway and Barbier 1990). Others are said to be overly centralized and hierarchical (Farrington 1988) or to neglect extension in favour of invention (Raina *et al.* 2006). No one disputes that there is plenty of need for improvement.[9] Moreover it may not be easy to reform these institutions, for numerous studies have shown that landed elites often resist (or capture) peasant-oriented development projects. Finally, despite the eagerness of administrators for short-term payoffs, even a well-organized and peasant-friendly research institution cannot be expected to produce results quickly. Staff at the Bavarian station took ten years to get the network of crop-improvement associations under way, and it was 20 years before the station's improved varieties were making a noticeable impact upon the seed market.

But how are NARS to be improved when the financial rug has been pulled from beneath them? During the 1980s and 1990s public-sector agricultural research was undermined as major donors embraced the 'Washington Consensus' and sought to roll back the state as a condition for countries receiving aid (Weis 2007; Rist 2008). The effects of this policy upon many countries were disastrous. Because Zambia, for example, could no longer afford to subsidize seed or fertilizer, the big production increases it had enjoyed into the mid 1980s collapsed, and it had to begin importing maize (Bowman 2011). A similar fate befell other countries in Africa, Mexico and the Philippines (Conway 1997; Bello 2009). By the late 1980s UNICEF was reporting that Structural Adjustment programmes were responsible for declining health, nutrition and education levels among large numbers of children in the developing world (Goldman 2005), and by the late 1990s it was clear that these countries had experienced zero growth of GDP per capita since the early 1980s (Wade 2001: 135). Arguably NARS were particularly hard hit by the Washington Consensus. They lost funding from national governments whose budgets were much reduced, but they also lost funding from donors such as the World Bank and USAID (Pardey *et al.* 2006) upon which many had come to rely from the 1950s to the 1970s (Flora and Flora 1989; Ruttan 1989; Dalrymple 2003).[10] Inevitably, this had a severe impact upon public-sector plant breeding; in some African NARS funding for it between 1985 and 2001 was slashed more than tenfold (Guimarães *et al.* 2006). Most observers agree that extension systems, for example, are now in poor condition, and most countries in the global South cannot afford to fix them (IFAD 2001: 10; Dalrymple 2006; Rivera 1996). Under these circumstances it becomes even more urgent that NARS is equipped to produce appropriate technology (Goodell 1984; Collinson 2000).

By 2007 even the World Bank was conceding that Structural Adjustment had too often failed to improve services, and where a private sector emerged, it had mainly served commercial farmers (World Bank 2007: 138; Zerbe 2001). The result, according to one analysis, is that the debate in mainstream circles over development strategy is no longer over 'market versus state' but rather over the form and degree of intervention (whether by state or by donor agency) which is necessary (Thomas 2000).[11] Recent discussion of the future of small farms points in the same direction. Since small farmers suffer differentially from market failures, there is a need for expansion of the public sector to compensate for this and as an effective way to promote poverty reduction (e.g. Wiggins *et al.* 2010; Hazell *et al.* 2010; Valdes and Foster 2010; Birner and Resnick 2010; Dorward *et al.* 2004). Noting the needs of many small farmers, for example, Robert Tripp has stressed the need for increased public investment in extension and education systems as well as in R&D, even though 'no one is yet willing to pay attention to the importance of building national agricultural research systems that are financially sustainable....' (Tripp 2001: 486; cf. Just and Rausser 1993; Falcon and Fowler 2002). A decade later, he reckons that little has changed in this respect (Tripp, pers. comm.).

Of course the World Bank itself is now calling for a major increase in investment in R&D for the global South – by states, donors and the private sector – citing evidence that the global community is 'massively underinvesting' in public goods for agriculture (World Bank 2007: 264). The UN's Hunger Task Force has followed suit, calling for a doubling of expenditure for agricultural research in the South (Hucho *et al.* 2005). But are these funders likely to pay up? As far as states in the global South are concerned, since the Bank accepts that many of their agricultural support services are in poor shape after being run down from the 1980s (e.g. World Bank 2007: 138), one wonders how these states will find the necessary funding (Woodhouse 2009). Similarly, since the Bank concedes that the private sector is not particularly interested in smallholder agriculture (World Bank 2007: 169), how much investment can one expect from that source? Lastly, as Tripp remarked, it is not clear whether donor agencies are willing to fund NARS on the required scale.

That donor willingness might be problematic becomes evident when one looks at the Bank's own development strategy, for its vision of the state's role and its support for an equitable and sustainable agriculture appear to be contradictory. The Bank, for example, is especially keen on states acting so as to strengthen the market economy. This is to be done by regulating competition, engaging in public–private partnerships, providing subsidies for privately provided agricultural insurance, facilitating 'market-based' land-reform (ibid.: 142–143), and providing 'market-smart' fertilizer subsidies (which are supposed to stimulate demand without displacing commercial sales and to encourage competition among fertilizer producers while being only temporary [ibid.: 152]). In this way rural poverty is to be alleviated not merely by improving the productivity of smallholders but rather by increasing their 'competitiveness' (ibid.: *passim*). But will such market-oriented state measures actually lead to the desired end? Although the Bank is keen on inputs to boost small-farm productivity, it recognizes that input subsidies are expensive (ibid.: 160). (When it comes to the Bank's environmental aims, it is not clear how championing the use of mineral fertilizer is consistent with an emphasis upon sustainability. Furthermore most of the technological and institutional solutions which the Bank proposes for environmental problems are not readily amenable to market-oriented solutions [ibid.: ch. 8].) Similarly, although the Bank thinks that governments should do more to encourage private investment in R&D (e.g. by strengthening intellectual property rights or lowering regulatory barriers to the release of new technologies), it argues at the same time that private breeders' improved varieties are primarily tailored to the needs of *large* commercial farmers (ibid.: 163ff.). One is reminded of Gilbert Rist's observation that consistency in recommendations is not a prominent feature of many development reports (Rist 2008). All in all, the *World Development Report 2008* seems to want it both ways: an equitable and sustainable agriculture but a state whose market-based solutions are unlikely to achieve it.

Perhaps the most radical challenge of late to the Washington Consensus is that of the economist, Ha-Joon Chang, who draws upon the history of development in order to argue that the state has played a far more important role than has so far been acknowledged (Chang 2002). He points out, for example, that Britain only embraced free trade in the nineteenth century (and the US in the twentieth) once their industries had achieved strong positions, thanks to tariffs and other forms of protection, while the governments of other now-developed countries invested heavily in railroads and other kinds of infrastructure and provided little or no protection of intellectual property. As far as agricultural policy is concerned, he shows that the governments of the developed countries often used land reform, policies supportive of small farmers (price stabilization, consolidation of smallholdings, input subsidies, cheap credit, tariffs), infrastructural improvements, and last not least research, education and extension (Chang 2009). It is a nasty irony, therefore, that since the 1980s the World Bank and IMF have been recommending policies to the developing world that bear little resemblance to those successfully pursued in the nineteenth and early twentieth centuries by the now industrialized countries. To insist upon market solutions, Chang concludes, is to wilfully ignore the lessons of the past.[12]

If the public sector's peasant-friendly potential is to be realized, however, strengthening IARCs and NARS in the global South is going to require far more funding than has been forthcoming over the last 30 years. Is the money available? The answer is 'yes' though at the moment it is divided between North and South in grotesque disproportion. For example, at the moment industrialized countries spend about one billion dollars per year on development assistance for the South's agriculture. The same countries, however, spend approximately one billion dollars *per day* on subsidies to their own farmers (Weis 2007: 25).[13] If rich countries were to scrap these subsidies (which protect their farmers from competition on world markets), the financial gain to the developing world would be *five times* the level of agricultural aid that they currently receive (World Bank 2007: 11).[14] Alternatively, since the North's subsidies to its farmers are over 30 times the amount currently spent worldwide upon agricultural research for the global South (Kiers *et al.* 2008: 320), the research budget for the South could be *doubled* if the North were prepared to reduce subsidies by a mere 3 per cent. If we look specifically at expenditure for agricultural research, we find that rich countries spend on average almost *70* times as much per farmer as do developing countries (Pardey *et al.* 2006: tab. 4). This means that were the rich countries to divert just 2 per cent of their agricultural research budgets to the developing world's NARS, the latter's budgets would more than double. There can be no doubt, therefore, that there is plenty of money available which could be used to strengthen the South's public-sector R&D. If rich countries choose not to spend it in this way, that is a political decision.

3 Bleak

The historical evidence is strong, therefore, that adequately resourced and organized state research institutions are capable of providing useful technical assistance to small farmers. And it would be nice to think that such evidence might be taken into account in development policy. But as we have seen, there has been an unfortunate tendency since 1945 to overlook such evidence. During the first generation of the Green Revolution programmes took no apparent notice of previous examples of successful state research while the second generation's experts ended up rediscovering many of the organizational principles and technical practices which were already part of early twentieth century development programmes, again largely unaware of their predecessors. And as I argued in Chapter 6, this neglect is not attributable to simple ignorance of the past. It appears to be due to indifference. In view of the endless declarations since 1945 of the need to alleviate rural poverty, why should the development industry take so little interest in the success or failure of its past interventions?

Accounting for this indifference would be a very large undertaking. There are several literatures upon which one could draw, including critiques of the development enterprise, works on the institutions and practices of the development industry, and studies of the operation of individual donor agencies. I certainly cannot attempt anything so ambitious here. Instead I offer a tentative interpretation which, though empirically grounded, is necessarily somewhat speculative. I hope that it might at least serve as a stimulus for debate.

One of the reasons why the development industry has difficulty in learning from its past may lie in the way in which it is structured. For one thing, since the rise of NGOs from the 1980s, the sheer number of these agencies has become enormous. Some estimates of the number of NGOs in the developing world are as high as 40,000. This often leads to 'fragmented, overlapping, discontinuous and sometimes contradictory' interventions (World Bank 2007: 257; cf. Farrington and Bebbington 1994). Lack of coordination among agencies also obstructs the learning process. As a consultant for a large NGO in Zambia complained, 'There is no institutional memory ... once a programme starts, it phases out, nobody collects the experience, nobody shares those experiences with others....' (cited in Bowman 2011: ch. 4). Moreover with a very large number of NGOs competing for donors' funding, a premium is placed on approaches that are perceived to be 'novel'. Projects that explicitly draw upon older methods, however effective, are less apt to be funded (Bellon *et al.* 2006).

It is almost certainly too crude, however, to say that 'development agencies' fail to learn from experience because that is to treat them as though they were homogeneous institutions. It seems more plausible to treat a development project as a differentiated hierarchy in which (at least) four groups of participants can be distinguished who have quite different interests and experience:

- Local employees (sometimes civil servants) in the host country who implement the programmes.
- Experts (e.g. academics, consultants, 'policy analysts') with a background in the natural or social sciences who have field experience and who provide advice on programme design.
- Planners (middle-ranking officers in foundations or aid agencies) who are responsible for putting together the project and overseeing it but who rarely leave their desks in New York, London or Washington.
- Decision-makers (high-ranking officials or ministers in the donor agencies) with the power to make or break a proposed project.[15]

Once one disaggregates projects in this way, it seems likely that some levels of the hierarchy may be more concerned with the question of efficacy (and thus seek to learn from past experience) than others but simply lack the power to implement necessary changes. Conversely, those who possess such power may be both too remote from the day-to-day operation of projects and too preoccupied with running an agency to be aware of which approaches were effective or even to care very much about outcomes.[16]

Some analysts of the development industry portray approaches to development as 'fashions' which come and go regardless of their efficacy and have been especially critical of *planners* (Porter *et al.* 1991). One question that ought be asked is to what extent planners are familiar with the society, culture and economy of the region for which they are responsible (including the history of previous development programmes there). There are indications that mid-level research staff in some agencies have been required to devise policy analyses for a particular region despite very limited knowledge or experience of it. The economists who staffed such research teams at the World Bank from the 1980s, for example, were expected to become experts on a region within a matter of months, and of those assigned to prepare the Bank's report on the East Asian development 'miracle', none had experience of living or working in Asia (Wade 1996: 33, 18).[17] Moreover, even where planners do possess the requisite background, the reward structure of agencies may not encourage them to look very carefully at the feasibility of proposed projects (*inter alia* by taking into account the history of development in their region). As Robert Tripp remarks, since donor agencies are under pressure to 'sell a distinct product', most projects are not thought through very carefully:

> The donor's urgency in moving funds and the contracting agency's priority in meeting payrolls conspire to extinguish any thoughts on long-term implications until reports are written [and] additional project phases are requested....
>
> (Tripp *et al.* 2006: 210–211)

The challenge, therefore, is to get donors and their projects 'to invest less in annual reports and self-congratulatory promotion and to pay more attention to careful, long-term monitoring of the course of agricultural change' (ibid.: 208).[18] Planners thus occupy a potentially awkward space between the (relatively well-informed) experts who report to them and the powerful who employ them. And that is probably why they spend more time on packaging and selling loans than on carefully preparing projects or keeping tabs on problems in the field (Goldman 2005).

As was clear from the Mexican Agricultural Program, however, it is almost certainly wrong to treat planners or other mid-level staff as a homogeneous group. There are often tensions and conflicts among different individuals and factions within the organization (Harwood 2009). That means that the reason why some views are taken into account in project design (or policy) and others are ignored presumably depends on the *decision-makers* at the top of the hierarchy. That group is yet further removed from the problems of rural poverty and how best to alleviate it. Nonetheless, when donors fail to take into account relevant information on previous development projects, it is not just the product of haste or ignorance. When evidence suggests that proposed schemes would not work or that consultants' conclusions are 'unacceptable' to decision-makers, such findings are often ignored or suppressed (Goldman 2005: 152, 164–169; Speich 2009b; Wood 1998; Porter *et al.* 1991: 3–4; Lepenies 2009). In describing the process whereby a regional research report was vetted by higher level officials within the World Bank, for example, Robert Wade remarks that

> It is not just that higher levels are more concerned with the Bank's and the system's integrity than with the integrity of the research. It is also that promotion criteria select people for the higher levels who make decisions quickly … using 'facts' selectively to support pre-conceived patterns and convictions.
>
> (Wade 1996: 31)

The same applies to assessing projects' success or failure. The World Bank's Operations Evaluation Department, for example, has a reputation for providing informed criticism of the Bank's work, but it is also said to be marginal within the Bank and thus 'largely ineffective' (Pincus 2009: 186). If the efficacy of projects is of so little concern within agencies, it is hardly surprising that planners take little notice of past experience.

By treating development projects (or agencies) as differentiated hierarchical organizations, therefore, we might gain a better understanding of why their capacity to learn from experience has been so limited. But is a lack of institutional memory attributable merely to the chaotic and uncoordinated state of the development industry, compounded by organizational flaws which divorce expertise from decision-making, so that agencies are run by careerist and ill-informed bureaucrats? When an institution fails to learn from experience, it has become a commonplace to cite Santayana's dictum that those who fail to learn from their own history are doomed to repeat it. To formulate the problem this way, however, is to portray the agencies of the development industry as rather hapless organizations who are well-meaning but haven't got round to doing their homework, as though they want to alleviate poverty but just haven't yet managed to tackle the problem systematically.

Over the last 20 years, however, critical analyses of the concept of 'development' as well as historical reflections upon the industry suggest that the problem is much more deep-seated (Ferguson 1990; Cooper and Packard 1997; Rist 2008; Mitchell 2002; Ziai 2007; Cullather 2000). For at the heart of the development concept is a contradiction. Despite abundant evidence that the social changes constituting development are impossible without political change, development is nonetheless portrayed as a process which can be fostered solely through technical measures. The history of the Green Revolution offers numerous examples where donor agencies have gone out of their way to avoid the political dimensions of development. This was evident, for example, in the early years of the Mexican Agricultural Program; by the 1950s a trustee of the Rockefeller Foundation, anticipating that the new technology would have political consequences, recommended that the Foundation withdraw from Mexico as soon as some scientific progress could be demonstrated and before political tension had become acute (Oasa and Jennings 1982). At the end of the decade the Foundation's head of agricultural research rejected an application from staff of the Program to attend a conference on land reform for fear that it could impair the Foundation's cooperation with the Mexican government. It was thought less controversial for the Program's economists to concentrate on marketing and production (Jennings 1988, 164). Mexico was not an isolated case. When a Peruvian agricultural scientist applied to the Foundation for support of a project on potato cultivation whose declared aim was to alleviate the poverty of the highland indigenous population, an official declined, saying that while the Foundation had an interest in the 'scientific side' of potato breeding, it could not support a proposal for work on 'the Indian problem' since that involved 'many complications on the political side' (cited in Shepherd 2005: 126). In Southeast Asia, too, the Foundation took an interest in those 'isolable technical problems' of rice cultivation that would not require changes in the social relations of production (Anderson 1991; cf. Goldsmith 1997).[19]

This aversion to politics is seen most clearly in discussions of land reform. The whole point of the Green Revolution, as one critic put it, was to avoid the need for land reform (Griffin 1999).[20] Since land reform is often cited as the single most effective measure in promoting increased productivity on small farms (Johnston and Kilby 1975; Ladejinksy 1973; Hayami and Ruttan 1985: 360), donors cannot entirely avoid the subject. It is illuminating, however, to examine how they have attempted to deal with this issue in recent reports. IFAD, for example, notes that although land reform has had considerable success in reducing poverty, 'it has run into many problems'. Of the problems cited, however, no mention is made of the *political* resistance to land reform. Rather than calling directly upon governments to pursue land reform, therefore, the authors appeal for poverty alleviation on the grounds that heavy biases against the rural poor are 'inefficient as well as unjust' (IFAD 2001: 230). The World Bank agrees that land reform can increase efficiency (World Bank 2007: 9) and goes further by acknowledging that desired agricultural reform can require political change. But the Bank's idea of how to promote such change reminds one more of the timid technocrat than the experienced politician. The *World Development Report 2008*, for example, declares that such changes 'always have gainers and losers', and these 'conflicts make decisions more difficult'. How, then, to ease the pain? The Bank outlines measures which it believes can increase the 'feasibility' of political change such as recommending that agencies use 'research-based evidence for information and debate' (as though this would persuade the opponents of reform to give way) (World Bank 2007: 242; cf. 249–250). As Ferguson (1990) points out, such limp recommendations have to be seen in context. The fact that donors must not be seen as intervening in the client country's politics means that if they are to sell development packages successfully, they must portray the client as likely to benefit from the kind of package they can provide. Thus donors' accounts typically exaggerate the power of government ('policy', 'planning') while discreetly ignoring the extent to which the economy is shaped by wider forces.

If development as a technical intervention is one fiction, development as essentially a philanthropic undertaking is another. Neither donor nor recipient governments have hesitated to pursue quite different aims. The beauty of development aid, as two historians conclude, '[lay] in the fact that [economic and political] aims could be camouflaged by humanitarian concerns' (Frey and Kunkel 2011: 228). As far as donors are concerned, it is clear from the history of the first generation of the Green Revolution that aid had a lot to do with politics and very little with poverty alleviation.[21] During the 1950s and 1960s, as we have seen, the Green Revolution was routinely supported as an anti-communist measure, and more recently some have argued that, in the US at least, the humanitarian aims of aid generally play second fiddle to foreign policy (Gow and Morss 1988). But aid has also served to expand overseas markets for the agribusiness sector. In the 1960s, for example, 90–100 per cent of development aid from the US, Japan, Canada and Italy was granted on the condition that it be spent on goods and services from the donor country (Müller 1969; cf. Mitchell 2002: 236–238; Hödl 2011). Worse, whenever such goods were not the cheapest or the highest quality available on the international market, the recipient country ended up in effect subsidizing the wealthier donor. The industrialized world's relative indifference to poverty alleviation in the global South is equally evident from the pattern of aid's distribution. Much of it goes to countries with relatively low poverty levels while little goes to India or China despite very high levels of poverty there (IFAD 2001: 40). Finally, the unfavourable terms of trade between North and South in recent years testify to a similar indifference: the North declines to open its markets completely to agricultural exports from the South while dumping heavily subsidized commodities in southern markets (FAO 2009b: 3; World Bank 2007: 10, 17). Despite their continuing rhetorical support for poverty alleviation, therefore, wealthy countries' actions instead hinder it (Rist 2008).[22]

Turning to client countries, their governments are often just as keen as the donors to use apparently neutral development programmes in order to pursue undeclared political ends (Ferguson 1990; Speich 2009a; Cooper 2010). Where peasant-oriented projects are involved, for example, client governments can be particularly uncooperative. For these run the risk of provoking a political backlash from estate owners who are invariably better organized than smallholders and are often part of the governing elite (Leonard 1986; Mitchell 2002; Birner and Resnick 2010; Bray 1986; Hazell *et al.* 2007).[23] And this is probably one of the reasons why, as we saw above, NARS in some countries make little effort to address the problems of the small farmer. More generally, whether because of corruption or for other reasons, host governments find numerous ways to divert funding away from intended beneficiaries (Porter *et al.* 1991: 212; Rist 2008: 151–155) or to put pressure on staff in order to influence a project's outcome.[24] The upshot, as Timothy Mitchell's history of US development policy in Egypt demonstrates, is that donors have relatively little control over the outcomes of their projects (Mitchell 2002). It is instead the political context in the host country which is decisive. Others agree:

> the receiving society and its historical momentum are much more powerful and important than the [development projects], and the latter only really have a chance to succeed if they can work with the flow and the momentum of the society's history to encourage the desired kinds of selective adaptations.
>
> (Woolcock *et al.* 2011: 87)

If we ask, therefore, why the development industry has taken so little interest, not just in what happened in early twentieth century Europe or Japan but even in the efficacy of recent projects, Gilbert Rist would seem to be correct. 'Development' is an ideological commitment and a vehicle for achieving a range of often political aims, not an empirically based learning process that seeks poverty alleviation (Rist 2008). And for all the emphasis placed in development discussions upon expertise, the role of expertise/evidence in policy-making or project design is actually quite limited. Each time that projects or policies are confronted with evidence, decision-makers (like politicians everywhere) take a straightforwardly instrumentalist view, ignoring evidence when it is inconvenient but invoking it whenever it can be used to justify the preferred course of action. Nor will they generally be held to account by client countries keen to see poverty alleviation succeed. From this perspective, there is little to be gained from going to the trouble to examine past projects, be they last century's or last year's.

Of course, if evidence plays such a small role in shaping policy, and donors and client countries alike are so little interested in poverty alleviation, the reader may well be wondering why I bothered to argue the case that peasant-friendly development can be fostered by state research institutions. For it looks as though what is feasible in principle may be unlikely in practice: 'bleak' trumps 'hopeful'. That would probably be the unavoidable conclusion if the development industry and client countries were monoliths. But there are indications that not all developing countries are indifferent to poverty alleviation or to the needs of smallholders (e.g. World Bank 2007: 258; Juma 1989: 189–190). As far as donors are concerned, my argument for a larger role for the state in R&D isn't likely to persuade the advocates of neo-liberal development strategies to change course. But I hope it will nonetheless be of use to those organizations which are committed to a more radical revision of the Washington Consensus than the World Bank and other major agencies are willing to countenance. And I take encouragement from the fact that not all donor agencies, past or present, have subscribed to a single orthodoxy (cf. Minogue and Kothari 2002). During the 1960s, for example, there were signs that West Germany promoted a more peasant-oriented route toward India's agricultural development than did the US (Unger 2010a). In the 1980s, as we have seen, Japan declined to embrace the World Bank's neo-liberal strategy and continued to emphasize the role of the state (Wade 1996). And more recently evidence from southern Africa suggests that some governments' aid agencies – for example, those of the Netherlands and some Scandinavian countries – have been particularly energetic in supporting peasant-friendly approaches (Bowman 2011). As long as the system incorporates some degree of pluralism, therefore, it may be possible for at least a small sector of the development industry to reflect upon the past while trying to devise more effective strategies for the future.

Notes

Introduction

1 Although the total number living in absolute poverty worldwide fell between 1990 and 2004, in 54 countries the levels of poverty actually increased (Rist 2008: 234–235).

2 Farmers' vulnerability through reliance upon agrochemicals was made all too clear in 2008–2009 when fertilizer prices rose threefold in some areas (Anon. 2008; Vidal 2008).

3 In the 1960s the International Rice Research Institute (IRRI) showed a film to visitors which declared that 'primitive methods' of cultivation in the Philippines were 'inefficient [and] wasteful of human energy' (cited in Cullather 2004: 237).

4 It bears saying that although these problems pose a particular threat to farmers in the global South, none of them is confined to agriculture in that region (e.g. Wittwer 1983). Farmers in some parts of Europe and North America have also been trying for decades to reduce excessive fertilizer-use and find safer agrochemicals, and in some regions of the US water-shortage has become acute.

5 At the moment this problem of distribution is aggravated by dietary preferences in the global North as well as increasingly in China and India. Half of all the cropland in the world is devoted to feeding animals rather than human beings (Weis 2007: 42), and rearing animals is a much less efficient way to convert food into calories than is growing plants (Lipton 2007: 37). Thanks also to recent incentives for growing biofuels, cropland that might have been dedicated to food production is being diverted to other uses.

6 In Malawi, for example, 2.6 million hectares of arable land remain uncultivated (by large landowners) at a time when smallholders are struggling and rural malnutrition is rife (Holmen 2005: 82). The situation in Zambia and elsewhere in sub-Saharan Africa is similar (Andrew Bowman, pers. comm.).

7 As Biggs *et al.* (2011) point out, however, there are still economists who continue to believe that large-scale mechanization on large farms is more efficient.

8 A few economists still adhere to this view, claiming that peasant agriculture is 'chronically ill-suited to innovation' (Collier 2009: 64).

9 On the shift of opinion among German-speaking historians, see Dipper 1993. A collection of German peasant sayings from earlier centuries includes more than a few that demonstrate a respect for learning and knowledge (Richey and Strich 1984: 7, 18, 24, 28–30, etc.).

10 Average farm size in Britain *c.*1870 was between 30 and 40 hectares, depending on the region, but over half of the total acreage was concentrated on much larger farms (Fletcher 1973: 40; Perkins 1986). In the US *c.*1900 the average farm was about 60 hectares (Hurt 2000: 100; Fite 1981: 20). Moreover US agricultural colleges have frequently been criticized for ignoring the needs of the country's smaller farmers (Hightower 1973; Paarlberg 1981; Busch and Lacy 1983).

11 The same point holds true for peasant-oriented research and development today. The rationale for 'participatory plant breeding' (Chapter 7), for example, is based upon population genetic theory.

12 The strongest evidence comes from Denmark, where before the First World War several experiment stations were carrying out varietal testing and breeding as well as instructing farmers in basic breeding techniques (Lang 1911c, 1912b, 1914b). Since Denmark at that time was internationally famous for its cooperative movement and its promotion of technical innovation among smallholders (Henriksen 1993), it is quite likely that the breeding work there was peasant-friendly. In Poland the experiment station in Sobieszyn began in the late 1890s to improve local rye varieties which were sold at reduced prices to peasant farmers (Sempolowsky 1903). Less clear is the picture in the Czech province of Bohemia, where from 1904 a unit at the agricultural college in Tetschen-Liebwerd (Libverda) was carrying out varietal development in ways similar to those of the German stations, though it is not clear whether the work was tailored to peasant needs (Lang 1914a). In searching for comparable institutions elsewhere, one needs to bear in mind that small farm size alone is not a reliable indicator of peasant-friendly breeding. In France during the 1920s, for example, 85 per cent of farms were smaller than 10 hectares, but public-sector plant breeding appears not to have been directed toward peasant needs, at least before 1945 (Bonneuil and Thomas 2009).

13 It is noticeable, too, that while many agricultural historians have used the generic term 'agricultural revolution' in describing the European transformation (e.g. Grigg 1984; Overton 1996; Mazoyer and Roudart 2006), few have attempted to apply the more specific concept of 'green revolution' to times and places other than the global South after 1945. Those that have done so have usually deployed the term as a convenient way of labelling twentieth century agricultural transformations in Europe but with little or no attempt at systematic comparison between transformations in Europe and those in the developing world since 1945 (Van Zanden 1991; Henriksen 1993; Dix and Langthaler 2006). Francesca Bray is a rare exception, using the term in order to draw attention to the similarities between the Green Revolution after 1945 and the state promotion of technological improvements in Chinese agriculture between the tenth and thirteenth centuries (Bray 1979).

14 A large market nearby often made such farms viable. If located near a city, a two-hectare farm growing vegetables or fruit under intensive cultivation could make a living (Henning 1978: 154). A study of Hessen around 1900 indicated that with good soil a farm of three hectares could still produce surplus grain for sale (Achilles 1993: 320).

15 Despite these similarities, it is possible that there were significant differences between 'peasant' farms in the two contexts which would merit systematic study. It is my impression – and it is no more than that – that a larger proportion of small south German farms were integrated into the capitalist economy than was the case in the global South after 1945 (or, for that matter, in Eastern Europe into the 1930s and beyond). That is, in Germany a significant proportion of smallholders seem to have taken decisions – for example, what crops to plant, what technologies to use – on grounds of expected profit (rather than on considering whether the increased family prosperity would be worth the additional labour required). Thus if one adheres strictly to Chayanov's definition of the peasant economy (Tschajanow 1923), a lower proportion of south German smallholders may have been 'peasants' than in the developing world.

16 At Wageningen University, for example, the Technology and Agrarian Development group incorporates perspectives from both genomics and history into its research on West African rice-growing strategies (e.g. paper by Harro Maat, summarized in Harwood and Sturdy 2010).

17 The Bank's Economic Development Institute (est. 1956) set up training programmes

for development officials whose courses included an introduction to 'past experience of economic development and ... the general principles which may be drawn from it' (Staples 2006: 38). Recently some of its staff co-edited (and the Bank funded) a collection of papers arguing history's relevance for development policy (Bayly *et al.* 2011). All of this is very welcome, but it remains to be seen whether this historical sensibility will feed back upon the Bank's own programme design and decision-making.

18 As an academic in development studies told me with some exasperation, 'Who reads programme reviews?' An American agricultural economist with intimate knowledge of the Indian GR programmes of the 1960s was equally blunt: 'The Ford program people and their Government of India colleagues were hell-bent on "doubling food production within five years" but had never examined whether that had been accomplished elsewhere and if so, how. They had no interest in the success/failure of past development efforts'.

19 As one observer put it, '...many Third World countries are repeating the mistakes of industrialized countries in the process of development' (B. S. Baviskar, cited by Warren 1989: 161). And another: 'In the history of development there are many good things, but the avoidable errors are appalling' (Chambers 2005: 84). When James Ferguson asked a development planner in the mid 1980s what he would recommend for Lesotho's future development, the planner replied that he would push the same crop-development plan which had failed ten years earlier (Ferguson 1990: 235).

20 'The function of history is to lend materials to politics....' (Santayana 1906: 66).

1 The origins of peasant-friendly research in Germany

1 Ironically, even where social historians have decried this one-sided political emphasis, the alternative approaches they have championed still remain pretty far removed from agricultural practice, institutions or in some cases even the economy (e.g. Evans and Lee 1986; Moeller 1986c; cf. Bessel 1989).

2 Although Aldenhoff-Hübinger makes an admirable start at remedying this deficiency, she acknowledges that her own book has much more to say about protectionist policies than about those aimed at restructuring agriculture. It bears saying that this preoccupation with trade policy may not be specific to the German-language historiography; there are signs of the same gap in recent English-language literature on agricultural policy (Ingersent and Rayner 1999).

3 In contrast with the history of American agriculture, where considerable attention has been paid to the role of the U.S. Department of Agriculture in promoting modernization (e.g. Marcus and Lowitt 1990), histories of German ministries of agriculture or similar agencies during this period scarcely exist.

4 Although local or regional 'economic societies' (*Oekonomische Gesellschaften*) had been established from the mid eighteenth century, for example, the initiative was taken by estate owners keen to exchange experience and explore new cultivation methods, though they enjoyed the support of Frederick the Great (Popplow 2010; Buchsteiner 1995).

5 Between 1816 and 1858 the German states' population increased by over 50 per cent while the proportion of the population working in agriculture declined from 80 per cent to 45 per cent (Teuteberg 1977: 17, 21).

6 In Prussia by about 1840 there were 108 Associations and by 1860, about 500 (Achilles 1993: 302; Ullmann 1988: 34).

7 In doing so, Prussia was following Württemberg, Hessen-Darmstadt and Baden which had established similar commissions a decade or more earlier (Henning 1996: 955–958).

8 In 1879, for example, the Agricultural Association of Rheinland-Prussia got 80 per cent of its finance from the state (Pyta 1991; cf. Pruns 1979: 141–166).

9 Designated as the *Landesculturrat* and *Zentralstelle für die Landwirtschaft*, respectively (Findbuch for Rep. E171, GLAK).

10 Sugar beets required intensive cultivation methods, and a substantial number of growers were peasants working small and medium-sized farms (Perkins 1981; Webb 1982). One estimate suggests that in the late nineteenth century 80 per cent of the shares of sugar refineries were held by peasant farmers (Dipper 1993: 32).

11 This undertaking represented a continuation of the work of the late eighteenth century economic societies which also sought to improve peasant cultivation practices (and even to recruit peasant members) as part of a general enlightenment project, though with little success (Popplow 2009).

12 Peasants' lack of enthusiasm for the Associations can be seen in Württemberg where in 1877 less than 15 per cent of the farming community were members (Lang 1971: 106).

13 In Austria-Hungary, too, peasants called for better representation, complaining that the existing agricultural societies were dominated by the middle classes and aristocratic landowners (Bruckmüller 1984).

14 In Switzerland cooperatives on the German model were also established in the 1860s and became very popular for much the same reasons (Baumann 1993: 57–62).

15 Between 1870 and 1880, for example, the proportion of wheat on British markets that was imported from Germany declined from 18 per cent to 3 per cent while that imported from the US increased from 23 to 65 per cent (Aldenhoff-Hübinger 2002: 31). And between the 1860s and 1890s US wheat exports to Germany increased fourfold (Goltz 1894: 125).

16 Triggered by the formation of the German Reich in 1871, the Council was set up the following year in order to provide farmers with representation at national level, its members being drawn from each state's agricultural commission (Gottwald 1984).

17 A recent dissertation on this topic offers an alternative hypothesis which supports the argument that I outline in section 4 below. Staus (2011) suggests that the movement for internal colonization was primarily an attempt to increase productivity by transferring land from inefficient estate owners to peasant farmers. The nationalist and domestic justifications officially offered for the policy, on the other hand, were only secondary, a form of political camouflage thought necessary in order to secure government support.

18 In Belgium about this time the socialist party's attempts to recruit in the countryside were also unsuccessful for many of the same reasons (Vanschoenbeek *et al.* 2002).

19 Over the period 1885–1900, 5,000 large estates went bankrupt (Berghahn 1987: 6).

20 Between 1850 and 1890 land prices in Germany increased threefold (Barkin 1970: 194).

21 The economist Georg Gothein was soon citing regional surveys since the late 1880s in support of this view (Gothein 1910/1911: 12–23), and it has since been endorsed by various social and agricultural historians, e.g. Rosenberg 1969; Henning 1996: 670ff.; Puhle 1986; Achilles 1994: 208–209).

22 On the shortage of agricultural workers, due largely to migration to the cities, see Anon. *Die Entwicklung*, 1914. Gustav Schmoller and other members of the Association for Social Policy (*Verein für Sozialpolitik*) took a similar view. Across-the-board tariffs or subsidies as demanded by the Agrarian League would just keep the weaker estates going; assistance needed instead to be targeted at competent farmers who were often peasants (e.g. Gothein 1910/1911; cf. Jatzlauk 1994; Steinbeck 1997; Frauendorfer 1957: 386ff.).

23 25 April 1895 – Johannes Miquel to Kaiser, Nr. 30170, Rep 89, I. HA, GStAPK.

24 One type of German cooperative grew fourfold between 1894 and 1904 (Peal 1988: 254). In the Rhineland the membership of dairy cooperatives increased sevenfold between 1895 and 1900 (Lichter 1994: 230) while in Silesia the number of cooperatives increased nearly tenfold between 1890 and 1896 (Merl 1994: 294).

25 Between 1882 and 1907 the total acreage nationwide occupied by most categories of farm was declining, with the exception of those in the mid-sized peasant category which grew by about 15 per cent (Henning 1996: 945; cf. Boelcke 1995). For the shift between 1895 and 1907, see Dade 1913: 562–564. The evidence from Switzerland for this period is very similar (Moser 2000).

26 One indicator of the scale of this shift was the fact that the number of pigs (a peasant speciality) reared in Germany increased nearly threefold from 1883 to 1913 (Bitterman 1956: tab. 10).

27 The same was true in Switzerland *c.*1900 (Laur 1910). In the case of animal husbandry this was because smallholders maintained a higher number of animals per hectare than did large farmers (Aereboe 1917: 530).

28 *Wochenblatt Baden 1900*: 42.

29 This is not to say that the League was indifferent toward technical modernization. On the contrary; certain of its facilities – for example, fertilizer factories, discounts on seed and machinery which it secured for members – were intended to help those farmers who sought to adopt intensive methods (Puhle 1966: 50–55). The point is that the League was not geared to *peasant* farmers' needs.

30 On this policy of *Sammlungspolitik* (*sammeln* = to gather) see Mommsen 1995: 148–149, 171–173; Lehmann 1970: 101–104.

31 The Association for Social Policy had been pushing this line since the 1880s (Jatzlauk 1994). On the political motivation behind state support for peasant farmers in other countries at this time see Van Molle 2008; Bonneuil 2006: 292; Koning 1994: 36, 104–105.

32 ('...revolutionäre Wühlereien an dem harten Schädel des deutschen Bauern scheitern müssen'). 10 November 1894 – Miquel to Kaiser, Nr. 30170, Rep 89, I. HA, GStAPK. The peasant's 'thick skull' was a fairly common trope in writings of the 1880s and 1890s, not least among socialists trying to explain their failure to recruit peasant voters (Lehmann 1970: 17–18, 61).

33 In Austria, too, cooperatives were perceived as a countermeasure against socialism (Bruckmüller 1977: 259).

34 The secondary literature on the chambers, especially for the period before 1914, is very small, partly because archival sources are sparse (cf. Flemming 1978). That which exists is useful in showing how the chambers were created in order to further agricultural modernization and in giving an account of their activities (Lichter 1994; Albers 1999; cf. Aldenhoff-Hübinger 2002: 197–210). But it usually misses the chambers' *political* dimension: the motives behind their establishment, the distribution of power within them, and the orientation of work they conducted. To be sure, the idea that the chambers were part of Miquel's anti-socialist programme was proposed many years ago by Heinrich-August Winkler (1979), but he approached the chambers as an instance of the bureaucratization of economic expertise without attempting to locate them within the history of agricultural policy. The one agricultural historian who has recognized the political intent behind the chambers' creation is Norbert Steinbeck, but his treatment of it is cursory (Steinbeck 1997: 102–103). For a fuller discussion, see Harwood 2008).

35 In Schleswig-Holstein, for example, the Agricultural Association's income was so small that three-quarters of it went toward administrative costs (Frentz 1995). One estimate in 1912 was that whereas the Associations commonly had an annual income of roughly 5000 marks, a large chamber's income, financed by a tax levied upon every farm in the province, could exceed 100,000 marks (Crone-Münzebrock 1912: 558). In addition, each chamber could reckon on sizeable state subsidies which often exceeded its income from the tax levy.

36 By the 1920s, for example, most provincial chambers in Prussia were employing 50 to 100 staff, and some large ones were employing several hundred (Altrock 1925; Albers 1999: 28).

37 Von Rümker 1913. In some regions such as East Prussia, many Associations seem to have been less concerned with technical innovation than with the defence of their traditional privileges (Puhle 1975: 56–58).
38 Before the First World War the chambers appear to have honoured this expectation (Wygodzinski 1916: 286). Winkler is thus probably right to say that the chambers allowed for the domestication of agricultural pressure groups through incorporating them into the state (Winkler 1979: 173).
39 Maintaining state control was clearly important. When the Agrarian League requested in 1894 that the chambers should be overseen, not by a government body but by an independent commission, the Minister declined (Steinbeck 1997: 96ff.).
40 In other countries at this time, estate owners expressed a similar unease, sensing their inability to dominate new chamber-like bodies (Koning 1994: 145; Aldenhoff-Hübinger 2002: 197ff.).
41 In the Rhineland, for example, the provincial president remarked that the owners of small farms (and those with close ties to the Peasants' Association) were generally more favourable to the new chambers than were 'the more influential landowners' since the latter were quite satisfied with the existing Agricultural Association (Steinbeck 1997: 44; in the same vein Lichter 1994: 145).

2 The movement for peasant-friendly plant breeding, 1880–1905

1 In what follows I have drawn particularly upon Bavarian sources, partly because it was the largest and most influential of the south German stations, but also because of the abundance of archival material. In Württemberg, by contrast, the destruction of the Agriculture Ministry during the Second World War means that archival sources for the period *c.*1880 to 1945 are very thin.
2 The best overview by far is Wieland's history of plant breeding as a discipline (Wieland 2004). 'Insider histories' exist since the early twentieth century, but they tend to be lists of achievements, are sometimes inaccurate, or do not bother to cite sources (e.g. Reitemeier 1904; Hillmann 1910b; Scheibe 1961; Bundesverb. Dt. Pflanzenzüchter 1987; Gäde 1993; von Rümker 1928b)
3 By 1908 one firm (Rabbethge und Giesecke) had further increased sugar content to 23 per cent and was producing one-third of the world market for sugar-beet seeds (p. 87, Shull, G.H., 'Notes on a European Tour undertaken by George H. Shull…' (1908)), Shull papers, American Philosophical Society.
4 In Sweden breeders also sought to improve Squarehead's winter hardiness by selection, though in vain (Weller 1953).
5 This strategy was hardly unique to Germany. At the agricultural college in Wageningen in the 1880s, for example, Dutch breeders also crossed Squarehead with local varieties for the same reason (Maat 2001). By 1914 one German breeder reckoned that probably all Swedish, Danish and German wheat varieties then on the market had been derived from Squarehead (Kiessling 1914: 24).
6 Hybridization was relatively easy with a naturally cross-fertilizing species like the potato. Rimpau's achievement, however, was in managing to construct *artificial* crosses with a normally self-fertilizing species like wheat.
7 In the US, too, the first wave of systematic plant breeding in the nineteenth century was undertaken by experimentally inclined farmers, some of whom later turned full time to breeding (Olmstead and Rhode 2008).
8 As late as 1960 various derivatives of the Petkus rye were still covering about 90 per cent of the rye acreage in several European countries (Scheibe 1961: 9).
9 By the mid 1870s there were over 20 experiment stations in Germany alone (Jahnel and Ludwig 1961; Finlay 1992: 250–254; Jas 2001: 169–171). In the US comparable seed testing does not seem to have begun until around 1900 (Kloppenburg 2004: 72), possibly because commercial breeding also took off relatively late there.

10 Seed adverts in the state's weekly agricultural newspaper, for example, typically referred to the type of grain for sale (rye, wheat, etc.) but neglected to mention the name of the variety.

11 On the other hand, to judge from a well-known handbook of the period, there was increased interest in this issue by the turn of the century (cf. *Schlipf's Handbuch*, 1880 and 1902).

12 Starting with 83 varieties in 1892, the programme had grown to 823 by 1914 (Haushofer 1960). Comparable organizations were set up during the 1880s in Denmark (Holtmeier-Schomburg 1908) and Austria (Tschermak 1936–1937).

13 'Allegedly' because in 1895 the Agrarian League established its own Hybred Seed Commission (*Originalsaatgutkommission*) which is said to have conducted both varietal testing and certification (Reich Min. Ldw., 'Begründung des Referentenentwurfes eines Saat- (Pflanz-) gutgesetzes' (1929), pp. 14–15, ML2040, BayHStA).

14 One mechanism by which certification procedures sought to combat fraud was by regulating names and claims. A new name, for example, could be used only if the 'new' variety was demonstrably different from an existing one. Similarly, a variety could be declared 'improved' only if it in fact performed better in field tests (Hillmann 1910b). That breeders were keen on certification is evident from the roughly eightfold increase in the number of varieties entered between 1901 and 1908 alone ('Geschäftsbericht', *Jahrbuch der DLG*, vol. 24 (1909), 179).

15 'Hybred' was common usage in the English-language literature at this time and presumably was shorthand for 'highly bred'. Of 62 varieties entered in the selection process for the Register in 1905, only 20 were chosen (Saatzuchtstelle der DLG 1905a). By 1913 there were 28 cereals varieties on the Register plus another dozen from other crops (Edler 1913). The Register was revised every four years (minutes of meeting of Saatzuchtabteilung, 22 February 1910, *Jahrbuch der DLG*, vol. 25 (1910), 33–60).

16 As one academic put it, the scale of the Division's varietal testing and certification needed to be expanded if its benefits were to be enjoyed 'by all farmers and not just members of the Agricultural Society' (Gisevius 1912: 9). Before the war and even into the 1920s academics continued to call for the expansion of varietal testing, in part to keep the seed market from being swamped by new varieties (Remy 1907b; von Rümker 1907b) but also so that the most economically effective geographical distribution of varieties could be established (von Rümker 1925a).

17 Commercially bred barleys, for example, quickly lost their desirable properties when planted in Bavaria, and in some cases their yield was not even as high as that of local barleys (Kraus 1903, 1914; Kiessling 1912).

18 Spanish rice planters in the 1940s were in a similar situation. Faced with possible shortages of water and mineral fertilizer, they often thought it not worth spending money on high-yielding varieties (Camprubi 2010).

19 At that time commercial varieties cost from 25 to 50 per cent more than a good quality local variety (Raum 1912: 934; Balster 1912; Kappen and Henkelmann 1932).

20 Unlike commercial varieties which typically bore the name of the individual breeder, local varieties bore the name of their region. This reflected their geographical origin but also the fact that they were not an individual's creation but a collective product: the end result of numerous decisions by many farmers who had exchanged seed among themselves over a long period.

21 This is my impression based on reports of the tests that were published in the *Jahrbuch der DLG*, 1890 to 1905. A notable exception appears to have been local oats varieties; many of these were tested, perhaps because this crop tended to be planted on poor soil for which few commercial varieties were available.

22 It is no accident that two academics who played a leading role in the Division from an early date, Kurt von Rümker and Wilhelm Edler, were also founder members of the Association of Commercial Breeders (*Gesell. zur Förderung der deutschen Pflanzenzucht*) (GFP 1933).

23 It was not until 1921 that the Division finally decided that the Seed Office would henceforth arrange sales to *any* farmer (1 March 1921 – minutes of the meeting of the Ausschuss der Saatzucht-Abteilung, R 8072/53, BAB).

24 The Seed Breeding Division seems to have sensed that this might be a delicate issue. In 1907 its head administrative officer concluded a survey of the Division's activities by emphasizing that although it might seem as though the Division existed merely for the benefit of breeders and seed multipliers, its work was in fact of general interest to all members of the Agricultural Society (Hillmann 1907).

25 The same might be said of the French wheat-seed trade which in 1925 declared that the term 'improved wheat' could only be used to describe varieties obtained via pedigree selection. Mass selection no longer counted as 'selection' (Bonneuil and Thomas 2010).

26 The most likely explanation for this dearth was that due to the enormous variation in soil and climate over quite short distances in the south, a variety bred in any one locality was unlikely to flourish across the region, much less elsewhere in Germany. The market for it would thus be very small.

27 M. Snoy, 'Geschichte des Instituts für Pflanzenzüchtung mit Landessaatzuchtanstalt', typescript, 1967; Anon. 'Entwicklung des Instituts für Acker- u Pflanzenbau', n.d., typescript both in UA-Hohen, 3/25.31.

28 Minutes of the Commission's plenary meetings, *Wochenblatt Württemberg*, 1890–1900.

29 Other kinds of agricultural technology were receiving far more attention at this time. Judging from the subjects discussed at meetings, presented at exhibitions or published in articles, even the German Agricultural Society paid far more attention to animal husbandry or mineral fertilizers than to the choice of variety (*Jahrbuch der DLG*, *Arbeiten der DLG*, 1890–1910).

30 Such cooperatives were not just a German phenomenon. In North Dakota, for example, experiment station staff also urged farmers to form cooperatives for seed growing and for the same reason: to make themselves more independent of commercial seed companies (Danbom 1989).

31 For example, seed-growing cooperatives were established in East Prussia in 1902 and in West Prussia in 1904 with support from their regional chambers of agriculture (Gisevius 1901; Krech 1904; Modrow 1904).

32 The rapid growth of this new kind of organization had interesting knock-on effects. With ever more organizations producing seed from a wider range of varieties, farmers once again needed guidance on quality. And that created an increased demand for both varietal testing and certification which the Agricultural Society was clearly unable to meet. Several of the newly established chambers of agriculture, therefore, soon began to get involved in both activities (Fruwirth 1911a; Hillmann 1910b; Merkel 1910; Steinbeck 1997: 125ff.). As a consequence, however, the Agricultural Society, despite its self-understanding as the leader of 'scientific agriculture', was beginning to lose control over seed quality and varietal testing. In an apparent attempt to regain control, the Seed Breeding Division decided in 1903 to establish its own administrative office (*Geschäftsstelle*) which would liaise with the new organizations in order to bring their testing and certification activities into line with those of the Society ('Aus den Verhandlungen der Ausschüsse', *Jahrbuch der DLG*, vol. 18 (1903), 392–400; 'Jahresbericht für 1903', *Jahrbuch der DLG*, vol. 19 (1904), 172; 'Bericht über die Tätigkeit der Abteilung im Jahre 1903', *Jahrbuch der DLG*, vol. 19 (1904), 232–236). There are indications, however, that not all of the newcomers were willing to play by the Society's rules (Gisevius 1907: 502; Remy 1907a: 297; von Rümker 1905: 18–19; Hansen and Fischer 1936: 194).

33 Schindler (1854–1937) was an agronomist, at the Polytechnic in Riga and from 1903 at the Technical College in Brno, with strong interests in plant physiology and plant geography/ecology who was particularly concerned to understand why imported

varieties did poorly in Moravia. Maintaining close relations to farmers, large and small, he was a consistent promoter of the value of local varieties through the 1920s (Proskowetz 1924; Schindler 1937; Harwood 2004).

34 Cf. *Ber. Bad. Ldw-Kammer*, 1920, 114.

35 Not all agreed. One eminent agronomist argued that because of widely varying growing conditions, intensive agriculture was not necessarily the most sensible strategy everywhere. In some places extensive cultivation methods would yield greater profits (Pommer 1900: 485–486).

36 During the 1890s only three Bavarian farmers took up breeding (Glaser 1902; Kiessling and Gassner 1927). In Württemberg no one did so until 1903 (Wacker 1933), and in Baden no commercial breeding took place before 1908.

37 The idea that plant breeding could be organized on a cooperative basis was not a radical departure since animal-breeding cooperatives were very common in south Germany by this time (e.g. *Jb.Bay. Ldw-Rats* 1899: 10–11; *Jahrbuch der DLG*, 17 (1902), 137; Häcker 1902).

38 In 1898, for example, the seed-growing association in the Probstei, facing the same problem, also established a breeding cooperative (Holtmeier-Schomburg 1908; Ramm 1926: 25). On Bavarian interest, see 25 July 1898 – Bay.Min.Ldw. to governments of Oberpfalz, Regensburg and Oberfranken, ML 309, BayHStA.

39 5 March 1900 – Georg Heim to Bay.Min.Ldw., ML 309, BayHStA.

40 9 March 1900 – Bay.Min.Ldw. to Bayerischer Landwirtschaftsrat, ML 309, BayHStA.

41 In 1899 the experiment station at Posen's chamber was providing farmers with instruction on individual selection (Steinbeck 1997: 130). And from 1898 staff at the experiment station in Lausanne had recruited local farmers as collaborators in a cereals-breeding programme (Moser 2003: 28ff.).

42 This success is presumably why the Association of German Chambers of Agriculture had recommended that varietal tests should routinely include a Svalöf variety among its standard test varieties (Ulander 1906). Another reason why Svalöf may have been an attractive model in south Germany is that the Swedes' experience showed that in general local varieties were a better starting-point for breeding than were imported ones (Anon. 'Die DLG' 1907; Kraus 1914).

43 Minutes of the meeting of the Seed Division, 21 June 1889, *Jahrbuch der DLG*, vol. 4 (1889), 59–62.

44 The Association's powers were widened so that it could henceforth make recommendations to government, not only on technical matters but also on economic and political ones, and an attempt was made to increase its peasant membership (Bay. Staatsmin. Innern 1897: 336–339; *Jb. Bay. Ldw-Rats* 1895: 1–5.

45 One of the problems with Haushofer's (1975) account of the establishment of the Bavarian station is that he portrays the Bavarian Agricultural Association (as well as the German Agricultural Society) as supporting this move. He thereby overlooks the fact that a major reason for the station's founding was that both organizations had *failed* to address peasant needs.

46 *Jb. Bay. Ldw-Rats* 1898: 2; cf. 15.

47 During the 1890s state spending on the breeding of small animals increased fivefold, on dairying sixfold and on secondary education over 20-fold ('Uebersicht der Aufwendungen aus öffentlichen Mittel zur Förderung der Landwirtschaft in Bayern, 1889 bis 1898', ML 204, BayHStA). In Württemberg, too, total government expenditure on agriculture roughly quadrupled between 1882 and 1907 (minutes of plenary meeting of 23rd DLG exhibition in Stuttgart, *Jahrbuch der DLG*, vol. 23 (1908), 264).

48 18 December 1896 – Bay.Min.Ldw. to Finanzministerium, MF 71660, BayHStA.

49 4 October 1897 – Bay.Min.Ldw. to Kgl. Regierungen u. Kammern des Innern, ML 204, BayHStA.

50 The emphasis of the 1897 volume was quite different from its predecessor in 1890.

When the possibility of a new volume was first raised in 1896, the top civil servant in the ministry suggested to his minister that simply to continue the older form of presentation – a compilation of statistics and analyses of the current state of various areas comprising agriculture – would be unwieldy as well as expensive and time-consuming to compile. Preferable, he argued, was a smaller volume which would not attempt to survey the state of agriculture but would focus primarily upon the measures taken by government bodies since 1890. The minister agreed (9 December 1896 – Ministerialrat Brettreich to Excellenz, ML 204, BayHStA). The successor volume in 1903 took the same form (7 August 1903 – Bay.Min.Ldw. to Finanzministerium, MF 71660, BayHStA).

51 4 Oct 1897 – Bay.Min.Ldw. to Kgl. Regierungen u. Kammern des Innern, ML 204, BayHStA.

52 378th session of the Kammer der Abgeordneten on 7 February 1898, *Stenograph. Bericht d. Kammer der Abgeordneten*, vol. 11 (1898), 16–22, ML 204, BayHStA.

53 18 February 1896 – cutting from *General-Anzeiger d. Kgl. Haupt- u Residenzstadt*, no. 41, ML 204, BayHStA.

54 *Mentzel und Lengerke*, vol. 67, pt 2 (1914), 310–318.

55 *Mitteilungen des KWIs für Landwirtschaft in Bromberg*, vol. 1 (1909) to 5 (1912).

56 On the Institute's mission see 16 July 1913 – draft of Preuss.Min.Ldw. to Director of KWI f. Landwirtschaft, Nr. 12498. On breeding work, see the Institute's annual reports for 1912–1916, Nr. 12489, both files in Rep 87B, GStAPK).

57 'Verhandlungen des kgl. Landesökonomie-Kollegiums', *Landwirtschaftliche Jahrbücher*, vol. 35, supplem vol. 1 (1906), 433. On plant breeding at chambers in the provinces of Saxony (1910) and Silesia (1913), see Anon. *Landwirtschaft* 1911b: 128 and Opitz 1913. Precisely what kind of work these chambers were conducting I do not know. They were certainly promoting the formation of seed-breeding cooperatives (*passim*, Nr. 10440 and 10441, Rep 87B, GStAPK; cf. Gisevius 1907).

58 As late as 1939 a student of Pammer's, again inspired by Proskowetz and Schindler, set up a very similar station at Rinn in the Tyrol (Siegl n.d.).

3 Research, development and extension at the south German stations

1 Württemberg's station was similarly attached to the college at Hohenheim, presumably because staff at both places had prior experience of breeding. Kraus had been breeding barley and oats at Weihenstephan since 1900 (*Jb Bayern 1903*: 7–8); Hohenheim's director, August Strebel, had experimented with mass selection in the early 1890s (Wacker 1933); and Fruwirth's breeding work at Hohenheim began in 1899 (Fruwirth 1907b).

2 10 January 1903, minutes of Beirat meeting, Nr. 2, Fach 86, Sekt II, UA-Weih. On staff perceptions see Steuert 1905. Ten years later, however, the station was separated from the College, coming under the auspices of the agricultural division of the Interior Ministry (Kiessling and Gassner 1927). From 1919 it was accountable to the newly established Ministry of Agriculture (Haushofer 1969: 16).

3 Said by contemporaries to be extremely gifted, he was short-listed after 1918 for a number of high profile chairs in agriculture but chose to remain in Bavaria, becoming professor of agronomy at the Technical College of Munich.

4 n.d. but January 1902, 'Statut der Saatzuchtanstalt', MK 14040, BayHStA.

5 Average annual rainfall in different regions of the state, for example, varied by fivefold (Weller and Lechner 1953), and soil types were so variable that a single farmer's holdings – often consisting of a collection of separated fields rather than one contiguous plot – sometimes included several different soil types (Kiessling 1910a).

6 10 January 1903, minutes of Beirat meeting, MK 14040, BayHStA; *Jb Bayern 1906*.

7 Kiessling 1906. Generally staff sought to visit each centre at least once a year, but

sometimes breeders came to the station in order to carry out procedures under supervision (*Jb Bayern 1904*: 52; Scharnagel 1929). On cost see *Jb Bayern 1905*: 60. In providing this service to small start-up firms, the station was playing a similar supportive role that public-sector breeders in recent years have provided to small private breeders in some developing countries (McIntyre *et al.* 2009b: 86).

8 *Jb Bayern 1910*; Kiessling and Gassner 1927. Initially the inspectors were funded by various agricultural organizations, but by 1921 Kiessling had persuaded the Ministry of Agriculture to take on all of the inspectors as state-funded employees of the station (Kiessling to Bay.Min.Ldw., 18 January 1920, MK 19697, BayHStA.

9 Seed-producing cooperatives that had begun to face stiff competition from non-Bavarian varieties sometimes sought the station's help in improving their local variety (e.g. 14 May 1904 – Kraus, 'Bericht über die Bestrebungen der Rübensamen-Züchter-Genossenschaft in Schweinfurt-Oberndorf', Nr. 2, Fach 86, Sekt II, UA-Weih). In such cases the actual breeding would usually be conducted by a single member under the supervision of the station while the other members would be responsible for growing up the improved variety in large quantities for sale (Kiessling and Gassner 1927; Fruwirth 1919a).

10 In 1905, for example, 10 per cent of the centres were run by local branches of the Agricultural Society, 37 per cent by staff at state institutions and 10 per cent by cooperatives (Kraus 1905b).

11 10 January 1903 – minutes of Beirat meeting, MK 14040, BayHStA; cf. Kraus and Kiessling 1918.

12 25 June 1907 – minutes of Beirat meeting, Nr. 2, Fach 86, Sekt II, UA-Weih.

13 26 June 1906 – minutes of Beirat meeting, MK 14040, BayHStA. The issue came up again in 1911 when rumours were circulating that the station had excessively promoted the varieties of two prominent private breeders (9 June 1911 – Kiessling to Herr Brunner, Nr. 1, Fach 86, Sekt II, UA-Weih). Kiessling evidently felt the rumours were not only false but damaging since he was prepared to sue for libel.

14 *Jb Bayern 1906*: 89. Perhaps this pressure explains why staff occasionally took pains to point out the successes of small breeders (Griesbeck 1924).

15 At this time the acreage planted in barley was much larger in Bavaria than in any other German state, and it also ranked among the largest states in terms of wheat acreage (Lange 1917: plates 19–33).

16 30 December 1931 – Scharnagel to Bay.Min.Ldw., ML 3697 and 18 July 1934 – Scharnagel to Krohn, Nr. 532, LABK, both in BayHStA. Beginning in the mid 1920s Theodor Scharnagel's work on baking quality in wheat demonstrated that, contrary to reigning opinion, it did not correlate with certain morphological properties of the kernel. Establishing more reliable indicators through laboratory studies, he was able to construct hybrid varieties which combined both high yield and baking quality, traits which had long been thought incompatible (Scharnagel 1936/37).

17 13 January 1914 – Grundregel der Geschäftsstelle des Bay. Saatzuchtvereins; and 17 January 1914 – Satzungen des Bay. Saatzuchtvereines', both in Nr. 2, Fach 86, Sekt II, UA-Weih.

18 Early evidence of this kind came from varietal tests on red clovers undertaken by the DLG's Seed Breeding Division from 1898 ('Ergebnisse d Anbauversuche im Klee verschiedener Herkunft', *Jahrbuch der DLG*, vol. 18 (1903), 26–27).

19 In Baden, for example, a local variety of alfalfa and a red clover grown in the Black Forest were seen as offering promising starting materials for breeding ('Denkschrift über die Errichtung einer Saatzuchtanstalt', 236/26751, GLAK).

20 That the private sector nationwide had taken little interest in these crops had been clear since the 1890s. The Seed Breeding Division of the DLG had established a prize competition for grass varieties, but it failed to attract many entries (*Jahrbuch der DLG*, vol. 12 (1897), 155–156). By the 1920s the number of clover and grass breeders was still dwarfed by the number breeding any of the cereals or beets (Evers 1924: 26).

And even today in the US grasses are of little interest to commercial breeders (Frey 1992).

21 Weller 1928; Weller and Lechner, 1953: 18; 10 January 1914 – Saatzuchtanstalt to Kgl. Bezirksämter, ML 4094, BayHStA; *Jb Bayern 1912–13*: 4.

22 23 September 1929 – Scharnagel to Bay.Min.Ldw., ML 3697, BayHStA; *Jb Bayern 1922–26*; Gassner 1927.

23 This is clear from numerous annual reports: e.g. *Jb Bayern 1903*: 28; *Jb Bayern 1904*: 50.

24 Kiessling and Gassner 1927; Pflug and Kiessling 1917; Kiessling 1912. For example, one variety of winter wheat had not been planted very widely in Bavaria, the station's director pointed out, because it yielded poorly in unfavourable growing conditions (29 October 1932 – Scharnagel to Bay.Min.Ldw., ML 2033, BayHStA).

25 Kiessling and Gassner 1927; 30 December 1931 – Scharnagel to Bay.Min.Ldw., ML 3697; 23 August 1941 – Scharnagel to Hergenröder, Nr. 532, LABK, both in BayHStA. The very successful 'Tassilo' wheat variety, for example, was a hybrid constructed at the station and handed over in its F4 and F5 generations to a breeder for finishing (18 July 1934 – Scharnagel to Krohn, Nr. 532, LABK, BayHStA).

26 30 December 1931 – Scharnagel to Bay.Min.Ldw., ML 3697, BayHStA.

27 Kraus n.d. In 1919–1921 of 39 varieties released by the station, one-third of them went to state farms (*Jb Bayern 1919–1921*: 9). In the only case known to me the fee charged was 5 per cent of the breeder's sales over the following five years (18 July 1934 – Scharnagel to Krohn, Nr. 532, LABK, BayHStA).

28 Kraus n.d. For example in 1906 the station agreed that each year it would provide the barley-growers' association in Moosburg with a small quantity of seed from an improved variety. The association's largest member would multiply it up and progressively share out the harvest with colleagues so that within four years it was expected that all members of the association would be able to grow it (*Jb Bayern 1906*: 81).

29 This claim seems to be borne out by the following: Scharnagel 1929: 496; Scharnagel 1928: 493; Scharnagel 1936/37: 39; 18 July 1934 – Scharnagel to Krohn, Nr. 532, LABK, BayHStA; .

30 This is noticeable in the revised statute of 1913 which obliged the station to promote the production and use of improved seed throughout the state ('Satzung d. K.B. Saatzuchtanstalt in Weihenstephan', n.d. but approved by the Interior Ministry on 8 June 1913, Nr. 2, Fach 86, Sekt II, UA-Weih).

31 Kiessling 1910b and Kiessling, n.d. but *c.*1913, 'Tätigkeit d. Saatzuchtanstalt nach aussen', MK 40158, BayHStA.

32 26 June 1906, minutes of Beirat meeting, MK 14040, BayHStA.

33 8 July 1914, minutes of Beirat meeting, Nr. 3, Fach 86, Sekt II, UA-Weih. By 1917 there were 247 potato-variety testing stations throughout Bavaria (Gassner 1927).

34 18 October 1930 – Saatzuchtanstalt to Bay.Min.Ldw., ML 2032, BayHStA.

35 The German Agricultural Society evidently thought this was a bad idea (Raum 1912: 924), but as we have seen, the Society was not especially interested in the problems faced by small farmers.

36 Kiessling 1913 and 1922b: 25–27; Müller 1927; Raum 1912; Kiessling and Gassner 1927. Between 1906 and 1917 the number of certified seed producers in the state rose eightfold (*Jb Bayern 1914–1918*: 119–120).

37 26 June 1906 – minutes of Beirat meeting, MK 14040, BayHStA; 25 June 1909 – minutes of Beirat meeting, Nr. 2, Fach 86, Sekt II, UA-Weih.

38 Virtually all of Kraus's and Kiessling's papers in the early years appeared in popular agricultural newspapers such as the weekly *Wochenblatt Bayern*. But the latter also established a series of pamphlets on agricultural topics – the *Landwirtschaftliche Hefte* – which were written so as to be accessible to small farmers (Kiessling 1913: 29).

39 Kiessling and Gassner 1927. By all accounts Kiessling was rushed off his feet. In

1905 he was on the road some 75 days, giving lectures, attending scientific meetings or visiting breeding centres (*Jb Bayern 1905*: 58–59). By the following year he found that he was getting far more requests for lectures than he could handle (*Jb Bayern 1906*).

40 Kiessling 1913: 22–25. Among many others, the station organized a large Bavarian Barley Exhibition in 1909 as well as the national barley exhibition in Munich in 1926 (Kiessling and Gassner 1927).

41 Kuhn 1900; Kraus 1897; Bay.Staatsmin. Innern 1897: 75. Uniformity of supply was particularly important in barley because brewers, unlike millers, did not get good results from heterogeneous grain (14 April 1932 – Scharnagel to Bay.Min.Ldw., ML 3697, BayHStA). Notice that, unlike the seed-producing and plant-breeding cooperatives discussed in the last chapter, the point of the crop-improvement associations was not to *produce* improved seed but to *consume* it.

42 This point seems not to have been grasped by those running the Mexican Agricultural Program in the 1940s with the result that the Program had relatively little impact upon peasant farming (Harwood 2009).

43 Kiessling appears to have invested an enormous amount of time and energy in promoting the movement. Indeed, one of his assistants felt that Kiessling had devoted far too much time to the movement, in the process neglecting matters of breeding which the assistant regarded as the proper focus of the station (16 February 1920 – Raum to Kultusministerium, MK35953, BayHStA.

44 Kraus 1909a; Kiessling 1911; Hecht 1912; 25 June 1907, minutes of Beirat meeting, Nr. 2, Fach 86, Sekt II, UA-Weih.

45 A similar relationship existed in Italy during the 1920s; some public-sector institutions provided seed-growing cooperatives with an appropriate variety and oversaw their work (Saraiva 2010).

46 Kiessling 1911; 31 July 1930 – Bay.Min.Ldw. to Saatzuchtanstalt, ML 2032, BayHStA.

47 Kiessling claimed that the crop-improvement associations were unprecedented, either in Germany or abroad (Kiessling 1913: 4). To some extent he may have been right. Olmstead and Rhode (2008: 164ff.), for example, have recently claimed that several countries attempted to copy an American institution whose aims and organization bear a close resemblance to the Bavarian associations (the 'one variety movement') though with little success (ibid.: 400). Bavaria was obviously not one of the imitators; its associations in fact preceded the US counterpart and, as we shall see, were very successful. The general features of the Bavarian associations, however, were not entirely new. As we saw in the previous chapter, the idea that small farmers would be more effective if they were organized in cooperatives and could draw upon the advice of experts from the public sector was being discussed during the 1890s.

48 6 June 1910 – Kiessling to Herr Ministerialrat, MK 40158, BayHStA; Kiessling 1913: 32. On the Minister's praise for the 'outstanding work' done at Weihenstephan, see 16 July 1910 – Bay.Min.Ldw. to Kultusministerium, MK 40157, BayHStA. Years later the Minister again congratulated the station for its accomplishments in this respect and called upon all districts in the state to ensure that the associations in their jurisdiction were adequately equipped to exert their maximal impact (16 June 1926 – Fehr to Landessaatzuchtanstalt, ML 2030; 31 August 1932 – Bay.Min.Ldw. to Landessaatzuchtanstalt, ML 2033, both in BayHStA).

49 In 1931, for example, three-quarters of the farmers in the region of Middle Franconia belonged to that region's crop-improvement association (21 December 1931 – Mittelfränk. Kreisackerbauverband to Staatsrat Dr Schäfter, MK 68193, BayHStA). In 1927 the Bavarian Federation of Crop Improvement Associations paid tribute to the station's role in supporting the movement (Schweiger and Balz 1927).

50 The Bavarian crop-improvement associations were quite similar to an institution set up in central Germany during the 1920s on the initiative of Theodor Roemer,

professor of plant breeding at the University of Halle (*Versuchsringe*). The difference between the two, however. was that the *Versuchsringe* were intended for estate owners (Uekötter 2010: 65–67), reflecting the different agrarian structures of central and south Germany.

51 According to one member of the station's staff, the barleys bred at Weihenstephan almost always won first prize at regional barley shows (Scharnagel 1914) and in 1909 Kraus and Kiessling were awarded a gold medal by the president of the German Brewers Exhibition for their work on barley improvement (*Jb Bayern 1909*: 8). Just why the brewers had been critical earlier is not clear (25 June 1907 – minutes of Beirat meeting, MK 14040, BayHStA). Kiessling felt the problem lay with the brewers' inability to specify which properties they wanted varieties to possess and complained that they had wrongly accused the station of neglecting brewing quality (Kiessling 1914; 8 July 1914 – minutes of Beirat meeting, MK 40158, BayHStA.

52 There are hints that this link may have hindered the development of Württemberg's station. The director had to devote part of his time to teaching agronomy and plant breeding at the college, and his colleagues at the college seem to have been rather unsupportive (Snoy, 'Geschichte des Instituts für Pflanzenzüchtung mit Landessaatzuchtanstalt', typescript, n.d. but *c.*1967, 3/25.31, UA-Hohen).

53 Fruwirth 1905. Others took a similar view (Schindler 1907a). Accordingly, the first course which the station held was not in breeding but in varietal testing and seed-growing (Fruwirth 1907b). Once the state established a chamber of agriculture in 1921, however, the station handed over this work and concentrated upon breeding.

54 Lang 1911a and 1914b: 20–21. The station's advisory service evidently met a need; in the first year it was bombarded with inquiries, requiring about a dozen written replies per day (Lang 1909b).

55 Kuhn 1902; Wacker 1918. Baden paid particular attention to maize and tobacco which were hardly grown in other states and were too difficult and expensive for the private sector (Anon. 'Von der neuen' 1908; Lang 1910a, 1914b: 15.

56 Lang 1910b; Bach 1914. Württemberg's station began breeding clover and fodder plants in 1907. The work seems to have expanded slowly until the late 1920s when the station established a department for fodder plants (Fruwirth 1907b; Weiss 1918; G. Baur 1928).

57 Lang, 'Einiges über die Züchtung von Landsorten', n.d. but 1910 or 1911, 236/26699, GLAK.

58 Wacker 1912; Anon. 1932; 25 June 1929 – minutes of Beirat meeting, 3/25.1.3(2), UA-Hohen; Lang, 'Einiges über die Züchtung' (as note 57).

59 Fruwirth 1919a; 7 July 1924 – 'Anmeldungen zur Teilnahme an dem Lehrgang …', 3/25.1.5, UA-Hohen

60 Entrup 2008; Lang 1909d and 'Einiges über die Züchtung' (as note 57); 'Denkschrift über die Errichtung einer Saatzuchtanstalt', 236/26751, GLAK.

61 Fruwirth 1911b and 1910: 649; *passim*, 236/26751, GLAK.

62 1 August 1901 – Bay.Min.Ldw. to Kultusministerium, MK 14040; 29 June 1909 – Welz to Aussenministerium, MK 40157; 'Finanzperiode 1912 und 1913', not dated, MK 40157; 27 July 1916 – 'Voranschlag', MK 40158; and 20 May 1914, von Knilling to Direktor der Akademie Weihenstephan, MK 40158; all in BayHStA. Fifteen years later when the economic crisis called for severe budget cutbacks, the Agriculture Minister argued forcefully that the station's staff budget should be maintained (27 November 1925 – Bay.Min.Ldw. to Finanzministerium, MF 68192, BayHStA). Growth of the other south German stations was a good deal more modest, though between 1908 and 1912 the budget for Baden's station was increased by 50 per cent (Anon. 'Zur Förderung' 1912), and when it moved to Rastatt after the First World War, the station acquired an experimental field 10 times the size of its previous one (Anon. 'Die Entwicklung' 1932a).

63 The phrases, respectively, are from 17 January 1930 – Kammer for Wiesbaden to Preuss.Min.Ldw., Nr. 10440; and 23 December 1924 [evidently a draft memo from a meeting of six top officials to Preuss.Min.Ldw., Nr. 10437, both in Rep 87B, 1. HA, GStAPK. On the Bavarian station's reputation see also Evers 1924: 27. In Dresden, similarly, Steglich cited the Bavarian station as a model for Saxony (Steglich 1907).

64 Personalakte Paul Kulisch, MK17473, BayHStA; Harwood 2005: 217–218.

4 Success breeds trouble: the controversy over public-sector breeding, 1902–1933

1 6 June 1910 – Kiessling to Herr Ministerialrat, MK 40158, BayHStA.

2 For example, Deborah Fitzgerald's important book on hybrid maize, for example, includes interesting material on the relations between the sectors, focusing on the state of Illinois, but does not examine the nature of this relationship nor how it changed over time (Fitzgerald 1990: 164–167, 199–208).

3 On several occasions before the war Kiessling complained about 'agitation' directed against the stations (Kiessling 1917: 29) and 'unfair' criticism (Kiessling 1913: 7, 9), but he doesn't reveal the critics' identity. Such suspicion was not peculiar to Germany. In Britain the seed trade voiced similar concern at the founding of the Plant Breeding Institute and the Scottish Plant Breeding Station (Palladino 1990).

4 One reason for the GFP's founding was that although the German Agricultural Society already had a Seed Breeding Division, the Society's statutes specified that as an organization concerned purely with technical matters, it would not get involved in lobbying on behalf of commercial breeders (von Rümker 1928a).

5 I have so far found little archival material on the deliberations leading to the founding of the GFP, but the timing of its formation suggests that it may have been in part a defensive response to the establishment over the preceding few years of the south German stations and other forms of public-sector breeding.

6 One reason for this may have been that its membership fee (100 marks) was very steep; even the GFP's chairman admitted that this was 'quite a lot' (19 May 1911 – Kühle to Minister, Nr. 20577, Rep 87B, GStAPK).

7 At the end of the Second World War virtually all of the central and north German firms were estates (i.e. more than 100 hectares), and a majority were larger than 300 hectares (Kratzsch 2008: 39–40).

8 In 1911, for example, of 49 members, only one was from south Germany, and two were cooperatives. Of 134 members in 1924, 13 were from the south, and three were cooperatives (membership lists in *Beiträge zur Pflanzenzucht*, vols 1 (1911) and 7 (1924).

9 The earliest varieties bred at Württemberg's station, he noted, included poppies and fodder peas which had not so far been bred in Germany (Fruwirth 1911b).

10 In their handbook of plant breeding Rudorf and Theodor Roemer listed a number of varieties which they deemed to be 'good' or which were planted on significant acreages. Public varieties were among the grass and clover varieties mentioned as were wheat varieties from Weihenstephan (Roemer and Rudorf 1941: vols 2 and 3).

11 Others have stressed the importance of the disease-resistant barleys bred at Weihenstephan (Scheibe 1987; Spennemann, n.d.; Scharnagel 1936/37). Recent reference books indicate the continuing importance of Weihenstephan's cereal varieties as parent lines (Lupton 1992).

12 'Denkschrift über die Neuorganisation der württembergischen Landespflanzenzüchtung' (*c.*1927), Nr. 3/25.02, UA-Hohen.

13 Some breeding cooperatives supported by the public sector were particularly successful. One wheat variety bred by a cooperative (*Rheinische Bauernverein*) gave the highest yield of any such variety in the Agricultural Society's varietal tests for three years running (12 January 1928 – Rhein. Bauernverein to Preuss.Min.Ldw., Nr. 10439, Rep 89B, GStAPK).

14 The criterion for inclusion in this volume was not as stringent as that required by the Hybred Varieties Register but required at a minimum that a variety be certified. Many of these varieties also received additional recognition, however, such as prizes at exhibitions and in several cases inclusion on the Hybred Varieties Register.

15 Unfortunately the author did not specify what criteria were used for inclusion, but the title of the handbook implies that the acreage planted in a variety was important.

16 Fifty per cent of the winter rye varieties, 30 per cent of the winter wheats, 50 per cent of the summer ryes, 18 per cent of the summer barleys, 10 per cent of the oats varieties, and 20 per cent of the maize, pea and flax varieties originated in south Germany (Anon. *Reichssortenliste*, n.d. but 1935).

17 The actual importance of public-sector support was undoubtedly greater, partly because a further 5 per cent were cooperatives (which almost certainly also drew upon technical assistance from a public institution), but also because the list includes about 50 breeders with south German addresses whose relationship to public-sector institutions is unknown to me.

18 The names of the substantial number of varieties originating during the 1920s and 1930s from Theodor Roemer's institute at the University of Halle, for example, did not reflect their public origins.

19 In that year 40 of the breeding centres supported by the station at Weihenstephan were selling seed on the market (*Jb Bayern 1914–18*: 96).

20 The public-sector competition was not coming just from south Germany. On complaints of competition from Svalöf, see 9 December 1925 – report of meeting of Abteilung f. Getreidezucht, Nr. 20578, Rep 87B, GStAPK.

21 For example, Nr. 10440–10442. In consequence only about 15 per cent of the cereals acreage was planted in certified seed (26 January 1926 – draft letter from Preuss.Min. Ldw. to all chambers of agriculture, Nr. 10438). This fall in demand also contributed to the collapse of several seed-breeding cooperatives (27 December 1927 – GFP to Preuss.Min.Ldw., Nr. 10439); all files in Rep 87B, GStAPK.

22 By the early 1930s sales of commercial seed of several cereals were 10–20 per cent of the level in 1925 (GFP 1933: 21; 4 December 1931 – Ackermann *et al.* to Bay.Min. Ldw., ML3697, BayHStA). The declining acreage planted in well-known commercial varieties was already evident in the decade before 1914 as farmers turned increasingly to local varieties (Böhmer 1914). Whether this can be attributed to the stations' work in varietal testing and extension is not clear.

23 Sometimes this proliferation of pseudo-varieties has been driven by attempts at product differentiation. In the US *c.*1935 all of the hybrid maize varieties commercially available were actually of public origin, so firms marketed them under aliases and kept quiet about their pedigrees (Kloppenburg 1988: 107).

24 Reich Min.Ldw., *Begründung des Referentenentwurfes eines Saat-(Pflanz-) gutgesetzes*, 7, Nr. ML 2040, BayHStA. Twenty-five of the varieties of yellow oats on the market in 1932 were indistinguishable from the Petkus yellow oat, a variety developed by a well-known breeder before the First World War (Laube 2004). Of 230 varieties of fodder beet on the market by 1934, 210 of them were indistinguishable (von der Schulenburg 2004). The French seed market at this time suffered from the same problem (Bonneuil and Thomas 2010).

25 Many chambers of agriculture did the same on behalf of their seed-breeding and seed-growing cooperatives who were equally in financial trouble (see Nr. 10440–10442, Rep 87B, GStAPK).

26 Following a lecture by von Rümker which called for a wide range of mechanisms by which the state could assist private breeders, the GFP sent an appeal for support to the Prussian Agricultural Ministry (18 December 1925 – GFP to Preuss.Min.Ldw., Nr. 10438, Rep 87B, GStAPK; This file indicates that during 1926 the GFP wrote to the Prussian Ministry about once a month.) and 18 December 1925 – Kiessling to Bay.Min.Ldw., ML4094, BayHStA. Lobbying continued into the early 1930s (e.g. GFP 1933: 36–38; Kühle 1933a).

27 17 March 1926 – memo to all ordinary members of the GFP, and 21 November 1931 – minutes of meeting of Hauptvorstand, both in R68IV/1, BAK. Cf. 30 January 1930 – Reich Min. Ldw. to Landesregierungen and Preuss.Min.Ldw., Nr. 10440, Rep 87B, GStAPK.

28 N.d. (probably March 1930) – Saatzuchtgenossenschaften to Ldw. Verein f d Fürstentum Lüneburg in Uelzen, Nr. 10440, Rep 87B, GStAPK.

29 For example, Kühle 1926; cf. Kühle's contributions to the discussion following Kiessling 1922b. In its remarkably hysterical annual report for 1931 the GFP declared 'We have had more than enough of the planned economy! On the contrary: we demand … unconditional and unrestricted restitution of the free market' (p. 3, Geschäftsbericht 1931, Nr. 10441, Rep 87B, GStAPK).

30 'Mitgliederliste', *Beiträge zur Pflanzenzucht*, vol. 2 (1912), 145–152; 'Mitgliederverzeichnis', *Beiträge zur Pflanzenzucht*, vol. 10 (1929), 114–122).

31 Minutes of Hauptversammlung of GFP, 29 June 1929 and 8 July 1929, both in R168/252, BAK.

32 11 March 1931 – minutes of meeting of the Verkaufskommission and the Verwaltungsausschuss, Nr. 2619, MPGA. In a letter to a south German commercial breeder, Baur was at pains to emphasize the differences between his institute and the south German stations, noting among other things that KWIB sold its promising varieties only to 'genuinely *recognized*' breeders (excerpt of 21 October 1932 – Baur to Ackermann, ML3697, BayHStA, emphasis in original).

33 Minutes of meeting of the Kuratorium, 7 July 1930 and 12 February 1931, both in Nr. 2616, MPGA.

34 In 1930, for example, the Association of German Seed Merchants (*Vereinigung der Samenhändler des Deutschen Reiches*) lodged objections with the Reich government and other public bodies against various chambers of agriculture whose commercial activity (i.e. selling seed themselves or promoting the seed of an associated cooperative) was said to be seriously damaging seed merchants. It was wrong, they argued, that cooperatives in financial difficulties should receive state subsidies while seed merchants did not (n.d. but probably 1930, printed form letter from *Vereinigung der Samenhändler des Deutschen Reiches* to Industrie- u. Handelskammer zu München, MWi 7984, BayHStA).

35 3 March 1926 – GFP to Deutscher Landwirtschaftsrat and Preuss. Hauptlandwirtschaftskammer, Nr. 10438, Rep 87B, GStAPK; May 1926 – 'Geschaftsbericht 1925/26', Nr. 53, R8073, BAB; 26 November 1931 – Kühle to AG für Saatenanerkennungswesen des Deutschen Landwirtschaftsrats, Nr. 11, R68IV, BAK. Other public-sector institutions such as the Kaiser-Wilhelm Institute for Agriculture were also targeted for criticism (10 March 1926 – Lembke to Bohne, Nr. 11, R68IV, BAK).

36 By 1922 there were signs of growing tension. The director of the Bavarian station, for example, voiced his hesitation in accepting an invitation to address the Seed Breeding Division of the Agricultural Society – an organization close to commercial breeders (cf. Chapter 2) – but said he had decided nonetheless to regard the invitation as a 'friendly' gesture toward the Bavarian movement (Kiessling 1922b).

37 19 February 1926 – 'Alarm!' (with a covering letter from Bohne), Nr. 22, R 168, BAK). Talk of the ecological limitations of individual varieties, the document maintained, was little more than propaganda on behalf of locally bred varieties.

38 The implication was that the problem of 'synonymous varieties' was attributable to varietal development at the south German stations. I have found no evidence to support this claim; indeed, evidence published a few years later suggested otherwise. A committee which had systematically examined the characteristics of all of the cereals varieties then on the market concluded that 'synonyms' were actually more common in north Germany than in the south (Appel *et al.* 1932).

39 Roemer, 'Die Organisation der Pflanzenzüchtung', typescript, n.d. but probably 1928, Nr. 29, Rep 34G, UA-Halle.

40 26 January 1928 – 'Ausschuss zur Untersuchung d. Erzeugungs- u. Umsatzbedingungen d. deutschen Wirtschaft (Enquete-Ausschuss)', Nr. 236/26260, GLAK.
41 17 March 1926 – memo to all ordinary members of GFP, R68IV/1, BAK.
42 9 December 1926 – minutes of meeting of Ausschuss V. of Baden's chamber, Nr. 236/26261; 13 February 1928 – minutes of meeting of Ausschuss A.V, Nr. 236/26260; and 6 December 1929 – copy of a proposal from the Verein badischer Pflanzenzüchter for consideration at a meeting of Ausschuss A.V., Nr. 236/26260, all in GLAK. Also 21 December 1928 – Graf Douglas to Reich Min.Ldw, ML 2040, BayHStA.
43 7 October 1930 – Deutsche Saatveredelung to GFP, R68IV/16, BAB. Similar organizations complained (2 March 1931 – memo to members of the GFP's Abteilung f. Futterpflanzenzüchtung, R68IV/6, BAK).
44 8 July 1929 – minutes of meeting of GFP Hauptversammlung, R168/252, BAK.
45 25 January 1929 – draft letter from Bay.Min.Ldw. to Reich Min.Ldw.; 18 October 1929 – Theodor Scharnagel to Bay.Min.Ldw.; 4 November 1929 – Generalsekretär des [illegible] christlichen Bauernvereins to Bay.Min.Ldw.; 24 December 1930 – Bay. Landesbauernkammer to Bay.Min.Ldw.; and 6 January 1931 – Verband deutscher Saatenanbauer to Bay.Min.Ldw.; and 18 March 1930 – Reichsratsausschüsse II und VII; all in ML 2040, BayHStA. In the Rhineland and Silesia, for example, some chambers declined to provide the GFP with information which would have been necessary for the GFP to levy licence fees upon chamber members ('Geschäftsbericht für 1931', Nr. 10441, Rep 87B, GStAPK).
46 For a nice illustration of the large breeders' assumption that the proper role of public-sector breeding was to serve private-sector breeders rather than farmers, see the discussion following Weller 1928.
47 8 July 1929 – minutes of meeting of Hauptversammlung der GFP, Nr. 252, R168, BAK; 'Neue Denkschrift zum Entwurf des Saat- und Pflanzgutgesetzes', enclosed with 29 March 1932 – Ackermann to Bay.Min.Ldw., ML3697, BayHStA; 'Geschäftsbericht für 1931', Nr. 10441, Rep 87B, GStAPK.
48 In 1924 only 4 per cent of the GFP's commercial breeders (5 of 126) were cooperatives (Anon. 'Verzeichnis d. Mitglieder . . .', *Beiträge zur Pflanzenzucht*, vol. 7 (1924), 102–111). By 1929 the proportion had fallen to 2 per cent (Anon. 'Mitgliederverzeichnis', *Beiträge zur Pflanzenzucht*, vol. 10 (1929), 114–122).
49 n.d. but *c.*30 March 1927 – letter from five seed-breeding cooperatives to Landwirt. Verein f. d. Fürstentum Lüneburg in Uelzen; see also 10 January 1927 – Verband Hannoversche Pflanzenzüchter to Reich Min.Ldw., both in Nr. 10440, Rep 87B, GStAPK).
50 18 October 1929 – Scharnagel to Bay.Min.Ldw.; 7 November 1929 – Bay.Min.Ldw. to Reichs Min.Ldw.; and 25 November 1929 – minutes of meeting in Reichs Agriculture Ministry with representatives of all state governments; all three in ML2040, BayHStA. The objection was accepted and the bill amended accordingly.
51 3 January 1929 – memo from Bohne to Vorstand of GFP, R168/252; and 'Geschäftsbericht für 1930', R168/212. On dissensus within the GFP over licence fees see R68IV/17 and 18, all in BAB. The number of breeders protesting was significant because it was about equal to the number of commercial breeders in the GFP at that time (Anon. 'Mitgliederverzeichnis', *Beiträge zur Pflanzenzucht*, vol. 10 (1929), 114–122).
52 This same 'breakaway' process was occurring elsewhere in south Germany. In Württemberg, for example, by 1908 two of the station's larger breeding centres had become independent (Fruwirth 1908). By the late 1920s in Baden 16 private breeders and four seed-growing associations were supported by the station, but two breeders had become independent (5 November 1929 – 'Verzeichnis der Zucht- und Aussenstellen', Nr. 236/26260, GLAK).
53 Dr Karl August von Schmieder (1867–1941) was a lawyer who inherited a 1000-hectare estate in 1901. With the help of his estate administrator, he organized in 1919

the Bavarian Society for the Promotion of Grassland Agriculture (*Verein zur Förderung der Grünlandwirtschaft in Bayern*) which subsequently expanded into a nationwide movement (Schlögl 1954: 96ff.; Röbbelen 2002). He was also a member of the GFP's fodder-plant division (R68IV/17, BAK).

54 2 September 1929 – Schmieder to Bay.Min.Ldw., ML 3697, BayHStA.

55 The way in which the station released varieties was not unique. The Welsh Plant Breeding Station was then doing much the same thing (Palladino 1990), as were maize breeders at the University of Illinois (Fitzgerald 1990: 199–208). In the latter case just as in Bavaria, large commercial breeders protested when the University supplied small breeders with inbred lines or hybrid varieties.

56 23 September 1929 – Scharnagel to Bay.Min. Ldw., ML 3697, BayHStA. Schmieder's estate-manager subsequently confirmed how valuable the station's help had been (Niggl 1953: 32ff.).

57 19 May 1928 – Schmieder to Lembke, R68IV/17, BAB.

58 24 June 1932 – Bay.Min.Ldw. to Ackermann, ML3697, BayHStA.

59 4 December 1931 – J. Ackermann *et al.* to Bay.Min.Ldw.; the complaints were further detailed in 29 March 1932 – Ackermann to Bay.Min.Ldw., both in ML 3697, BayHStA.

60 30 January 1925 – Kulisch to Kultusministerium, MK 19682, BayHStA. Chairman of the GFP's cereals-breeding division during the 1920s, by 1931 Ackermann was deputy chair of its executive committee (von Rümker 1928b; 24 June 1931 – GFP to all members, Nr. 10436, Rep 87B, GStAPK). In 1923 he received an honorary doctorate from the Technical College at Munich (Kiessling 1938).

61 30 December 1931 – Scharnagel to Bay.Min.Ldw., ML3697, BayHStA. In 1917 Ackermann himself had received a *finished* barley variety from the station (Kiessling and Gassner 1927).

62 Memorandum dated 15 January and signed by Ministerialrat Lang and two others, appended to the end of a draft letter dated 'January 32' – Bay.Min. Ldw. to Ackermann, ML3697, BayHStA. Ironically, only a few years earlier Ackermann had been regarded as a friend of the station (Hiltner 1911); he was happy to acknowledge how much he had benefited from the station's help (Ackermann 1928) and paid tribute to the station's achievements which were 'recognized throughout all of Bavaria and beyond' (Ackermann 1927). Since that time the rapidly deteriorating agricultural economy had apparently changed his mind.

63 23 February 1932 – Dr Stützel to Ackermann, ML 3697, BayHStA.

64 25 April 1927 – Lembke to GFP, Nr. 1, R68IV, BAK.

65 14 April 1932 – Scharnagel to Bay.Min.Ldw., ML 3697, BayHStA.

66 24 June 1932 – Dr Stützel to Ackermann, ML3697, BayHStA, emphasis in original.

67 2 January 1933 – Philipp Lichti to Ministerialrat Lang, ML3697, BayHStA.

68 The farmers' representative was to be the chairman of the Bavarian Federation of Crop Improvement Associations (*Landesverband bayerischer Ackerbauvereine*), the state-wide coordinating body for the local crop-improvement organizations whose establishment Kiessling had done so much to encourage. The breeders' representative was to be the chairman of the Bavarian Seed Breeding Association (cf. Chapter 3) of which Kiessling was also a founding member. By the 1920s all Bavarian breeders and most cooperatives were members, totalling 243 in all (Müller 1927; Kiessling and Gassner 1927; Kiessling 1922a).

69 27 September 1933 – Bay.Min.Ldw. to Ackermann with attached draft minutes of meeting on 9 January 1933, signed by Scharnagel, ML3697, BayHStA.

70 Cf. a forthcoming collection of papers on the history of agricultural policy, edited by Peter Moser (Bern) and Tony Varley (Galway).

71 The dependence can be mutual. Public-sector breeding institutions potentially also gain through a cooperative relationship with firms, for example through obtaining information as well as access to facilities and breeding material (Kittlitz 2005: 119–120).

72 Competition may also be useful in other ways. In the US one history of agricultural research has concluded that the public sector has an important 'watchdog' role to play in testing the private sector's claims about research and extension (Huffman and Evenson 1993: 253).

5 The fate of peasant-friendly breeding under National Socialism

1 7 April 1933 – Bohne to Lembke, R68IV/11. On breeders' rights see 17 March 1933 – minutes of meeting of Hauptvorstand of GFP, R68IV/1, both in BAB.

2 20 April 1933 – Ludwig Kühle to Hitler, R68 IV/11, BAK. Later that year, in his capacity as representative of the GFP on the executive committee of the Kaiser-Wilhelm Institute for Breeding Research, Kühle wrote to Richard Walther Darré, the top Nazi agricultural official who had by then installed himself as chair of KWIB's executive committee. Kühle requested that no change be made to the existing arrangements whereby the Institute turned over any promising results to the private sector at an early stage for finishing and sale (26 October 1933 – Ludwig Kühle to Herr Vorsitzenden des Kuratoriums, Nr. 2612, Abt. I, Rep 1A, MPGA). Kühle himself may have felt a certain affinity for the new regime, judging from his annual report to the GFP in 1931 in which he railed against the then government and declared his belief in the need for a 'new economic and social order' ('Geschäftsbericht 1931', Nr. 10441, Rep 87B, GStAPK; cf. Kühle 1933b).

3 18 December 1933 – Kreisbauernschaft Regensburg to Bay.Min.Ldw., ML 3697, BayHStA.

4 These issues are discussed at greater length in Harwood 2010.

5 The huge and festive annual conferences which the party held each autumn in Goslar were not named 'agricultural' or 'farmers' conferences but 'peasants' conferences. Accordingly, with the reform of higher agricultural education in 1935, 'the German peasantry' became one of the five main subjects in the curriculum (Klemm 1994: 40).

6 One source estimates that between about 1929 and 1938 cereal yields increased 15 per cent while those for root crops rose 20 per cent (Tornow 1972: 14; cf. Berthold 1988).

7 Between 1933 and 1939 1.4 million agricultural workers left the countryside for industrial jobs (Kluge 1989, vol. 1: 36).

8 By the late 1930s this shift of policy away from the peasantry is noticeable, in the writings of Konrad Meyer (e.g. Meyer 1939b). The task of agricultural policy, he noted, was to strengthen what he chose to call at several points, not the 'peasantry' (*Bauerntum*) but the 'rural population' (*Landvolk*) (Meyer 1939a: 195–196, 202–203). In the first years of the Third Reich, agricultural policy had been entirely devoted to the rescue of the peasantry, he acknowledged, but this was essential as a matter of population policy rather than for reasons of production. There was no single optimal farm size, he emphasized; what was needed was a healthy ratio of small to large farms.

9 The reductions ranged from about threefold in maize to 15-fold in potatoes (Anon. *Reichssortenliste* n.d. but 1935; cf. Anon. *Ratgeber* 1938a). And although seed-merchants and breeders were still allowed to advertise their approved varieties, advertisements required the prior approval of an RNS agency (Spennemann n.d.: 49).

10 *Passim*, R68IV/1, BAB. Membership of the RdP was compulsory and thus extended – unlike the GFP's – to all breeders, including the smallest (who paid a smaller subscription fee) as well as to seed growers. This gave the RdP an initial membership nearly three times the size of the former GFP's, and by 1942 it was over five times the GFP's (RdP *Ratgeber*, 1942).

11 On the regime's concern with prices, see Backe 1938: 84; 22 October 1935 – minutes of meeting of Hauptvorstand der RdP; and 22 and 23 October 1935 – transcripts of

talks given at a meeting in Goslar. For GFP complaints, see various correspondence, October to December 1935; all three in R68IV/2, BAB.

12 I have found no systematic data on private breeders' profits during this period, but one well-known firm's profits jumped from 1935 and remained healthy from 1936 on (Laube 2004). One breeder who worked in the private sector from 1936 claimed that the Decree led to an upswing of breeding operations in many firms (Kuckuck 1951: 30, 73).

13 The percentage may have varied by crop, since one RNS official claimed that in 1934–1935 as much as 18 per cent of the winter-rye acreage was planted in improved varieties (Krohn, talk to RdP in Goslar, 22–23 October 1935, R68IV/2, BAB). For comparison, in fascist Italy by 1940 the new high-yielding wheat varieties had been planted on over 50 per cent of the national wheat acreage while in several northern provinces the proportion was over 90 per cent (Saraiva 2010: 472–473). In fascist Spain one high-yielding hybrid rice variety obtained in 1936 was accounting for 75 per cent of Spanish rice production 15 years later (Camprubi 2010: 530).

14 27 October 1936 – minutes of meeting of Mitgliederversammlung d. Reichsverband, R68IV/3; 21 December 1936 – minutes of meeting of Abteilungsvorsitzenden d. Reichsverband, R68IV/4; both in BAB.

15 17 October 1934 – Müller to Bay.Min.Ldw., ML 4094, BayHStA.

16 4 May 1935 – Bayerische Landesanstalt für Moorwirtschaft to Reichsbauernführer, and latter's reply on 21 May 1935, both ML 4094, BayHStA.

17 23 August 1941 – Scharnagel to Herr Hergenröder (Landesbauernschaft Bayern), Nr. 532, LABK, BayHStA.

18 Baden's station, unlike those in Bavaria or Württemberg, had been incorporated in 1914 into the state's chamber of agriculture. Since all chambers became part of the RNS after 1933 (Haushofer 1975), Baden's station was thenceforth at the disposal of Nazi officials.

19 28 January 1941 – Scharnagel to Bay.Min.Ldw., ML 3698, BayHStA.

20 For example, 18 July 1934 – Scharnagel to Krohn; 2 August 1934 – Krohn to Scharnagel; and 7 September 1934 – Scharnagel to Krohn. A few weeks later Theodor Roemer, professor of agronomy and plant breeding at the University of Halle, was also asked by the RNS to submit such proposals (24 August 1934 – Roemer to Scharnagel); all four letters are in Nr. 532, LABK, BayHStA.

21 Darré's letter of 25 January 1937 to Reich Erziehungsministerium and Reich Min. Ldw. was incorporated into 11 March 1937 – Reich Min.Ldw. to Württemberg. Kultusministerium, Nr. 3/25.1.3(2), UA-Hohen. Why it took him so long to address this issue is not clear. The fact that the Four Year Plan had been declared a few months earlier was possibly crucial. Until then the peasantry had enjoyed considerable political support so that it may have been inopportune for the RNS to try to curb the activities of peasant-friendly institutions.

22 That Darré's proposals bore a marked resemblance to the handover arrangements which had prevailed at the Kaiser-Wilhelm Institute for Breeding Research was probably no coincidence. As chair of that Institute's overseeing body (*Kuratorium*) from 1933 to 1936 (30 June 1933 – Max Planck to Darré, Nr. 2987, Abt. I, Rep 1A, MPGA), he was familiar with the particular kind of public–private sector relationship favoured there, a very different kind of relationship from that which operated in south Germany.

23 19 April 1937 – Kiessling to Rektor of TH München, Sign = V, 33, Bd. 2, TUM.

24 24 June 1937 – Roemer to Scharnagel, no. 532, LABK, BayHStA; in a similar vein 3 November 1937 – Roemer to Reichs Erziehungsministerium, Nr. 975, R4901, BAB. From his reply (25 June 1937 – Scharnagel to Roemer (no. 532, LABK, BayHStA) Scharnagel appears to have agreed, as did the director of the breeding station in Württemberg (24 April 1937 – Lowig to Rektor of Hohenheim, 3/25.1.3.(2), UA-Hohen).

25 Copy of 29 April 1937 – Roemer to Reichs Erziehungsministerium and to Kurator,

Nr. 59, Rep 33, UA-Halle. Most of these objections were also voiced by other public breeders; see correspondence with the Reich Erziehungsministerium for 1937–1938 in Nr. 975, R4901, BAB.

26 Moreover, compared to Darré's 1937 proposals, the revised 1939 version of the Guidelines granted public-sector institutions the freedom to hand over unfinished varieties to private breeders of their choice, subject to the Handover Committee's agreement (Anon. 'Umschau' 1939).

27 To judge from the few handover agreements I have seen, it looks as though the institutes' wishes were at least sometimes taken into account. At Halle, for example, one agreement honoured Roemer's requests that he be granted the right to keep an eye on the subsequent breeding work undertaken by the recipient and that the name given to the variety should reflect its origins at Halle ('Bedingungen für die Abgabe', attached to 16 February 1939 – Roemer to Scharnagel, no. 532, LABK, BayHStA). And a year later the naming-right was formally incorporated into the revised edition of the Guidelines of 5 August 1939 (Anon. 'Umschau' 1939).

28 The RNS seems to have recognized that even when deadlock within the Committee was broken in this way, there might still be conflict at some later date between the donating institute and the designated recipient. So a further revision of the Guidelines added a clause stating that in the event of such conflict, it would be resolved by an RNS official (Anon. 'Richtlinien' 1940a).

29 According to a note in the archive of the RNS, the files of the department concerned with agronomy (IIC1) were 'lost' in January 1945 (R16/I, Nr. 155, BAB).

30 Enforcing this regulation may have taken a certain amount of effort, for the public sector appears not always to have played by the rules. At one meeting of the RdP, for example, public-sector representatives were reminded that they were not to enter into agreements with a breeder of their own choice (31 March 1941 – minutes of meeting of RdP, no. 531, LABK, BayHStA). And when on one occasion this nonetheless occurred, the Committee intervened, negating the contract and reassigning the variety to another breeder (2 July 1941 – Scharnagel to Lang, ML 4094, BayHStA).

31 In this case other Committee members doubted whether a crop-improvement association made up of small peasant farmers in northeast Bavaria was qualified to finish the breeding of a potato variety developed by the Bavarian station (31 March 1941 – minutes of meeting of RdP, no. 531, LABK, BayHStA).

32 2 July 1941 – Scharnagel to Lang, ML 4094, BayHStA. Scharnagel also complained that he was sometimes not informed who potential recipients were until the meeting at which the decision was to be taken, by which time it was too late for him to establish which recipient was the most suitable (11 November 1941 – Scharnagel to Lang, ML 4094, BayHStA).

33 11 November 1941 – Scharnagel to Lang, ML 4094; 23 August 1941 – Scharnagel to Herr Hergenröder, Nr. 532, LABK; cf. 2 July 1941 – Scharnagel to Lang, ML 4094; all in BayHStA. The director of Württemberg's station had earlier made the same point; the station's varieties were bred for very specific growing conditions in Württemberg and would not thrive elsewhere. Moreover, if the state's small breeders were to continue to develop such varieties suitable for the region, he argued, their close ties with the station must be maintained since they could not survive without it (24 April 1937 – Lowig to Rektor of Hohenheim, 3/25.1.3.(2), UA-Hohen). Attempts on the part of Württemberg's breeders to strengthen their link to the station by founding a cooperative were dissolved a few months later (von Kittlitz 2005: 53–54).

34 1 November 1941 – Görlach to Lang and 20 November 1941 – Lang to Görlach, both ML 4094. See also 31 October 1941 – RNS to Landesbauernschaft Bayern, Nr. 532, LABK; all in BayHStA.

35 Just why this was the case is not clear, but I discuss possible explanations in Harwood 2010.

36 3 October 1960 – 'Abgabe von Zuchtmaterial d. Bay. Landessaatzuchtanstalt'; and 4

October 1960 – 'Begründung zu den Vorschlägen über die Abgabe '; both in Nr. 532, LABK, BayHStA.

37 *Passim*, Nr. 532, LABK, BayHStA. Deliberations continued into the summer of 1962, but there is no record in the archive of when the guidelines were eventually agreed nor precisely what form they took.

38 8 March 1966 – Simon to Direktor, Nr. 533, LABK, BayHStA.

39 n.d. but probably autumn 1968 – 'Entwurf: Die zukünftige Abgabe von Zuchtmaterial durch die Bayer. Landessaatzuchtanstalt', Nr. 533, LABK, BayHStA.

40 1 October 1968 – Ulonska to Direktor; 30 September 1968 – Simon to Direktor; 18 March 1969 – Zscheischler, 'Stellungnahme'; all in Nr. 533, LABK, BayHStA.

41 Precisely what provisions the eventual agreed contract contained is not clear because no copy of that contract exists in the archive. What follows here, therefore, is based on what looks like the latest draft: 8pp. typescript, n.d. but probably after November 1971, Nr. 533, LABK, BayHStA

6 The Green Revolution and its critics

1 The naivety of some claims now seems astounding. In 1970 a former director of the FAO told the *New York Times* that agricultural technology could eliminate hunger from the world within a decade (Cotter 2003: 253).

2 Harrar 1963: 84. According to World Bank insiders, Robert McNamara had a similar faith in the power of technology to alleviate poverty (Hodge 2010: 39).

3 In the early 1950s experts on Asia, in government as well as in the Council on Foreign Relations often advanced this view (Anderson 1991; Oasa and Jennings 1982; Goldman 2005: 59, 69; Cullather 2007). On the political rationale for US agricultural assistance to China see Stross 1986: 213. In 1949 India's Ministry of Agriculture warned that India would go communist within five years if no progress were made on food production (Perkins 1997). Two years later Rockefeller Foundation trustees enthusiastically recommended Warren Weaver's report on the 'World Food Problem' which linked food shortages with the dangers of communist infiltration (Harwood 2009).

4 In the late 1950s, for example, the US government declined to support the UN Food and Agricultural Organization's 'Freedom from Hunger Campaign' (Staples 2006).

5 Officials at the Rockefeller Foundation appear to have hoped for an additional trickle-down effect whereby the resulting food surplus might help to fund improved welfare programmes (Anderson 1991).

6 Harrar 1963: 60, cf. 8. The Foundation's Mexican Agricultural Program took the same view, sharing with the Mexican government of the day a desire to foster industrial growth through agricultural modernization (Jennings 1988; Perkins 1997: 114–115).

7 Goldman 2005: 298, cf. 54–55. As one agricultural economist noted, development programmes were sometimes driven less by the needs of the recipient country than by the export interests of the donor country (Johnston 1970). In the case of the Green Revolution, for example, the expectation was that by boosting production, food imports would be reduced so that Third World governments could afford to import manufactured products from the West (Anderson *et al.* 1991: ch. 2).

8 This pattern is not peculiar to the developing world. In the US, too, large farmers generally gain a greater advantage from agricultural research (Paarlberg 1981; White 1986).

9 In Mexico wheat prices fell by as much as 50 per cent between about 1950 and 1970 (Dalrymple and Jones 1973). On world markets grain prices fell steadily between 1950 and the mid 1980s: about 50 per cent for rice and wheat and more for maize (Mazoyer and Roudart 2006: fig. 11.3).

10 Griffin 1974: 241; Dahlberg 1979: 177; and the discussion following Stone 2002. The

argument for India has been challenged (Dalrymple 1979: 719–720), but similar shifts occurred in Guatemala and Bangladesh (Francis 1990; Alauddin and Tisdell 1991).

11 Dalrymple 1977. The problem is particularly acute in Africa where in recent years farmers have been paying the highest fertilizer prices in the world (Conway and Toennissen 2003). Even in coastal cities the cost of nitrogen fertilizer there is nearly five times that in Europe (Conway and Waage 2010: 28). One result is that in Uganda only 3 per cent of farms use fertilizer (Bahiigwa *et al.* 2005).

12 Because of its expensive requirements, the rice variety IR-8 was often referred to as a 'Cadillac' variety (Barker 1970).

13 It is not just the absolute level of these costs but also their unpredictability that discourages smallholders. Input prices fluctuate, depending on the price of oil and – where they have to be imported – on currency valuations (Pinstrup-Andersen and Hazell 1985). For a large farmer this means fluctuations in profit margins, but for a small farmer it may mean a drastic drop in income or even in food for the family.

14 Nor have subsidies always been available to smallholders; in one development scheme in Malawi only the largest 25 per cent of farms qualified for assistance (Hodge 2010: 40).

15 Not all small farmers' circumstances are identical, of course. For those who rely on non-farm work to supplement their income, labour-saving technology is very welcome (Tripp 2001) as it is in regions where labour shortage is a problem for small farmers (Woodhouse 2009). But for most farmers in the global South mechanization is simply too expensive. Approximately one-half of farmers in Latin America and Asia (and a much higher proportion in Africa) cannot afford even animals to pull their ploughs (Mazoyer and Roudart: 449).

16 Hayami 1971; Oasa 1987; Lipton with Longhurst 1989; Oasa and Jennings 1982; Lipton 1978; Pearse 1980: 142ff. Even those studies that find that small farmers gained from the new varieties acknowledged that tractor use did not actually increase farm income per hectare and was only economic on large farms (Blyn 1983). Some authors, recognizing that mechanization has been bad for rural areas, have tried to defend the Green Revolution by arguing that mechanization was not really a part of the GR package (Ruttan 1977). While this may be strictly correct, it hardly makes much difference in practice since mechanization was so often accelerated by the Green Revolution.

17 That 'all-rounders' are better suited than 'high achievers' to the needs of small farms is equally evident from animal breeding. A breed of German cows introduced to one region in India as part of a development programme gave much higher milk yields than local breeds. But the cows turned out to be too large to climb the region's steep hills or to drag a plough, and the costs of maintaining them were greater than the increased income from its milk (Unger 2010a).

18 The International Centre for Maize and Wheat Improvement (CIMMYT) set out to breed a similar wheat variety (Fischbeck 1988). One observer at the time warned against an infatuation with simple solutions: 'What we have in hand seems to many people to approach magic; let us hope that it does not become the source of deception' (Wharton 1969: 475).

19 This is probably why in 1980 Bangladeshi breeders were able to identify at least 4500 different rice varieties cultivated there (Biggs 2008); the number in India was estimated at 40,000 (Farmer 1979).

20 Something similar happened with maize. The maize varieties developed for Kenya by the CIMMYT were designed for prime land, not for the agro-ecological conditions in which most Kenyan farms were located (Smith 2009: 472–73).

21 At IRRI in 1970, for example, trustees felt obliged to refute the critics and discussed various ways in which this might be done (Oasa and Jennings 1982).

22 The criticism did not come exclusively from social scientists; even some plant breeders felt that the Green Revolution had failed (e.g. Simmonds 1979: 362).

23 'Plan Puebla', a programme developed by CIMMYT to extend the benefits of improved maize varieties to smallholders, illustrates this approach (Redclift 1983). Planners chose a region for the project which was highly atypical of those occupied by most Mexican small farmers in that it enjoyed abundant rainfall, good soils, cheap credit, strong political support and a high rate of literacy among the farmers. The technological package that was provided, however, was little different from that which large farmers had been offered.

24 'Many of the things the GR has been criticized for are reflections of more general agricultural or social problems.... Technology by itself is unlikely to provide the answers to rural inequities.' (Dalrymple and Jones 1973: 78, 80). As the Rockefeller Foundation's advisors said with reference to the Mexican Agricultural Program, 'There were some disappointments, but they were due less to scientific than to political causes' (Stakman *et al.* 1967: 71). Pearse (1980: 216) and Freebairn (1995) have also drawn attention to this view of technology's relation to society.

25 Some agricultural economists seem to have been aware that in their enthusiasm for the GR package they had left social and political matters out of account. Commenting upon a paper by a fellow economist, Randolph Barker observed: '... as Falcon suggests, we are scarcely prepared by background and training to undertake research in some of the basic socioeconomic issues of the developing countries. Perhaps there is a need for institutional reform in the agricultural economics profession'. (Falcon 1970: 711).

26 During the 1970s, one agricultural economist has reflected, although many in his discipline claimed to analyse poverty issues, they did not actually address questions of power, the control of markets or the distribution of assets (Biggs 2008: 490).

27 Defenders of the package also sometimes made the point that although the Green Revolution was imperfect, had it not occurred, the food situation in the developing world would have been far worse than it was (e.g. Evenson and Gollin 2003). This is, of course, a false dichotomy. The alternative to the Green Revolution was not to continue with traditional cultivation practices but rather to develop a set of improved cultivation practices that were explicitly tailored to the needs of smallholders.

28 The task, as Robert Chambers saw it, was '... not only to criticise [the Green Revolution] but also to construct, not only to analyse what has gone wrong but also to learn from it how to create something better' (Chambers 1977: 398).

29 In the view of an experienced colonial plant breeder, '... [it is an] illusion that agricultural problems can be solved by massive centrally planned research' (Hutchinson, foreword to Farmer 1977: x). Nevertheless, it was just such a centralizing tendency that impaired work at the Bangladeshi Rice Research Institute through the 1970s (Anderson *et al.* 1982; cf. Sisler and Colman 1979: 8–12). In West Africa after 10 years of varietal testing in which some 2000 imported rice varieties were tried in the mangrove swamps, only 2 were found which performed as well as local varieties (Spencer 1986).

30 At IRRI scientists reportedly found extension activity boring (Anderson *et al.* 1991: 117).

31 One advocate of integrated pest management, for example, argued that her fellow entomologists had too often handed over a package of IPM practices that were either too expensive for farmers or too difficult for them to grasp, thus putting the burden upon an extension service which was usually not up to the job. A solution, she felt, was to develop the package further, making it so cheap and easy to use that adoption no longer depended upon the extension service as intermediary (Goodell 1984).

32 In the mid 1960s scientists at IRRI were worried that the new variety they were developing, IR-8, was still vulnerable to pests and of inadequate quality, but they were under pressure nonetheless to release it, stemming in part from the Rockefeller and Ford foundations' concern with public relations (Anderson *et al.* 1991: 67–71; Cullather 2004).

33 The Rockefeller-funded agricultural programme in China during the 1930s is said to have failed because the problem of extension was huge but the facilities were inadequate. The staff on the insect-control programme, for example, found that it was 'easier to deal with locusts themselves than with farmers. . . .' (Stross 1986: 211).

34 For example, IRRI attempted to curb the work of two Indian rice research institutes whose work took a different though productive approach (Anderson *et al.* 1991: 92; Juma 1989).

35 Dahlberg argued that this policy of 'walking on two legs' was part of the reason for the doubling of rural income in China between the early 1950s and late 1960s (Dahlberg 1979: 207–209), and similar claims were made for Sri Lanka and India (Anderson *et al.* 1982: part 2). On the need for farmer participation more generally, see Chambers 1983; Pearse 1980: 234–236; Millikan and Hapgood 1967; Porter *et al.* 1991: 72–77.

36 On elite resistance to land reform in India, see Griffin 1974: 220–229; Frankel 1971: 4, cf. 205–206; Bose 1997. Anticipating the theme of the next section of this chapter, it bears saying that landlord power did not begin in 1945. In Chapter 1 we saw that estate owners in Prussia tried to block reforms aimed at peasant farmers, but the same happened elsewhere in Europe (Vivier 2008). Similarly, in colonial Kenya planters – who primarily grew export crops – objected to local government attempts to develop research on domestic food production (Clarke 2005). On landlords' resistance to the Rockefeller Foundation's peasant-oriented development programme in China in the 1930s, see Stross 1986: 211–212. Why, then, did it take until the 1970s before green revolutionaries grasped this point?

37 In considering the success of the stations' outreach work, it should not be overlooked that they were able to count upon the services of both agricultural school teachers and plant-breeding inspectors in each region, constituting an extension network a good deal more luxurious than anything available in most developing countries.

38 Cited in Maat 2007: 54. For general discussion of these issues see Hodge 2007; van Beusekom 2000; Moon 2004 and 2007; Tilley 2003; Maat 2001. For a first-hand account from British India, see Howard 1940.

39 Hodge 2007: ch. 7; Richards 1985: ch. 1; Bonneuil 2001; Cooper 1997; Bowman 2010. Systematic research on inter-cropping in sub-Saharan Africa, for example, did not resume until the 1970s (Spencer 1986).

40 Significantly, extension workers were encouraged to live in the villages they served rather than spending most of their time in district headquarters (Ogura 1967: 640). On the importance of cooperatives see Dore 1960 and Ishikawa 1981: 270–272. In view of the central role played by extension in both Central Europe and Japan, it is remarkable that advisors and staff on the Mexican Agricultural Program only began to think seriously about this issue when the Program had been under way for a few years (Harwood 2009).

41 Most of the lecturers employed at the newly established agricultural schools were also European, but within a few years the Japanese had concluded that European (and especially English) agricultural science was largely useless in Japanese conditions (Ogura 1967: 643; Iinuma 1973).

42 Several dwarf rice varieties were brought to the US and grown extensively before the First World War (Jones 1936; Dalrymple 1986: 16).

43 As in Japan, the initial strategy was simply to improve local varieties by selecting and distributing the best of them. Later the hybrid variety Ponlai, once adapted to Taiwanese soil and climate, was distributed free to farmers. The result was a two-and-a-half-fold increase in yield over the period (Carr and Myers 1973).

44 Ogura 1967: part 4; Hayami *et al.* 1975: ch. 8. The Japanese experience also demonstrated, however, that double cropping and heavy use of mineral fertilizer reduced soil fertility (Sawada 1965: 333), a lesson that had to be relearned by green revolutionaries a generation later.

45 Ruttan implies that this might have been part of the problem when he writes that it became clear only in the 1960s that agricultural expertise based on experience in the temperate zone was not very useful in tropical zones (Ruttan 1989: 174, 186).

46 In 1919 one commentator in rural Pennsylvania noted that local poor farmers refused to have anything to do with the state agricultural experiment station 'which they considered as an institution having unlimited resources with which to experiment and one carrying on farming under conditions which the ordinary farmer, for lack of money, could not hope to imitate' (A. B. Ross, cited by Carlson 2005: 103).

47 That heavy use of mineral fertilizer led to soil acidification and decreased fertility was well known to western agricultural scientists by the 1950s at the latest (Uekötter 2010).

48 One World Bank report on Lesotho is said to have been oblivious to the country's colonial history although it had clearly shaped the country's economy (Ferguson 1990).

49 Contact did not depend just upon formal institutional mechanisms. Before and after the First World War a number of German breeders visited the US for extended periods (von Rümker 1911a; von Rümker and Tschermak 1910; Roemer 1926). On his visit to the University of Minnesota in 1925, for example, Theodor Roemer met Elvin Stakman who later became one of the key figures in the design of the Mexican Agricultural Program. As a result of the visit, the two men set up a student exchange between their universities, and Stakman was visiting professor at Halle in 1930 (Harwood 2009). Similarly, in 1908 the American Mendelian George Shull visited Europe in order to meet the major plant breeders and geneticists, several of whom would have been well informed about the activities of the south German stations ('Notes on a European tour undertaken by George H. Shull …', Shull papers, American Philosophical Society).

50 Albert Mann, 'Report of Trip to England, Sweden and other European Countries for special Barley Investigations', E. S. Beaven papers, Institute for Agricultural History, University of Reading. I thank Paolo Palladino for giving me a copy of this report.

7 Reforming the revolution: peasant-friendly innovation, 1970–2010

1 Just why it took so long to reach this conclusion is not clear. Already in the 1930s several highly publicized surveys had shown that malnutrition was common in the wealthiest countries despite food surpluses (Staples 2006: 72–74).

2 When he first came to northern Nigeria in the 1960s, David Norman was struck by this phenomenon. 'Since the experiment station had been there for almost 50 years, the question that immediately came to mind was why none of the results or lessons from the research station had "rubbed off" on neighbouring farms' (Norman 2000: 30). Elsewhere the same phenomenon was in evidence as green revolutionaries claimed that their research would alleviate the 'burden of labor' faced by rice farmers in the Far East (Stakman *et al.* 1967: 289), evidently unaware that such regions tended to suffer instead from surplus labour.

3 Nevertheless, some felt that not enough effort was being devoted to developing varieties suited to intercropping (Maurya 1989).

4 One evaluation of extension systems published during the 1980s, for example, concluded that simply spending more on extension was not enough. It was essential, not only that farmers were given more control over the extension process, but that researchers '[got] the technology right in the first instance' (Antholt 1998: 360; in a similar vein see Berg and Whitaker 1986: part 3).

5 One should not exaggerate the degree of continuity. If one takes south Germany as a benchmark, for example, both FSR and agroecology were quite novel. Environmental issues were not an issue for the south German stations, nor did they express any interest in improving upon indigenous cultivation practices. As far as participation is

concerned, the fact that the director of the Bavarian station saw the experts' task as overcoming the suspicions of peasants and making sure that they implemented the recommended practices suggests that the hierarchical relation between expert and farmer was largely taken for granted there (Kiessling 1913: 27–28).

6 Experts differ over precisely how FSR is to be defined. Biggs, for example, sees FSR as one form of participatory technology development (Biggs 2008). Others do, too, but regard it as an essentially 'extractive' form of participation in which the scientist tries to understand the system in question, improves it via on-station research, and then returns the improved practices to the farmer. While the farmer is consulted at various points, it is the scientist who remains in control throughout the process (Chambers *et al.* 1989: 167–169; Scoones and Thompson 1994b).

7 In these cases the breeding is collaborative. The initial crosses and selection are conducted at the experiment station, resulting in a mixture of lines which the breeder hands over to farmers who carry out the subsequent selection (Conway 1997: ch. 10).

8 For example, in the Philippines during the 1980s small farmers were worried about health damage from pesticide use and began to look for alternatives to the Green Revolution (Frossard 2002). On related concerns at that time in Bangladesh, see Alauddin and Tisdell 1991.

9 In Zambia FSR was institutionalized within the national agricultural research system, but as at IRRI it was segregated within a single team – which was mistrusted by crop specialists – rather than dispersed throughout the system (Bowman 2011).

10 To be sure, in 1996 the CGIAR established a 'System-wide Program for Participatory Research and Gender Analysis' which was co-sponsored by four of its research centres. According to one well-placed source, however, over the last decade the CGIAR's Science Council has consistently sought to marginalize this Program.

11 As one academic in this field wrote to me, PPB 'has served to energize some … staff and alienate many others'.

12 In his classic analysis of centralized top-down state-led schemes of improvement Scott remarks that 'What is perhaps most striking about high-modernist schemes … is how little confidence they repose in the skills, intelligence and experience of ordinary people' (Scott 1998: 346).

13 Similar distinctions have been made before (e.g. W. F. Whyte cited by Redclift 1983: 570). For an analysis of why the latter paradigm has remained marginal to mainstream agricultural research, see Hogg 2000.

14 According to IFAD, biotechnology is a scale-neutral technology whose scientific prospects were excellent (IFAD 2001: 127ff.). An issues brief for FAO was similarly enthusiastic but worried that public fears about the technology would prevent it helping all farmers (FAO 2009c). The United Nations Development Programme has also expressed its support for biotechnology as a potential solution to the problem of food security (McMichael 2004). Only the *World Development Report 2008* acknowledged in passing that the technology's private-sector concentration meant that it was reaching only commercial farmers (World Bank 2007).

15 According to one insider account, drafts of the various reports received a large number of critical 'and sometimes rude' reviews during the last few months prior to publication, especially from the US (Feldman *et al.* 2010: 69). The only such review I have seen (from an agricultural scientist at USAID) was indeed severely critical of Chapter Two of the *Global Report*, dismissing parts of it as 'very biased', 'combative', 'inflammatory' and a 'harangue' which represented only 'one school of thought'.

16 So common is the emphasis upon smallholder agriculture by now that some authors refer to it as 'the current model favoured by donors' (Collier and Dercon 2009).

17 The claim that *only* biotech-based breeding (rather than conventional breeding) can adequately maintain the food supply has been disputed by a number of high-profile breeders (Murphy 2007: 235–237). My concern here is merely to examine the claims for what GM can do.

18 For a general discussion of the industry's stance, see Janssen and Gupta 2009; for Monsanto's view, see Combes 2006 and Krueger 2001.

19 A decade ago the US minister of agriculture insisted that 'There is no way to feed a hungry world … without embracing the future. We must make this choice [for biotechnology]' (Cited in Lappé and Bailey 1999: 20).

20 The president of the Boyce Thompson Institute for Plant Research at Cornell University has declared that 'We can't afford to walk away from technology. [GMOs are] the best way to feed the world' (*Cornell Magazine*, January/February 2000, 31).

21 Although I will focus only upon the use of biotechnological methods to construct transgenic organisms (also known as 'genetic modification' or 'genetic engineering'), the term can also be used in a more general sense to include tissue culture as well as knowledge and techniques from genomics (e.g. 'marker-assisted breeding'). These two forms of 'biotechnology' in a wider sense have been uncontroversial and are already widely used in conventional plant breeding because they make the process considerably more efficient.

22 By 'moderate' I mean those authors who portray biotech, not as a panacea but as a useful addition to the breeder's toolkit, and who also acknowledge the potential limitations of the technology (e.g. Conway 1997; Lipton 2001; Fukuda-Parr 2007a; Tripp 1997; Pimentel *et al.* 1989).

23 In sub-Saharan Africa, for example, about 40 per cent of farmers work in arid or semi-arid conditions, and only 4 per cent of the land is irrigated (Paarlberg 2008: 149–150).

24 Pray 2002; Gouse 2007. On Pioneer Hi-Bred's early establishment of breeding facilities in Latin America and Asia, see Buttel *et al.* 1985. If one focuses upon sub-Saharan Africa where farms are overwhelmingly small, less than 2 per cent of all agricultural R&D is funded by the private sector (Paarlberg 2008: 106). Of R&D on biotechnology in the region, only 10–15 per cent is funded by the private sector (Ndiritu 2000: 113) and most of that has been aimed at commercial farmers (Nwalozie *et al.* 2007).

25 Sorghum and millet constitute over half of the total cereals acreage in sub-Saharan Africa, partly because they will grow on poor soil and require relatively little water, as does cassava. But in 2003 less than one-tenth of 1 per cent of the total GM acreage was planted in any major food staple (Lipton 2007: 52).

26 To be sure, in India a few biotech firms have developed hybrids of millet and sorghum (Pray and Naseem 2007), but the general tendency in the Indian private sector is said to be much the same as in the multinationals (Spielman *et al.* 2006; Scoones 2006). Others have noted, similarly, that during the 1980s the amount of commercial biotech-breeding devoted to animals was far larger than that on cereals, despite the fact that most peasants own few animals and can only rarely afford to eat meat (Buttel *et al.* 1985). The private sector's neglect of certain commodities is not confined to agriculture; pharmaceutical companies have been similarly uninterested in developing drugs for diseases like malaria or schistosomiasis since a large proportion of the sufferers have no money (Spillane 2002). As a result, 90 per cent of the world's medical research is targeted at just 10 per cent of the population (Smith, J. 2009: 84).

27 The private sector is little interested in cassava for related reasons; as a perennial crop, there is no need for farmers to replant annually (Stone 2002). Much more profitable is maize. Unlike rice or wheat, maize is a naturally cross-fertilizing plant, so it is relatively easy for the breeder to construct F1 hybrids. Since the hybrid's desirable properties are rapidly lost in subsequent generations, the farmer is forced to purchase fresh seed annually. This is why private-sector investment in maize is about ten times higher than in wheat (Swanson 2002: 15–16) as well as high in other crops where hybrids have been developed (Murphy 2007: n. 231), and it is why multinational companies supply nearly 90 per cent of maize seed in Brazil (da Silveira and Borges 2007:113).

28 The best-known varieties of this kind are Monsanto's 'Roundup-ready' varieties. Sales of the herbicide, Roundup, are said to generate 17 per cent of Monsanto's total annual sales (Middendorf *et al.* 2000: 110), though some estimates are as high as 45 per cent (*Guardian*, 20 August 2002).

29 As we saw above, developing F1-hybrid varieties is one way to force farmers to buy seed regularly, but this is difficult to do in some crops (e.g. wheat). In recent years, therefore, the industry has been trying to develop alternative methods ('genetic use-restriction technologies') which in effect make the seed from a GM plant sterile (Middendorf *et al.* 2000). Faced with an avalanche of criticism, however, Monsanto pledged in 1999 not to commercialize its own version of this technology – dubbed by critics the 'terminator technology' – for the next five years (Pinstrup-Andersen and Schiöler 2001: 136). Nevertheless, most biotechnology firms are believed to hold patents on similar technologies (Pingali and Traxler 2002; Crouch 2001; Swanson and Goeschl 2002).

30 The price differential depends upon the crop and the country. In Brazil GM soybeans cost 60–90 per cent more than non-GM seed (Da Silveira and Borges 2007: 118–119) while estimates for other crops and countries vary from 30 per cent more expensive (Benbrook 2003) to nearly 100 per cent (Seitzer 2008: 571). In India estimates of the cost of Bt cottonseed vary from only 10 per cent more than non-Bt seed (Ramaswami and Pray 2007:166) to four times as much (Scoones 2006: 28).

31 Initially, for example, subsistence farmers in one region of South Africa rapidly adopted Bt cotton (thanks to the availability of input credit) because it brought increased yield and reduced the need for pesticide. Once credit was withdrawn, however, the acreage in Bt cotton plummeted since most of these farmers could not afford the seed. Smallholders there were similarly keen on Monsanto's Bt maize because of its large yield increases, but in this case levels of adoption were low; since maize was classified as a subsistence crop, no credit was available (Gouse 2007).

32 Not all public-sector work is peasant-friendly. In 1998 the U.S. Department of Agriculture joined with a commercial seed company to obtain a patent on a form of 'terminator' technology whose intention is to prevent farmers from reusing seed (Stone 2002).

33 *Advances* (newsletter of the John Innes and Sainsbury Laboratory), autumn 2006 and spring 2007 issues; Herrmann 2009; Conway 1999.

34 Although in 1980 the United States devoted 25 per cent of its development assistance to agriculture, by 2003 the proportion was 1 per cent, and the US was spending far more on military aid than on food aid and much more on the latter than on agricultural development (Paarlberg 2008: 93; Falcon and Naylor 2005: figs. 1 and 2).

35 Although it serves several billion peasant farmers throughout the developing world, the CGIAR's entire budget is only about the size of Brazil's agricultural research budget (Falcon and Fowler 2002: 216; Falcon and Naylor 2005: 1120).

36 In 2000 Aventis's R&D budget was ten times what the entire CGIAR system spent on biotechnology (Kloppenburg 2004: 334). Monsanto's was 20 times greater (Spillane 2002: 73–74).

37 The roll-back of the public sector has not just affected research, of course. Structural Adjustment programmes have also hurt small farmers through cutbacks in credit facilities, price subsidies and marketing support (Rosset 2002) as well as seed production and distribution systems (Morris and Ekasingh 2002).

38 For example, in 2000 sub-Saharan Africa was spending only 0.7 per cent of its agricultural GDP on agricultural research (and India 0.3 per cent) vs. 2.6 per cent in the US and 3.6 per cent in Japan (World Bank 2007: tab. 7.1).

39 To be sure, neither funding nor reward systems in the North's public sector encourage plant scientists to address problems of the developing world (Murphy 2007: 295–296). One survey of plant research journals, for example, showed that only about 4 per cent of the articles were devoted to the mechanisms of plant-growth in unfavourable

environments (Coffman 1983). Nor has public-sector biotechnology research in the North paid much attention to Third World concerns (Byerlee and Fischer 2002).

40 Between about 1980 and 2000 the number of patents issued on genes as well as the major cereals rose over 40-fold (Fowler and Lower 2005: 39).

41 Public-sector breeders keen to place new varieties in the hands of resource-poor farmers are said to be frustrated, in part by the obstacles posed by patent rights (Manning 2000) but also because they feel that policy formulation on agricultural biotechnology has so far been dominated by the private sector and by environmental organizations (Spillane 2002).

42 Pringle 2003: 35; Hucho *et al.* 2005: 306–307. Monsanto has similarly offered to license its drought-resistant maize varieties free for use in Africa, though only after sales to American farmers have been under way for five years (Pollack 2008). African farmers, it seems, will just have to wait.

43 Toenniessen and Delmer 2005; Graff *et al.* 2003; Doyle 1985. In the US a law passed in 1980 allowed universities to patent work which had been federally funded. Since then the number of patent applications from academia has increased about 20-fold. In 1998 Stanford University, for example, earned $61 million from this source. (Press and Washburn 2000).

44 Recognizing the absurdity that public-sector breeders were having difficulty obtaining access, even to technologies that had been developed in their own institution (and thence licensed to firms), university and foundation presidents in the US proposed the formation of a network through which public-sector institutions in future would manage their patents collectively (Atkinson *et al.* 2003).

45 An instance of this is an agreement between the Japanese corporation, Plantek, and some CGIAR centres who are developing an insect-resistant cassava variety (Rausser *et al.* 2000).

46 As one USAID official remarked, ‘no USAID collaborative projects have been a fully philanthropic exercise for the private-sector’ (Lewis 2000: 199). This may be one reason why more than one public-sector institution has not been keen to collaborate, at least with multinationals (Paarlberg 2008: 165).

47 Morris and Ekasingh 2002; Just and Rausser 1993. On one Indian agricultural university’s partnership with Monsanto, see Scoones 2006: 100 and 123. A recent workshop, which brought representatives of the CGIAR and the private sector together, suggests that such concerns are justified (CGIAR Science Council, ‘Workshop on public–private partnerships and associated needs for product stewardship and liability – Final Report’, 11–13 November 2009).

48 Some of these have been pretty naive; for example, the suggestion – by a director of the CGIAR no less – that what is needed are ‘alliances of the caring’ (Serageldin and Persley 2000: 34).

49 The list of ‘obstacles’ varies somewhat from one author to the next. Some include the intellectual property barriers faced by public-sector breeders or unnecessarily provocative moves by the biotech industry, and many cite the rancorous public debate over the safety of GM (e.g. Conway 1999 and 2003).

50 In contrast with many developing countries, levels of pesticide use in China have been very high – comparable to those in rich countries – because of state subsidies (Qaim and Zilberman 2003).

51 The net gain to farmers of Bt cotton has been estimated at about US$300–400 per hectare, and the reported incidence of pesticide poisoning is much lower (Pray *et al.* 2002).

52 Interestingly, there are signs that a vigorous Chinese public sector is also curbing the power of multinationals through mechanisms very similar to those used by its south German counterparts a century ago: by making technical support available to small firms, by allowing seed bred in the public sector to be sold by numerous small locally oriented companies (Huang *et al.* 2007), and by forcing down the cost of seed through competition (Morris and Ekasingh 2002; FAO 2004: 49–51).

53 After this section was finished, I was pleased to discover that Dominic Hogg, though using evidence very different from mine, had earlier drawn a similar conclusion (Hogg 2000: 209).

8 Three conclusions

1 I thank Andy Bowman for helpful discussion of this issue as well as a stimulating paper by Gerald Hödl: 'Prologue or first act? Historical continuities and discontinuities of colonial development', given at the winter-school, 'History of Development', organized by the ETH Zürich and held at Ascona, Switzerland, 24–28 January 2011.

2 On the British colonial authorities' decision in the late 1930s to replace the previous 'exploitation' policy for Kenya with one of 'development', see Cowan and Shenton 1996: ch. 6.

3 Keeley and Scoones' (2000) analysis of the changing fortunes of productivist and environmentalist policies since the 1960s in Ethiopia offers one way in which one might make sense of shifting policy emphases (even if their focus is at regional and national rather than global levels).

4 Between the 1960s and the mid 1980s in the US, for example, the public sector's share of R&D in plant breeding fell by three-quarters while private-sector funding increased over tenfold (Murphy 2007: 104–108).

5 More properly, that part of the Institute which engaged in varietal development and other work of direct commercial interest was sold. The division of the PBI which worked on basic plant genetics remained in the public sector.

6 In the US, similarly, public-sector breeders are concerned that declining funding is available for the improvement of subsistence crops grown by resource-poor farmers (Weebadde and Mensah 2006).

7 Recent studies of rates of return for projects in developing countries show an average of 43 per cent (World Bank 2007: 14; cf. Ruttan 1980). Conservative estimates of the economic payoff from research at the CGIAR institutes for 1960 to 2001 indicate that the gain has been twice the cost (Conway and Waage 2010: 68). India's gain from the rice research at IRRI is estimated to be 20 times the size of IRRI's entire budget (Fan *et al.* 2005: 375, 378). The US's benefit from research at CIMMYT is estimated to be nearly 200 times the size of its expenditure on that institution (Puente-Rodríguez 2010: 39). Over the period 1988–2002 the total annual benefit worldwide of CIMMYT's wheat varieties has been estimated to be 100 times the annual cost of the research (Murphy 2007: 245).

8 Furthermore, while IARCs are important centres of innovation, it should be noted that in most cases the improved varieties developed at places like IRRI and CIMMYT have not been entirely new but rather based on generations of earlier breeding work at NARS and regional public-sector institutions (Dalrymple 1977: 181).

9 Given these criticisms, one might ask whether the needs of smallholders might be met more effectively by NGOs rather than NARS, as has been the preferred strategy of many donors since the 1980s. While NGOs have enjoyed a certain success in taking over responsibility for extension (Chambers 1989; Morris *et al.* 2006), however, few observers see them as credible institutions for research (Tripp *et al.* 1990). Moreover, NGOs have their own weaknesses. Given their small size and limited resources, they are not in a position to take on all the tasks assumed by NARS (Farrington and Bebbington 1994), and since they are dependent (in most cases) upon funding from donors, some see NGOs as more responsive to donors' priorities than to farmers' (Shrum 2000; Shepherd 2006). Finally, the very large numbers of NGOs acting independently of one another in some countries means that they are less likely than state institutions to develop coordinated long-term solutions. As one development expert in Zambia put it, when the 2002 food crisis hit, '[the NGOs] are suddenly experts; when

the disaster is over, they just leave. ... Nobody is here for more than two years' (Bowman 2011: ch. 4).

10 This is not to say that most NARS were *well* funded even during the earlier period. Even before the Washington Consensus began to bite, some agricultural economists were calling attention to the underfunding of NARS (Hayami and Ruttan 1985).

11 Even those who continue to endorse modified forms of the Washington Consensus now argue that since markets do not always work effectively in poor countries, more attention needs to be paid to the role of the state (Ashley and Maxwell 2001) and to forms of intervention used in the past by successful public-sector programmes (Kydd and Dorward 2001).

12 Chang is not alone in pointing to the role of state intervention in previous economic transformations. On the importance of the U.S. Department of Agriculture's wide-ranging promotion of American farming (*inter alia* via R&D) since the late nineteenth century see Ferleger and Lazonick 1993. On the crucial importance of the state in the post-war growth of East Asian agriculture see Djurfeldt *et al.* 2005; Kydd and Dorward 2001.

13 Ironically, during the 1980s when the World Bank and IMF were requesting aid recipients in the global South to reduce public expenditure, farm subsidies in the global North were increasing sharply (Weis 2007: 124).

14 Some rich countries are remarkably tight-fisted when it comes to foreign aid. In 1995, for example, the U.S. ranked last among the industrialized countries in foreign aid as a percentage of GDP. Even in terms of absolute amounts, US development aid was less than half that of Japan (Mellor 1998: 56).

15 In a similar vein, Porter *et al.* (1991: 59–61) distinguish between those who 'talk' development and those who actually 'do' it.

16 At the same time one must be careful not to assume that those working at the grass-roots level are necessarily dedicated to learning from experience and conveying hard truths to project managers or planners. As van Ufford (1993) points out, it can be in the interests of local project staff and local government officials to maintain certain fictions about a project where these are thought to be important for its survival.

17 The same criticism has been levelled at consultants, sent into the field for a few months at a time and poorly briefed (Hayter 2005; Bowman 2011). Even where field staff are experienced and knowledgeable, however, it is not clear how much influence they can exert over the design of programmes. One study of the World Bank argues that researchers and consultants are under pressure to produce analyses that fit smoothly with policy directives from above (Goldman 2005: 126ff.).

18 For the pressure on operations staff at the World Bank to 'move money', see Pincus 2009: 186–190; the same pressure exists elsewhere (Porter *et al.* 1991: 33ff.). One middle-ranking officer at the World Bank is reputed to have said in the late 1970s 'I don't know what the hell the goals are, but I'm moving ahead with projects' (Porter *et al.* 1991: 97).

19 This phenomenon was hardly confined to the Rockefeller Foundation. FAO staff are said to have been similarly uneasy with the political dimensions of development (Staples 2006).

20 A World Bank official is said to have remarked that one of the most attractive things about the Philippines' GR programme for smallholders was that it didn't antagonize landlords the way land reform did (Bello 2009; in a similar vein Freebairn 1995: 269). On the Bank's reticence to acknowledge the role of class power, see Pincus 2009.

21 By 'donors' I would emphasize that I am here referring to decision-makers within donor agencies, not to staff in the field whose commitment to poverty alleviation is often evident (e.g. Porter *et al.* 1991: xvii; Anderson *et al.* 1991).

22 Some observers have even suggested that the North reduced aid for agriculture in the developing world from the 1980s for fear that the latter's development might harm the North's export markets (Falcon and Naylor 2005).

23 Conversely, the reason why huge amounts are spent in some countries on fertilizer subsidies – rather than on R&D or infrastructure which offer a much greater return in the longer term – is thought to be their appeal to wealthier farmers (World Bank 2007: 115–116).

24 While working as an agricultural economist in Malawi in the early 1970s David Colman found that officials from that country's Ministry of Agriculture did not want to hear that their favoured approach to development was not working (Colman, pers. comm.). In cases like this some donor agencies have been known to knuckle under, turning a blind eye to the host country's preference for dubious development policies (Goldman 2005: 173–174). As Pincus points out, agencies like the World Bank have an interest in maintaining good relations with client countries. Drawing attention to the latter's corruption or other shortcomings might jeopardize future projects (Pincus 2009: 201–204).

Abbreviations

(Abbreviations of publication-titles are listed under 'Printed Primary Sources'.)

Bay.Min.Ldw – Bayerisches Ministerium für Landwirtschaft (I use this abbreviation for the period before 1918 when agricultural matters were handled by a division of the Interior Ministry as well as after 1918 when they were dealt with by the Wirtschaftsministerium/Abteilung Landwirtschaft).

CGIAR	Consultative Group on International Agricultural Research
CIMMYT	International Centre for Maize and Wheat Improvement
DLG	Deutsche Landwirtschafts-Gesellschaft (German Agricultural Society)
FAO	Food and Agricultural Organization of the UN
FSR	farming systems research
GFP	Gesellschaft zur Förderung der deutschen Pflanzenzucht (Association of Commercial Breeders)
IARC	international agricultural research centre
IFPRI	International Food Policy Research Institute
IRRI	International Rice Research Institute
NARS	national agricultural research system
NGO	non-governmental organization
PPB	participatory plant breeding
Preuss.Min.Ldw.	Prussian Agricultural Ministry
Reich Min.Ldw.	Reich Agricultural Ministry
USAID	U.S. Agency for International Development

Archival sources (and abbreviations used)

BAB (Bundesarchiv, Berlin)
R8073 – Deutscher Landwirtschaftsrat
R3602/alt R168 – Biologische Reichsanstalt
R68IV – Gesellschaft zur Förderung deutscher Pflanzenzucht

BAK (Bundesarchiv, Koblenz)

BayHStA (Bayerisches Hauptstaatsarchiv, Munich)
LABK – Landesanstalt für Bodenkultur u. Pflanzenbau
ML – Landwirtschaftsministerium
MK – Kultusministerium
MF – Finanzministerium

GLAK (Generallandesarchiv Baden-Württemberg, Karlsruhe)
Bestand 236 – Badisches Innenministerium
Bestand 237 – Badisches Finanzministerium

GStAPK (Geheimes Staatsarchiv Preussischer Kulturbesitz)
R87B – Landwirtschaftsministerium
R 89 – Finanzministerium

George H. Shull papers, American Philosophical Society, Philadelphia

MPGA (Archiv zur Geschichte der Max-Planck-Gesellschaft, Berlin)

TUM (Archiv der Technischen Universität München)

UA-Halle (Universitätsarchiv, Halle)

UA-Hohen (Universitätsarchiv, Stuttgart-Hohenheim)

UA-Weih (Archiv der Technischen Universität München, Zweigstelle Weihenstephan)

Printed primary sources (with abbreviations)

Arbeiten der DLG = *Arbeiten der Deutschen Landwirtschafts-Gesellschaft*

Ber. Bad. Ldw-Kammer = *Bericht über die Tätigkeit der Badischen Landwirtschaftskammer im Jahre...* for 1920, 1925, 1930 (Karlsruhe: Badische Landwirtschaftskammer)

Deutsche Landwirtschaftliche Presse, 1900–1914, 1919–1920

Illustrierte Landwirtschaftliche Zeitung, 1905–1914, 1919–1920

Jahrbuch der DLG = *Jahrbuch der Deutschen Landwirtschafts-Gesellschaft*

Jb. Baden 19xx = *Jahresbericht der Grossherzoglichen Badischen Saatzuchtanstalt für das Jahr 19xx*, 1910–1913 (Karlsruhe: Macklot'sche Buchdruckerei); the reports for 1920, 1925 and 1930 were published in *Bericht über die Tätigkeit der Badischen Landwirtschaftskammer im Jahre 19xx*

Jb. Bayern 19xx = *Bericht der Königlichen Bayerischen Saatzuchtanstalt, Weihenstephan*, no. 1 (1903) – no. 7 (1909) (Munich or Freising, various publishers). Numbers 8 (1910) to 14 (1922–1926) were published in the journal, *Landwirtschaftliches Jahrbuch für Bayern*

Jb. Württ. 19xx = *Bericht über die Tätigkeit der Württembergischen Saatzuchtanstalt Hohenheim im Jahre 19xx* (Plieningen: Friedrich Find, 1906–07 – 1913–14)

Jb. General-Comites = *Jahresberichte des General-Comites des landwirtschaftlichen Vereins in Bayern für das Jahr ...*, 1890–1894

Jb. Bay. Ldw-Rats = *Jahresberichte des Bayerischen Landwirtschaftsrats für das Jahr ...*, 1895–1901

Mentzel u. Lengerke = *Mentzel und von Lengerke's landwirtschaftlicher Hülfs- und Schreib-Kalender* (Berlin: Paul Parey)

Mitteilungen der DLG = *Mitteilungen der Deutschen Landwirtschafts-Gesellschaft*, 1886–1930

Schlipfs Handbuch = *Schlipfs populäres* [from 1920: *praktisches*] *Handbuch der Landwirtschaft* (Berlin: Parey, 9th edn 1880, 14th edn 1902, 24th edn 1922)

Wochenblatt Baden = *Wochenblatt des Landwirtschaftlichen Vereins im Grossherzogtum Baden* (from 1911: *Badisches Landwirtschaftliches Wochenblatt für das Jahr...)*, 1895–1920

Wochenblatt Bayern = *Wochenblatt des Landwirtschaftlichen Vereins in Bayern*, 1890–1905

Wochenblatt Württemberg = *Württembergisches Wochenblatt für Landwirtschaft*, 1890–1905

Bibliography

(Abbreviations of journal titles are listed under 'Printed Primary Sources'.)

Abel, W. (1962) *Die drei Epochen der deutschen Agrargeschichte*, Hannover, M. and H. Schaper.

Achilles, W. (1993) *Deutsche Agrargeschichte im Zeitalter der Reformen und der Industrialisierung*, Stuttgart, Ulmer.

Achilles, W. (1994) Betriebswirtschaftliche Leitbilder in der ostdeutschen Gutswirtschaft seit dem Ende des 19. Jahrhunderts. In Reif, H. (ed.) *Ostelbische Agrargesellschaft im Kaiserreich und in der Weimarer Republik*, Berlin, Akademie Verlag.

Ackermann, J. (1924) Aus zwanzigjähriger Züchtererfahrung. *Beiträge zur Pflanzenzucht*, 7, 33–39.

Ackermann, J. (1927) [Zum 25 ?] -jährigen Bestehen der Bayerischen Landessaatzuchtanstalt *Bayerische Ackerbauzeitung* (= Sonderbeilage z. *Wochenblatt Bayern*, Nr. 46).

Ackermann, J. (1928) *25 Jahre Pflanzenzucht seit 1903*, Niederbayern, Saatzuchtwirtschaft Gut Irlbach.

Ackermann, J. (1929) Bayerische Saatzucht. In Dörfler, H. (ed.) *Die Landwirtschaft in Bayern: eine Sammlung von Darstellungen über Entwicklung und Stand der bayerischen Landwirtschaft*, Freising, Datterer.

Aereboe, F. (1917) *Allgemeine landwirtschaftliche Betriebslehre, 2nd edn*, Berlin, Parey.

Agriculture & Environment Biotechnology Commission (2005) *What Shapes the Research Agenda in Agricultural Biotechnology? Plant Breeding Case Study*. London, Agriculture and Environment Biotechnology Commission.

Äkerman, A. *et al.* (eds) (1948) *Svalof, 1886–1946: History and Present Problems*, Lund, Carl Bloms.

Alauddin, M. and Tisdell, C. (1991) *The 'Green Revolution' and Economic Development: the Process and its Impact in Bangladesh*, London, Macmillan.

Albers, H. (1999) *Die stille Revolution auf dem Lande. Landwirtschaft und Landwirtschaftskammer in Westfalen-Lippe, 1899–1999*, Münster, Landwirtschaftskammer Westfalen-Lippe.

de Alcántara, Cynthia Hewitt (1976) *Modernizing Mexican Agriculture: Socioeconomic Implications of Technological Change 1940–1970*, Geneva, U.N. Research Institute for Social Development.

Aldenhoff, R. (1996) Agriculture. In Chickering, R. (ed.) *Imperial Germany: a Historiographical Companion*, Westport, CN, Greenwood Press.

Aldenhoff-Hübinger, R. (2002) *Agrarpolitik und Protektionismus. Deutschland und Frankreich im Vergleich, 1879–1914*, Göttingen, Vandenhoeck & Ruprecht.

Allen, T. (1987) Updating 'development policies': the Berg Report as the dust settles. In Shanin, T. (ed.) *Peasants and Peasant Societies*, Oxford, Blackwell.

Altieri, M. A. and Hecht, S. B. (eds) (1990) *Agroecology and Small Farm Development*, Boca Raton, CRC Press.

von Altrock, W. (1917) Zur Erinnerung an das 75jährige Bestehen des Königlich Preussischen Landes-Oekonomie-Kollegiums. *Deutsche Landwirtschaftliche Presse*, 44, 447–451.

von Altrock, W. (1925) Landwirtschaftskammern. In Elster, L. *et al.* (eds) *Handwörterbuch der Staatswissenschaften.* Jena, G. Fischer.

Anderson, R. S. (1991) The origins of the International Rice Research Institute. *Minerva*, 29, 61–89.

Anderson, R. S. *et al.* (eds) (1982) *Science, Politics and the Agricultural Revolution in Asia*, Boulder, Colorado, Westview Press.

Anderson, R. S., Levy, E. and Morrison, B. M. (1991) *Rice Science and Development Politics: Research Strategies and IRRI's Technologies Confront Asian Diversity (1950–1980)*, Oxford, Clarendon Press.

Andersson, G. (1963) The Swedish Seed Association: its history and organization. *Recent Plant-Breeding Research*, ?, 9–23.

Anon. (1892) Die Entwicklung der DLG vom 1.10.91 bis dahin 1892. *Jahrbuch der DLG*, 7, x.

Anon. (1894) *Die Agrarkonferenz vom 28. Mai bis 2. Juni 1894. Bericht über die Verhandlungen … zur Erörterung agrarpolitischer Massnahmen*, Berlin, Paul Parey.

Anon. (1900) Ueber den Anbauwerth verschiedener Weizensorten. *Allgemeine Norddeutsche Zeitung.* 24 March edn.

Anon. (1907) Die D.L.G. und die Saatzuchtbestrebungen einzelner Bezirke. *Mitteilungen der DLG*, 22, 168.

Anon. (1908) Von der neuen Saatzuchtanstalt. *Wochenblatt Baden*, 791–793.

Anon. (1911a) Förderung der Saatzucht in Hessen. *Deutsche Landwirtschaftliche Presse*, 38, 61.

Anon. (1911b) *Landwirtschaft und Landwirtschaftskammer in der Provinz Sachsen 1906–1910*, Halle, Otto Thiele.

Anon. (1912) Zur Förderung des Pflanzenbaus. *Wochenblatt Baden*, 361–363.

Anon. (1913) Eröffnungssitzung. *Beiträge zur Pflanzenzucht*, 3, 1–7.

Anon. (ed.) (1914) *Die Entwicklung der Landwirtschaft in der Provinz Sachsen während der letzten 25 Jahre (= Arbeiten der Landwirtschaftskammer für die Provinz Sachsen, no. 31)*, Halle, Verlag der Landwirtschaftskammer.

Anon. (1922) Geschäftlicher Teil. *Beiträge zur Pflanzenzucht*, 6, 5–10.

Anon. (1925) Pflanzenzüchterkurs in Rastatt. *Deutsche Landwirtschaftliche Presse*, 52, 673.

Anon. (1927) Eröffnungsansprache des Vorsitzenden. *Beiträge zur Pflanzenzucht*, 9, 5–24.

Anon. (1932a) Die Entwicklung der Pflanzenzüchtung in Baden. In Anon. (ed.) *25 Jahre Badische Landwirtschaftskammer 1907–1932.* [no city or publisher cited].

Anon. (1932b) 25 Jahre Badische Landwirtschaftskammer 1907–1932. In Anon. (ed.) *25 Jahre Badische Landwirtschaftskammer 1907–1932.* [no city or publisher cited]

Anon. (1934a) Professor Dr. Wacker. *Wochenblatt der Landesbauernschaft Württemberg*, 101, 1207.

Anon. (1934b) Verordnung über Saatgut vom 26. März 1934. *Reichsgesetzblatt*, 1934, part 1, 248.

Anon. (n.d. but 1935) *Reichssortenliste der zugelassenen Kulturarten. Nach dem Stand vom 1. Juni 1935*, Berlin, Reichsnährstand.

Anon. (1938a) *Ratgeber für Saatgutbeschaffung, Saatgutwechsel und Sortenwahl*, Berlin, Reichsverband der deutschen Pflanzenzuchtbetriebe.

Anon. (1938b) Richtlinien über die Abgabe von Pflanzenzuchtmaterial vom 5. März 1938. *Reichsministerialblatt der Landwirtschaftlichen Verwaltung*, 10, 174–175.

Anon. (1939) Umschau. *Der Forschungsdienst*, 8, 279–280.

Anon. (1940a) Richtlinien über die Abgabe vom Pflanzenzuchtmaterial vom 5. August 1939. *Reichsministerialblatt der Landwirtschaftlichen Verwaltung*, 47, 1193–1195.

Anon. (1940b) *Sortenliste. Liste der zugelassenen Sorten deutscher Kulturpflanzen*, Berlin, Reichsnährstand.

Anon. (1948) *Sortenliste der in der sowjetisch-besetzten Zone zugelassenen Sorten von Kulturpflanzen*, Berlin, Deutscher Zentralverlag.

Anon. (1949) *Sortenliste der im Bundesgebiet zugelassenen Sorten von landwirtschaftlichen Arten, Gemüsearten, Beerenobstarten*, Bonn, Sortenamt für Nutzpflanzen.

Anon. (1987) Der Sortenschutz in Deutschland. In Bundesverband Deutscher Pflanzenzüchter (ed.) *Landwirtschaftliche Pflanzenzüchtung in Deutschland. Geschichte, Gegenwart und Ausblick*, Gelsenkirchen-Buer, Verlag Th. Mann.

Anon. (2008) Fertilizer prices threaten decades of gains. *International Herald Tribune*, 30 April.

Anon. (2010) Gentechnik für die Welternährung. *Frankfurter Allgemeine Zeitung*, 10 July, p. 4.

Antholt, C. (1998) Agricultural extension in the 21st century. In Eicher, C. K. and Staatz, J. M. (eds) *International Agricultural Development*, Baltimore, Johns Hopkins University Press.

Appel, O. *et al.* (1932) Ergebnisse der Arbeiten der Getreidesorten-Registerkommission. *Der Züchter*, 4, 245–254.

Araghi, F. (2000) The great global enclosure of our times: peasants and the agrarian question at the end of the 20th century. In Magdoff, F., Foster, J. B. and Buttel, F. H. (eds) *Hungry for Profit: the Agribusiness Threat to Farmers, Food and the Environment*, New York, Monthly Review Press.

Arndt, T. M., Dalrymple, D. and Ruttan, V. W. (eds) (1977) *Resource Allocation and Productivity in National and International Agricultural Research*, Minneapolis, MN, University of Minnesota Press.

von Arnim, [Graf] (1925) Ein halbes Jahrhundert märkischer Weizenbau und Weizenzüchtung. *Illustrierte Landwirtschaftliche Zeitung*, 479–480.

von Arnim-Schlagenthin, Graf (1901) Die schwedische Saatzuchtanstalt Svalöf. *Deutsche Landwirtschaftliche Presse*, 28, 138–140.

Ashby, J. (1990) Small farmers' participation in the design of technologies. In Altieri, M. A. and Hecht, S. B. (eds) *Agroecology and Small Farm Development*, Boca Raton, CRC Press.

Ashley, C. and Maxwell, S. (2001) Rethinking rural development. *Development Policy Review*, 19, 395–425.

Atkinson, R. C. *et al.* (2003) Public sector collaboration for agricultural IP management. *Science*, 301, 174–175.

Aufhammer, G. (1970) Ferdinand von Lochow (1849–1924). In Franz, G. and Haushofer, H. (eds) *Grosse Landwirte*, Frankfurt a.M., DLG-Verlag.

Bach, W. (1914) Die amtliche Futterbauberatung in Baden, bisherige Tätigkeit und Erfolge. *Wochenblatt Baden*, 82, 269–272.

Backe, H. (1936a) Bauerntum und Vierjahresplan. *Archiv des Reichsnährstandes*, 4, 75–96.

Backe, H. (1936b) Rede anlässlich der ersten Sitzung des Forschungsrates. *Der Forschungsdienst*, 1, 3–8.

Backe, H. (ed.) (1943) *Gesetzgebung auf dem Gebiet der Ernährungswirtschaft*, Munich, C.H. Beck.

Bahiigwa, G., Rigby, D. and Woodhouse, P. (2005) Right target, wrong mechanism? Agricultural modernization and poverty reduction in Uganda. *World Development*, 33, 481–496.

Baier, T. (2009) Ein Reis für alle Fälle. *Süddeutsche Zeitung*, 21 April.

Baker, D. C. and Norman, D. W. (1990) The farming systems research and extension approach to small farm development. In Altieri, M. A. and Hecht, S. B. (eds) *Agroecology and Small Farm Development*, Boca Raton, CRC Press.

Ballwanz, I. (1978) Der Zusammenhang zwischen der Produktionsentwicklung und der Betriebsgrösse in der deutschen Landwirtschaft von 1871 bis 1914. *Jahrbuch für Wirtschaftsgeschichte*, 3, 77–99.

Balster, O. (1912) Einrichtung von Saatgut-Anbaustellen durch landwirtschaftliche Vereine. *Deutsche Landwirtschaftliche Presse*, 39, 631.

Barker, R. (1970) Green Revolution. *Current Affairs Bulletin*, 45, 66–79.

Barkin, D. (1987) The end of food self-sufficiency in Mexico. *Latin American Perspectives*, 14, 271–297.

Barkin, K. (1970) *The Controversy over German Industrialisation, 1890–1902*, Chicago and London, University of Chicago Press.

Barton, J. and Berger, P. (2001) Patenting agriculture. *Issues in Science and Technology*, online. Available at: www.issues.org/17.4/barton.htm.

Bauer, J. (1993) 100 Jahre Pflanzenzüchtung auf dem Gutshof Niedertraubling.... *Die Oberpfalz*, 81, 230–237.

Baumann, W. (1993) *Bauernstand und Bürgerblock. Ernst Laur und der Schweizerische Bauernverband 1897–1918*, Zürich, Orell Füssli.

Baur, E. (1914) Die Bedeutung der primitiven Kulturrassen und der wilden Verwandten unserer Kulturpflanzen für die Pflanzenzüchtung. *Jahrbuch der DLG*, 29, 104–110.

Baur, E. (1933) Private oder staatliche Pflanzenzucht? *Deutsche Tageszeitung*, 14 March.

Baur, G. (1928) Ziele der Futterpflanzenzüchtung. *Mitteilungen der DLG*, 43, 74–78.

Bayerisches Staatsministerium des Innern (1890) *Die Landwirtschaft in Bayern.* Munich, R. Oldenbourg.

Bayerisches Staatsministerium des Innern (ed.) (1897) *Die Massnahmen auf dem Gebiete der landwirt. Verwaltung in Bayern, 1890–1897.* Munich, R. Oldenbourg.

Bayerisches Staatsministerium des Innern (ed.) (1905) *Die Massnahmen auf dem Gebiete der landwirt. Verwaltung in Bayern, 1897–1903.*, Munich, R. Oldenbourg.

Bayliss-Smith, T. P. (1982) *The Ecology of Agricultural Systems*, Cambridge, Cambridge University Press.

Bayly, C. A., Rao, V., Szreter, S. and Woolcock, M. (eds) (2011) *History, Historians and Development Policy: a Necessary Dialogue*, Manchester, Manchester University Press.

Bundesverband Deutscher Pflanzenzüchter (ed.) (1987) *Landwirtschaftliche Pflanzenzüchtung in Deutschland. Geschichte, Gegenwart und Ausblick*, Gelsenkirchen, Th. Mann.

Becker-Dillingen, J. (1931) Abriss über die Leistungen der deutschen Zucker- und Futterrübenzüchter. *Der Züchter*, 3, 172–181.

Bello, W. (2009) *The Food Wars*, London, Verso.

Bellon, M. R. *et al.* (2006) Humanizing technology development in the Green Revolution's home. In Cernea, M. M. and Kassam, A. H. (eds) *Researching the Culture in Agri-Culture: Social Research for International Development.* Wallingford, CABI Publishing.

Benbrook, C. (2003) Sowing seeds of destruction. *New York Times*, 11 July, A17.

Bentley, J. W. (1994) Fact, fantasy and failures of farmers' participatory research. *Agriculture and Human Values*, 11, 140–150.

Berg, R. J. and Whitaker, J. S. (eds) (1986) *Strategies for African Development*, Berkeley and London, University of California Press.

Berghahn, V. (1987) *Modern Germany: Society, Economy and Politics in the 20th Century*, Cambridge, Cambridge University Press.

Bergmann, H. (1986) *Der Bayerische Bauernbund und der Bayerische Christliche Bauernverein, 1919–1928*, Munich, C.H. Beck.

Bernstein, H. (2006) Studying development/development studies. *African Studies*, 65, 45–62.

Berthold, R. (ed.) (1988) *Produktivkräfte in Deutschland. 1917/18 bis 1945*, Berlin, Akademie-Verlag.

Beseler, O. (1909) Erfahrungen in der Getreidezüchtung. *Jahrbuch der DLG*, 24, 189–196.

Beseler, O. (1911) Die landwirtschaftliche Pflanzenzüchtung in Deutschland in den letzten 25 Jahren. *Jahrbuch der DLG*, 26, 89–101.

Bessel, R. (1989) Making sense of the German countryside: some recent writing on rural life and politics in Germany. *European History Quarterly*, 19, 115–128.

Beusekom, M. V. (2000) Disjunctures in theory and practice: making sense of change in agricultural development at the Office du Niger, 1920–1960. *Journal of African History*, 41, 79–99.

Biggs, S. (1980) The failure of farmers to adopt new technological 'packages' entirely may be a sign of creativity rather than backwardness. *Ceres*, 13, 23–26.

Biggs, S. (2007) Reflections on the social embeddedness of science and technology in rural and agricultural transformations: Learning from positive experiences of poverty reduction and social inclusion in Nepal. *Studies in Nepali History and Society*, 12, 251–282.

Biggs, S. (2008) The lost 1990s? Personal reflections on a history of participatory technology development. *Development in Practice*, 18, 489–505.

Biggs, S. and Farrington, J. (1991) *Agricultural Research and the Rural Poor: a Review of Social Science Analysis*, Ottawa, International Development Research Centre.

Biggs, S. and Gauchan, D. (2001) The broader institutional context of participatory plant breeding in the changing agricultural and natural resources R&D system in Nepal. In Anon. (ed.) *An Exchange of Experiences from South and Southeast Asia: Proceedings of an International Symposium on Participatory Plant Breeding and Participatory Plant Genetic Resource Enhancement.* Cali, Colombia, CGIAR.

Biggs, S., Justice, S. and Lewis, D. (2011) Patterns of rural mechanization, energy and employment in South Asia: Reopening the Debate. *Economic & Political Weekly*, 46, 78–82.

Birner, R. and Resnick, D. (2010) The political economy of policies for smallholder agriculture. *World Development*, 38, 1442–52.

Bischoff, A. D. (1922) Saatzuchtgenossenschaften. *Journal für Landwirtschaft*, 70, 81–100.

Bittermann, E. (1956) Die landwirtschaftliche Produktion in Deutschland, 1800–1950. *Kühn-Archiv*, 70, 1–149.

Blackbourn, D. (1984) Peasants and politics in Germany, 1871–1914. *European History Quarterly*, 14, 47–75.
Blackbourn, D. (1997) *The Fontana History of Germany, 1780–1918: the Long Nineteenth Century*, London, Fontana.
Blackbourn, D. and Eley, G. (1984) *The Peculiarities of German History*, Oxford, Oxford University Press.
Blyn, G. (1983) The Green Revolution revisited. *Economic Development and Cultural Change*, 31, 705–725.
Boelcke, W. (1995) Ueber die säkulare Strukturentwicklung der klein- und mittelbäuerlichen Landwirtschaft in Deutschland während des 19./20. Jahrhunderts. In Anon. (ed.) *Entwicklungstendenzen in der agrargeschichtlichen Lehre und Forschung.* Berlin, Institut für Agrarpolitik, Marktlehre u. Agrarentwicklung, Humboldt-Universität zu Berlin, and Fördergesellschaft Albrecht Daniel Thaer.
Böhmer, G. (1914) Die Entwicklung der Sortenfrage und ihre Lösung durch Sortenprüfung. *Kühn-Archiv*, 5, 191–206.
Bohne, W. (1924) Wirtschaft und Wissenschaft. *Beiträge zur Pflanzenzucht*, 7, 96–101.
Bohne, W. (1928) Organisation und Aufgaben der deutschen Pflanzenzüchtung. *Mitteilungen der DLG*, 43, 1080–1082.
Bonjean, A. P. and Angus, W. J. (eds) (2001) *The World Wheat Book: a History of Wheat Breeding*, London, Intercept/Editions TEC & DOC/Lavoisier.
Bonneuil, C. (1999) Penetrating the natives: peanut-breeding, peasants and the colonial state in Senegal (1900–1950). *Science, Technology, and Society*, 4, 273–302.
Bonneuil, C. (2001) Development as experiment: Science and state building in late colonial and postcolonial Africa, 1930–1970. *Osiris*, 15, 258–281.
Bonneuil, C. (2006) Mendelism, plant-breeding and experimental cultures: agriculture and the development of genetics in France. *Journal of the History of Biology*, 39, 281–308.
Bonneuil, C. and Thomas, F. (2009) *Gènes, Pouvoirs et Profits: Recherche publique et régimes de production des savoirs de Mendel aux OGM*, Versailles and Lausanne, Editions Quæ and Fondation pour le Progrès de l'Homme.
Bonneuil, C. and Thomas, F. (2010) Purifying landscapes: the Vichy regime and the genetic modernization of France. *Historical Studies in the Natural Sciences*, 40, 532–568.
Borlaug, N. (2000) Ending world hunger. The promise of biotechnology and the threat of antiscience zealotry. *Plant Physiology Online*, 124, 487–490.
Borscheid, P. (1976) *Naturwissenschaft, Staat und Industrie in Baden (1848–1914)*, Stuttgart, Ernst Klett Verlag.
Bose, S. (1997) Instruments and idioms of colonial and national development: India's historical experience in comparative perspective. In Cooper, F. and Packard, R. (eds) *International Development and the Social Sciences: Essays on the History and Politics of Knowledge.* Berkeley, University of California Press.
Bowman, A. (2011) *Disruptive Technologies, Divided Experts: Scientific Knowledge, Development and the Problem of Stabilising Change, Zambia 1945–2010* (Ph.D. diss., Centre for the History of Science, Technology & Medicine, University of Manchester).
Brassley, P. (2000) Agricultural science and education. In Collins, E. J. T. (ed.) *The Agrarian History of England and Wales; vol. 7 (part 1), 1850–1914.* Cambridge, Cambridge University Press.
Bray, F. (1979) Green revolution: a new perspective. *Modern Asian Studies*, 13, 681–688.

Bray, F. (1986) *The Rice Economies: Technology and Development in Asian Societies*, Berkeley, University of California Press.

Bray, F. (1998) Technics and civilization in late Imperial China: an essay in the cultural history of technology. *Osiris*, 13, 11–33.

Broili, J. (1899) Welche Sorten von unseren Getreidearten werden jetzt in Süddeutschland angebaut...? *Jahrbuch der DLG*, 14, 214–219.

Broili, J. (1907) Ein Beitrag zu den 'Gedanken über die Gefahren und Nachteile des modernen Pflanzenzuchtbetriebes' in d. Nr. 87... *Deutsche Landwirtschaftliche Presse*, 34, 732–733.

Bruckmüller, E. (1977) *Landwirtschaftliche Organisationen und gesellschaftliche Modernisierung. Vereine, Genossenschaften und politische Mobilisierung der Landwirtschaft Österreichs vom Vormärz bis 1914*, Salzburg, Wolfgang Neugebauer.

Bruckmüller, E. (1984) Strukturwandel der österreichischen Landwirtschaftgesellschaften im 19. Jahrhundert. *Zeitschrift fur Agrargeschichte und Agrarsoziologie*, 32, 1–30.

Brummenbaum, A. (1936) Unsere Arbeit am Hof. *Archiv des Reichsnährstandes*, 4, 123–135.

Buchinger, A. (1935) Wie sucht das Ausland praktisch wichtige Fragen auf dem Gebiete der Pflanzenzüchtung zu lösen? *Zeitschrift für Züchtung, Reihe A*, 20, 503–513.

Buchsteiner, I. (1995) Von der Elitegesellschaft zur sozial orientierten Agrarorganisation. Gründung und Entwicklung des Mecklenburgischen Patriotischen Vereins. In Institut Für Agrarpolitik, Marktlehre u. Agrarentwicklung d. Humboldt-Univ. Zu Berlin. (ed.) *Entwicklungstendenzen in der agrargeschichtlichen Lehre und Forschung (Festschrift für Prof. Dr. habil. Volker Klemm).* Berlin, Fachgebiet Sozialgeschichte der Agrarentwicklung der Humboldt-Universität zu Berlin and Fördergesell. Albrecht Daniel Thaer.

Bundesverband Deutscher Pflanzenzüchter (ed.) (1987) *Landwirtschaftliche Pflanzenzüchtung in Deutschland. Geschichte, Gegenwart und Ausblick.* Gelsenkirchen-Buer, Verlag Th. Mann.

Busch, L. and Lacey, W. (1983) *Science, Agriculture, and the Politics of Research*, Boulder, Westview Press.

Busch, L. *et al.* (1991) *Plants, Power and Profit: Social, Economic and Ethical Consequences of the New Biotechnologies*, Cambridge, MA/Oxford, Blackwell.

Büschel, H. and Speich, D. (2009) Einleitung – Konjunkturen, Probleme und Perspektiven der Globalgeschichte von Entwicklungszusammenarbeit. In Büschel, H. and Speich, D. (eds) *Entwicklungswelten. Globalgeschichte der Entwicklungszusammenarbeit.* Frankfurt a.M., Campus Verlag.

Buss, H. (1925) *Stand und Bedeutung der Maiszuchten in Deutschland*, Karlsruhe, G. Braun.

Buttel, F. H. (2003) The global politics of GEOs: the Achilles' heel of the globalization regime? In Schurman, R. A., Kelso, D., and Doyle, T. (eds) *Engineering Trouble: Biotechnology and its Discontents.* Berkeley, University of California Press.

Buttel, F. and Busch, L. (1988) The public agricultural research system at the crossroads. *Agricultural History*, 62, 303–324.

Buttel, F., Kenney, M. and Kloppenburg, J. (1985) From green revolution to biorevolution: some observations on the changing technological bases of economic transformation in the Third World. *Economic Development and Cultural Change*, 34, 31–55.

Büttner, W. (1954) *Die Saatgutordnung. Eine Einführung in das Saaten- und Sortenrecht*, Hannover, Alfred Strothe.

Byerlee, D. (1998) The search for a new paradigm for the development of national agricultural research systems. *World Development*, 26, 1049–1055.

Byerlee, D. and Fischer, K. (2002) Assessing modern science: Policy and institutional options for agricultural biotechnology in developing countries. *World Development*, 30, 931–948.

Byres, T. J. (1972) The dialectic of India's green revolution. *South Asian Review*, 5, 99–116.

Cameron, J. (2005) Journeying in radical development studies: a reflection on thirty years of researching pro-poor development. In Kothari, U. (ed.) *A Radical History of Development Studies: Individuals, Institutions and Ideologies.* London, Zed.

Camprubi, L. (2010) One grain, one nation: Rice genetics and the corporate state in early Francoist Spain (1939–1952). *Historical Studies in the Natural Sciences*, 40, 499–531.

Cantrell, R. and Reeves, T. (2002) The cereal of the world's poor takes center stage. *Science*, 296, 53.

Carlson, L. W. (2005) *William J. Spillman and the Birth of Agricultural Economics*, Columbia, Missouri, University of Missouri Press.

Carney, D. (1995) The changing public role in services to agriculture: a framework for analysis. *Food Policy*, 20, 521–528.

Carr, C. and Myers, R. H. (1973) The agricultural transformation of Taiwan: the case of Ponlai rice, 1922–42. In Shand, R. T. (ed.) *Technical Change in Asian Agriculture.* Canberra, Australian National University Press.

Ceccarelli, S. (1989) Wide adaptation: how wide? *Euphytica*, 40, 197–205.

Ceccarelli, S. (2009) Evolution, plant breeding and biodiversity. *Journal of Agriculture and Environment for International Development*, 103, 131–145.

Ceccarelli, S. (n.d.) Decentralized-participatory plant breeding: adapting crops to environments and clients. Int. Center for Agricultural Research in the Dry Areas, Aleppo/Syria.

Ceccarelli, S. *et al.* (1994) Genotype by environment interaction and international breeding programmes. *Experimental Agriculture*, 30, 177–187.

Cernea, M. M. and Kassam, A. H. (2006) Stock taking and new challenges to social research; editors' preface. In Cernea, M. M. and Kassam, A. H. (eds) *Researching the Culture in Agri-Culture: Social Research for International Development.* Wallingford, CABI Publishing.

Chambers, R. (1977) Challenges for rural research and development. In Farmer, B. H. (ed.) *Green Revolution? Technology and Change in Rice-Growing Areas of Tamil Nadu and Sri Lanka.* London, Macmillan.

Chambers, R. (1983) *Rural Development: Putting the Last First*, London, Longman.

Chambers, R. (1984) Beyond the Green Revolution: a selective essay. In Bayliss-Smith, T. P. and Wanmali, S. (eds) *Understanding Green Revolutions: Agrarian Change and Development Planning in South Asia.* Cambridge, Cambridge University Press.

Chambers, R. (1989) Reversals, institutions and change. In Chambers, R., Pacey, A. and Thrupp, L. A. (eds) *Farmer First: Farmer Innovation and Agricultural Research.* London, Intermediate Technology Publications.

Chambers, R. (1990) Farmer-first: a practical paradigm for the third agriculture. In Altieri, M. A. and Hecht, S. B. (eds) *Agroecology and Small Farm Development.* Boca Raton, CRC Press.

Chambers, R. (2005) Critical reflections of a development nomad. In Kothari, U. (ed.) *A Radical History of Development Studies: Individuals, Institutions and Ideologies.* London, Zed.

Chambers, R. and Farmer, B. H. (1977) Perceptions, technology and the future. In Farmer, B. H. (ed.) *Green Revolution? Technology and Change in Rice-Growing Areas of Tamil Nadu and Sri Lanka.* London, Macmillan.

Chambers, R., Pacey, A. and Thrupp, L. A. (eds) (1989) *Farmer First: Farmer Innovation and Agricultural Research*, London, Intermediate Technology Publications.

Chambers, R. and Wickremanayake, B. W. E. (1977) Agricultural extension: myth, reality and challenge. In Farmer, B. H. (ed.) *Green Revolution? Technology and Change in Rice-Growing Areas of Tamil Nadu and Sri Lanka.* London, Macmillan.

Chang, H.-J. (2002) *Kicking Away the Ladder: Development Strategy in Historical Perspective*, London, Anthem.

Chang, H.-J. (2009) Rethinking public policy in agriculture: lessons from history, distant and recent. *Journal of Peasant Studies*, 36, 477–515.

Chataway, J. (2005) Introduction: Is it possible to create pro-poor agriculture-related biotechnology? *Journal of International Development*, 17, 597–610.

Chrispeels, M. (2000) Biotechnology and the poor. *Plant Physiology*, 124, 3–6.

Clarke, S. (2005) *Experts, Empire and the Development of Fundamental Research for the British Colonies, 1940–1960* (Ph.D. diss., Imperial College London).

Cleveland, D. (2006) What kind of social science does the CGIAR, and the world, need? *Culture and Agriculture*, 28, 4–9.

Cleveland, D. and Soleri, D. (2002) Introduction: Farmers, scientists and plant breeding: knowledge, practice and the possibilities for collaboration. In Cleveland, D. and Soleri, D. (eds) *Farmers, Scientists and Plant Breeding.* Wallingford, CAB International.

Cleveland, D. and Soleri, D. (2007) Extending Darwin's analogy: bridging differences in concepts of selection between farmers, biologists and plant breeders. *Economic Botany*, 61, 121–136.

Cochrane, W. (1979) *The Development of American Agriculture*, Minneapolis, University of Minnesota Press.

Coffman, W. R. (1983) Plant research and technology. In Rosenblum, J. W. (ed.) *Agriculture in the 21st Century.* New York, John Wiley.

Cohen, J. (2005) Poorer nations turn to publicly developed GM crops. *Nature Biotechnology*, 23, 27–33.

Collier, P. (2009) Africa's organic peasantry: beyond romanticism. *Harvard International Review*, 31, 62–65.

Collier, P. and Dercon, S. (2009) African agriculture in 50 years: smallholders in a rapidly changing world? *How to Feed the World in 2050.* Rome, Economic and Social Development Department, FAO.

Collinson, M. (2000) My FSR origins. In Collinson, M. (ed.) *A History of Farming Systems Research.* Wallingford and Rome, CABI Publishing and FAO.

Collinson, M. and Lightfoot, C. (2000) The future of Farming Systems Research. In Collinson, M. (ed.) *A History of Farming Systems Research.* Wallingford and Rome, CABI Publishing and FAO.

Combes, T. (2006) Response: There's no chance that this technology will replace GM. *The Guardian.* London. 7 November.

Compton, J. L. (ed.) (1989a) *The Transformation of International Agricultural Research and Development*, Boulder, Lynne Rienner.

Compton, J. L. (1989b) The integration of research and extension. In Compton, J. L. (ed.) *The Transformation of International Agricultural Research and Development.* Boulder, Lynne Rienner.

Conroy, C. (2009) Use and institutionalisation of process innovations: participatory crop improvement processes in India and Nepal. Unpublished manuscript, University of Greenwich.

Conway, G. (1997) *The Doubly Green Revolution: Food for All in the 21st Century*, London, Penguin.

Conway, G. (1999) The Rockefeller Foundation and plant biotechnology, online. Available at: http://biotech-info.net/gordon_conway.html (Accessed: 1 March 2011).

Conway, G. (2003) Biotechnology and hunger. Lecture to House of Lords, London, 8 May, online. Available at: www.rockfound.org.

Conway, G. and Barbier, E. (1990) *After the Green Revolution: Sustainable Agriculture for Development*, London, Earthscan.

Conway, G. and Toenniessen, G. (2003) Science for African food security. *Science*, 299, 1187–88.

Conway, G. *et al.* (2006) The Rockefeller Foundation and social research in agriculture. In Cernea, M. M. and Kassam, A. H. (eds) *Researching the Culture in Agri-Culture: Social Research for International Development.* Wallingford, CABI Publishing.

Conway, G., and Waage, J. K. with Delaney, S. (2010) *Science and Innovation for Development*, London, UK Collaborative on Development Sciences.

Cooper, F. (1997) Modernizing bureaucrats, backward Africans, and the development concept. In Cooper, F. and Packard, R. (eds) *International Development and the Social Sciences: Essays on the History and Politics of Knowledge.* Berkeley, University of California Press.

Cooper, F. (2005) *Colonialism in Question: Theory, Knowledge, History*, Berkeley, University of California Press.

Cooper, F. (2010) Writing the history of development. *Journal of Modern European History*, 8, 5–23.

Cooper, F. and Packard, R. (1997) Introduction. In Cooper, F. and Packard, R. (eds) *International Development and the Social Sciences: Essays on the History and Politics of Knowledge.* Berkeley, University of California Press.

Corni, G. (1990) *Hitler and the Peasants: Agrarian Policy of the Third Reich, 1930–1939*, New York, Berg.

Corni, G. and Gies, H. (1994) *'Blut und Boden': Rassenideologie und Agrarpolitik im Staat Hitlers*, Idstein, Schulz-Kirchner Verlag.

Corni, G. and Gies, H. (1997) *Brot – Butter – Kanonen. Die Ernährungswirtschaft in Deutschland unter der Diktatur Hitlers*, Berlin, Akademie Verlag.

Cotter, J. (1994) The Rockefeller Foundation's Mexican Agricultural Project: a cross-cultural encounter, 1943–1949. In Cueto, M. (ed.) *Missionaries of Science: the Rockefeller Foundation and Latin America.* Bloomington, Indiana University Press.

Cotter, J. (2003) *Troubled Harvest: Agronomy and Revolution in Mexico, 1880–2002*, Westport, CN, Praeger.

Cowan, M. and Shenton, R. (1996) *Doctrines of Development*, London, Routledge.

Crone-Münzebrock, A. (1912) Landwirtschaftliches Vereinswesen und Landwirtschaftskammern. In von Kerckerinck Zur Borg, H. (ed.) *Beiträge zur Geschichte des westfälischen Bauernstandes.*

Crouch, M. L. (2001) From golden rice to terminator technology: Agricultural biotechnology will not feed the world or save the environment. In Tokar, B. (ed.) *Redesigning Life? The Worldwide Challenge to Genetic Engineering.* Montreal and Kingston, McGill-Queen's University Press.

Crush, J. (1995) Introduction: Imagining development. In Crush, J. (ed.) *Power of Development.* London, Routledge.

Cullather, N. (2000) Development? It's history. *Diplomatic History*, 24, 641–653.

Cullather, N. (2003) Endless villages and rat cities: Imagining the food-population race at the Rockefeller Foundation. Paper to conference *Foundations of Globalisation*, University of Manchester, November 2003.

Cullather, N. (2004) Miracles of modernization: the Green Revolution and the apotheosis of technology. *Diplomatic History*, 28, 227–254.

Cullather, N. (2007) The foreign policy of the calorie. *American Historical Review*, 112, 337–364.

Dade, H. (1913) Die Gesamtentwicklung der deutschen Landwirtschaft 1888–1913. In Dade, H. (ed.) *Die Deutsche Landwirtschaft unter Kaiser Wilhelm II, vol. 2*. Halle, Carl Marhold.

Dahlberg, K. (1979) *Beyond the Green Revolution: The Ecology and Politics of Global Agricultural Development*, New York/London, Plenum.

Dalrymple, D. (1969) *Technological Change in Agriculture: Effects and Implications for the Developing Nations*, Washington, DC, Foreign Agricultural Service of the U.S. Dept. of Agriculture and Agency for International Development.

Dalrymple, D. (1977) Evaluating the impact of international research on wheat and rice production in the developing nations. In Arndt, T. M., Dalrymple, D. and Ruttan, V. W. (eds) *Resource Allocation and Productivity in National and International Agricultural Research*. Minneapolis, Minnesota University Press.

Dalrymple, D. (1979) The adoption of high-yielding grain varieties in developing nations. *Agricultural History*, 53, 704–726.

Dalrymple, D. (1985) The development and adoption of high-yielding varieties of wheat and rice in developing countries. *American Journal of Agricultural Economics*, 67, 1067–1073.

Dalrymple, D. (1986) *Development and Spread of High-Yielding Rice Varieties in Developing Countries*, Washington, DC, U.S. Agency for International Development.

Dalrymple, D. (2003) Scientific knowledge as a global public good: contributions to innovation and the economy. In Esanu, J. and Uhlir, P. (eds) *The Role of Scientific and Technical Data and Information in the Public Domain: Proceedings of a Symposium*. Washington, DC, The National Academies Press.

Dalrymple, D. (2006) Setting the agenda for science and technology in the public sector: the case of international agricultural research. *Science and Public Policy*, 33, 277–290.

Dalrymple, D. (2008) International agricultural research as a global public good: concepts, the CGIAR experience, and policy issues. *Journal of International Development*, 20, 347–379.

Dalrymple, D. and Jones, W. I. (1973) Evaluating the 'Green Revolution'. Paper for session on 'Nutrition and New Food Technology', joint meeting of American Assoc. for Advancement of Science and Consejo Nacional de Ciencia y Tecnología. Mexico City, 20 June.

Danbom, D. (1979) *The Resisted Revolution: Urban America and the Industrialization of Agriculture, 1900–1930*, Ames, Iowa State University Press.

Danbom, D. (1989) Politics, science and the changing nature of research at the North Dakota Agricultural Experiment Station, 1900–1930. *North Dakota History*, 56, 16–29.

Danbom, D. (1990) *'Our Purpose is to Serve': the First Century of the North Dakota Agricultural Experiment Station*, Fargo, North Dakota Institute for Regional Studies.

Daniel, P. (1985) *Breaking the Land: the Transformation of Cotton, Tobacco and Rice Cultures since 1880*, Urbana, IL, University of Illinois Press.

Darré, R. W. (1936) Ueber die Aufgaben der Landbauwissenschaft. *Der Forschungsdienst*, 1, 247–249.
Darré, R. W. (1938) Die Erzeugungsschlacht 1938. *Der Forschungsdienst*, 5, 1–7.
Davis, M. (2006) *Planet of Slums*, London, Verso.
von Dietze, C. (1934) Agrarpolitik im deutschen Reich. Anfang 1933 bis Mitte 1934. *Jahrbücher für Nationalökonomie und Statistik*, 140, 427–459.
Dipper, C. (1987) Bauern als Gegenstand der Sozialgeschichte. In Schieder, W. and Sellin, V. (eds) *Sozialgeschichte in Deutschland....Bd. 4 – Soziale Gruppen in der Geschichte*. Göttingen, Vandenhoeck & Ruprecht.
Dipper, C. (1993) Landwirtschaft im Wandel: Neue Perspektiven der preussisch-deutschen Agrargeschichte im 19. Jahrhundert. *Neue politische Literatur*, 38, 29–42.
Dix, A. and Langthaler, E. (eds) (2006) *Grüne Revolutionen. Agrarsysteme und Umwelt im 19. und 20. Jahrhundert (= Jahrbuch für Geschichte des ländlichen Raumes, vol. 3)*, Innsbrück, StudienVerlag.
Dix, W. (1911) Forderung des staatlichen Eingreifens zum Schutz des Saatguthandels und zur Förderung der Pflanzenzüchtung. *Illustrierte Landwirtschaftliche Zeitung*, 31, 102–103.
Djurfeldt, G. *et al.* (eds) (2005) *The African Food Crisis: Lessons from the Asian Green Revolution*, Wallingford, CABI Publishing.
Deutsche Landwirt.-Gesellschaft (1907) Der schwedische Saatzuchtverein Svalöf. *Illustrierte Landwirtschaftliche Zeitung*, 27, 332–334 and 340–342.
Dolezal, E. (1931) Hofrat Prof.Dr. agrar., Dr. agrar. h.c. Carl Fruwirth. Ein Lebensbild. *Oesterreichische Zeitschrift für Vermessungswesen*, 1–8.
Donald, C. M. (1968) The breeding of crop ideotypes. *Euphytica*, 17, 385–403.
Dore, R. P. (1960) Agricultural improvement in Japan, 1870–1900. *Economic Development and Cultural Change*, 9, 69–91.
Dorward, A. *et al.* (2004) A policy agenda for pro-poor agricultural growth. *World Development*, 32, 73–89.
Doyle, J. (1985) *Altered Harvest: Agriculture, Genetics and the Fate of the World's Food Supply*, New York, Viking.
Duvick, D. (1996) Plant breeding, an evolutionary concept. *Crop Science*, 36, 539–548.
Duvick, D. (2002) Theory, empiricism and intuition in professional plant breeding. In Cleveland, D. A. and Soleri, D. (eds) *Farmers, Scientists and Plant Breeding: Integrating Knowledge and Practice*. Wallingford, CAB International.
Eastman, C. and Grieshop, J. (1989) Technology development and diffusion: potatoes in Peru. In Compton, J. L. (ed.) *The Transformation of International Agricultural Research and Development*. Boulder, Lynne Rienner.
Edler, W. (1898) Die Besichtigung der Wirtschaften der Zucht- und Verkaufs-Genossenschaft für Pirnaer Saatroggen. *Jahrbuch der DLG*, 13, (370)-(386).
Edler, W. (1908) Deutsche Gesellschaft zur Förderung der Pflanzenzucht. *Fühlings Landwirtschaftliche Zeitung*, 57, 225–228.
Edler, W. (1913) Der gegenwärtige Stand der landwirtschaftlichen Pflanzenzüchtung in Deutschland. *Internationale Agrartechnische Rundschau*, 3–10.
Eicher, C. (1995) Zimbabwe's maize-based green revolution: preconditions for replication. *World Development*, 23, 805–818.
Elliott, L. (2011) World Bank warns of threat from soaring food prices. *Guardian*, 16 February, 23.
Elofson, A. (1905) Mitteilungen über die Pflanzenzüchtungsarbeiten des schwedischen Saatbauvereins in Svalöf. *Bericht der Züricher botanischen Gesellschaft*, 25–41.

Entrup, E. L. (2008) Futter- und Rasengräser, Gramineae. Eine breite Artenpalette für die verschiedensten Zwecke. In Röbbelen, G. (ed.) *Die Entwicklung der Pflanzenzüchtung in Deutschland (1908–2008). Hundert Jahre GFP, e.V. – eine Dokumentation.* Göttingen, Gesellschaft für Pflanzenzüchtung.

Esteva, G. (2010) Development. In Sachs, W. (ed.) *The Development Dictionary: a Guide to Knowledge as Power.* London, Zed.

Evans, R. J. and Lee, W. R. (eds) (1986) *The German Peasantry: Conflict and community in Rural Society from the 18th to the 20th Centuries*, London, Croom Helm.

Evenson, R. E. and Gollin, D. (2003) Assessing the impact of the Green Revolution, 1960 to 2000. *Science*, 300, 758–762.

Evers, K. H. (1924) *Die Organisation der Pflanzenzucht und des Saatbaus in der deutschen Landwirtschaft*, Berlin, Walter de Gruyter.

Fahlbusch, L. and Hartwig, E. (1984) Vereinigung der deutschen Bauernvereine (VdB) 1900–1934. In Fricke, D. *et al.* (eds) *Lexikon zur Parteiengeschichte. Die bürgerlichen Parteien und Verbände in Deutschland (1789–1945).* Leipzig, VEB Bibliographisches Institut.

Falcon, W. P. (1970) The Green Revolution: generations of problems. *American Journal of Agricultural Economics*, 52, 698–712.

Falcon, W. P. and Fowler, C. (2002) Carving up the commons – emergence of a new international regime for germplasm development and transfer. *Food Policy*, 27, 197–222.

Falcon, W. P. and Naylor, R. L. (2005) Rethinking food security for the 21st century. *American Journal of Agricultural Economics*, 87, 1113–1127.

Fan, S. *et al.* (2005) National and international agricultural research and rural poverty: the case of rice research in India and China. *Agricultural Economics*, 33, 369–379.

Fan, S., Zhang, L. and Zhang, X. (2002) *Growth, Inequality and Poverty in Rural China: the Role of Public Investments*, Washington, DC, IFPRI.

FAO (1966) *Agricultural Development in Modern Japan: the Significance of the Japanese Experience (Agricultural Planning Studies no. 6)*, Rome, Food and Agricultural Organization of the U.N.

FAO (2004) *The State of Food and Agriculture, 2003–04: Agricultural Biotechnology – Meeting the Needs of the Poor?*, Rome, Food and Agricultural Organization of the U.N.

FAO (2009a) *The State of Food Insecurity in the World 2009: Economic Crises – Impacts and Lessons Learned*, Rome, Food & Agriculture Organization of the U.N.

FAO (2009b) How to Feed the World in 2050 (briefing paper for High-Level Expert Forum, 12–13 October 2009). Rome, Food and Agricultural Organization of the U.N.

FAO (2009c) The technology challenge (issues brief for the High-Level Expert Forum 'How to Feed the World in 2050', 12–13 October). Rome, Food and Agricultural Organization of the U.N.

Farmer, B. H. (ed.) (1977) *Green Revolution? Technology and Change in Rice-Growing Areas of Tamil Nadu and Sri Lanka*, London, Macmillan.

Farmer, B. H. (1979) The 'green revolution' in South Asian rice fields: environment and production. *Journal of Development Studies*, 15, 304–319.

Farmer, B. H. (1986) Perspectives on the green revolution in South Asia. *Modern Asian Studies*, 20, 175–199.

Farquharson, J. E. (1976) *The Plough and the Swastika: the NSDAP and Agriculture in Germany, 1928–1945*, London, Sage.

Farquharson, J. E. (1986) The agrarian policy of National Socialist Germany. In Moeller, R. G. (ed.) *Peasants and Lords in Modern German History: Recent Studies in Agricultural History.* London, Allen & Unwin.

Farr, I. (1978) Populism in the countryside: the Peasant Leagues in Bavaria in the 1890s. In Evans, R. J. (ed.) *Society and Politics in Wilhelminian Germany.* London, Croom Helm.

Farr, I. (1986) Tradition and the peasantry: on the modern historiography of rural Germany. In Evans, R. and Lee, W. R. (eds) *The German Peasantry.* London, Croom Helm.

Farrington, J. (1988) Farmer participatory research: editorial introduction. *Experimental Agriculture*, 24, 269–279.

Farrington, J. and Bebbington, A. (1994) From research to innovation: getting the most from interaction with NGOs. In Scoones, I. and Thompson, J. (eds) *Beyond Farmer First: Rural People's Knowledge, Agricultural Research and Extension Practice.* London, Intermediate Technology Publications.

Fehse, B. (1980) Die Entwicklung der Zuckerrübenzüchtung von 1919–1939 in Deutschland. Diplom diss., Humboldt-Universität zu Berlin.

Feldman, S. and Biggs, S. (2012) The politics of international assessments: the IAASTD process, *Journal of Agrarian Change*.

Feldman, S., Biggs, S. and Raina, R. (2010) A messy confrontation of a crisis in agricultural science. *Economic and Political Weekly*, 45, 66–71.

Fensch (1936) Betriebsgrössen, Erzeugungsschlacht und Ertragsreserven. *Archiv des Reichsnährstandes*, 4, 303–320.

Ferguson, J. (1990) *The Anti-Politics Machine: 'Development', Depoliticization and Bureaucratic Power in Lesotho*, Cambridge, Cambridge University Press.

Ferleger, L. and Lazonick, W. (1993) The managerial revolution and the developmental state: the case of U.S. agriculture. *Business and Economic History*, 22, 67–98.

Finlay, M. (1992) *Science, Practice and Politics: German Agricultural Experiment Stations in the 19th Century*. Ph.D. diss., Iowa State University.

Finlay, M. (2001) New sources, new theses, and new organizations in the new Germany: recent research on the history of German agriculture. *Agricultural History*, 75, 279–307.

Finnemore, M. (1997) Redefining Development at the World Bank. In Cooper, F. and Packard, R. (eds) *International Development and the Social Sciences: Essays on the History and Politics of Knowledge.* Berkeley, University of California Press.

Fischbeck, G. (1988) Pflanzenzüchtung in Bayern – Zukunftsperspektiven für mittelständische Pflanzenzuchtbetriebe. In Verband Bayerischer Pflanzenzüchter, e.V. (ed.) *75 Jahre Verband Bayerischer Pflanzenzüchter e.V. (Festschrift).* Munich, Verband Bayerischer Pflanzenzüchter e.V.

Fischbeck, G. (2004) Private und staatliche Pflanzenzüchtung in Süddeutschland während der NS-Zeit. *Vorträge für Pflanzenzüchtung*, 65, 61–67.

Fite, G. C. (1981) *American Farmers: the New Minority*, Bloomington, Indiana University Press.

Fitzgerald, D. (1986) Exporting American agriculture: the Rockefeller Foundation in Mexico, 1943–53. *Social Studies of Science*, 16, 457–483.

Fitzgerald, D. (1990) *The Business of Breeding: Hybrid Corn in Illinois, 1890–1940*, Ithaca, Cornell University Press.

Flemming, J. (1978) *Landwirtschaftliche Interessen und Demokratie. Ländliche Gesellschaft, Agrarverbände und Staat 1890–1925*, Bonn, Verlag Neue Gesellschaft.

Fletcher, T. W. (1973) The great depression of English agriculture, 1873–1896. In Perry, P. J. (ed.) *British Agriculture, 1875–1914.* London, Methuen.

Flitner, M. (1995) *Sammler, Räuber und Gelehrte. Die politischen Interessen an pflanzengenetischen Ressourcen 1895–1995*, Frankfurt a.M., Campus.

Flora, C. B. and Flora, J. L. (1989) An historical perspective on institutional transfer. In Compton, J. L. (ed.) *The Transformation of International Agricultural Research and Development.* Boulder, Lynne Rienner.

Föhr, S. (1999) *Modernization for the Honor of the Estate: the Deutsche Landwirtschafts-Gesellschaft, 1884–1914.* Ph.D. diss., Georgetown University.

Fowler, C. (1994) *Unnatural Selection: Technology, Politics and Plant Evolution*, Yverdon, Switz., Gordon and Breach.

Fowler, C. and Lower, R. (2005) The politics of plant breeding. *Plant Breeding Reviews*, 25, 21–55.

Francis, C. A. (1990) Potential of multiple cropping systems. In Altieri, M. A. and Hecht, S. B. (eds) *Agroecology and Small Farm Development.* Boca Raton, CRC Press.

Franck, S. (2007) Wer züchtet in Zukunft noch Getreide? Paper for conference *80th Anniversary of I.G. Pflanzenzucht.* Würzburg, 15 June.

Francks, P. (1984) *Technology and Agricultural Development in Pre-War Japan*, New Haven, Yale University Press.

Frankel, F. (1971) *India's Green Revolution: Economic Gains and Political Costs*, Princeton, Princeton University Press.

Franz, G. (ed.) (1963) *Quellen zur Geschichte des Deutschen Bauernstandes in der Neuzeit*, Munich, R. Oldenbourg.

von Frauendorfer, S. (1957) *Ideengeschichte der Agrarwirtschaft und Agrarpolitik im deutschen Sprachgebiet. Teil 1 – von den Anfangen bis zum ersten Weltkrieg*, Munich, Bayerischer Landwirtschafts-Verlag.

Freebairn, D. (1995) Did the Green Revolution concentrate incomes? A quantitative study of research reports. *World Development*, 23, 265–279.

Frentz, T. (1995) 30.6.94: Preussisches Gesetz über die Landwirtschaftskammern. Chancen und Probleme aus Schleswig-Holsteinischer Sicht. In Institut für Agrarpolitik, Marktlehre u. Agrarentwicklung d. Humboldt-Univ. Zu Berlin. (ed.) *Entwicklungstendenzen in der agrargeschichtlichen Lehre und Forschung: Ehrensymposium anlässlich der Verabschiedung von Prof. Dr. Volker Klemm.* Berlin, Institut für Agrarpolitik, Marktlehre u. Agrarentwicklung und Fördergesellschaft Albrecht Daniel Thaer.

Frey, K. J. (1992) Plant breeding perspectives for the 1990s. In Stalker, H. T. and Murphy, J. P. (eds) *Plant Breeding in the 1990s.* Wallingford, CAB International.

Frey, K. J. (2000) A national strategy for plant breeding in the United States. In Fuglie, K. and Schimmelpfennig, D. E. (eds) *Public-Private Collaboration in Agricultural Research: New Institutional Arrangements and Economic Implications.* Ames, Iowa State University Press.

Frey, M. and Kunkel, S. (2011) Writing the history of development: a review of the recent literature. *Contemporary European History*, 20, 215–232.

Fricke, D. and Hartwig, E. (1984) Bund der Landwirte (BdL) 1893–1920. In Fricke, D. *et al.* (eds) *Lexikon zur Parteiengeschichte. Die bürgerlichen Parteien und Verbände in Deutschland (1789–1945).* Leipzig, VEB Bibliographisches Institut.

Friedeburg, R. V. (2004) Brach liegende Felder: Grundzüge der deutschen Agrargeschichtsschreibung. In Bruckmüller, E., Langthaler, E. and Redl, J. (eds) *Agrargeschichte Schreiben: Traditionen und Innovationen im internationalen Vergleich.* Innsbruck, StudienVerlag.

Frölich, G. (1922) Die wirtschaftliche Bedeutung der Pflanzenzucht und ihre Förderung durch Staat und Korporationen. *Beiträge zur Pflanzenzucht*, 5, 30–45.

Frossard, D. (2002) How farmer-scientist cooperation is devalued and revalued: a Philippine example. In Cleveland, D. and Soleri, D. (eds) *Farmers, Scientists and Plant Breeding: Integrating Knowledge and Practice.* Wallingford, CAB International.

Fruwirth, C. (1896) *Landwirtschaftliche Pflanzenzüchtung und ihre Stätten in Oesterreich*, Leipzig, Franz Deuticke.

Fruwirth, C. (1905) Was soll eine Anstalt für Prüfung und Züchtung von Saatsorten in Württemberg? *Wochenblatt Württemberg*, 28, 177–179.

Fruwirth, C. (1906) *Wie kann sich der Landwirt Pflanzenzüchtung, Sortenversuche und Saatgutbau zu Nutze machen?*, Berlin, Parey.

Fruwirth, C. (1907a) Die Förderung des Saatgutbaues durch Saatbauvereine. *Illustrierte Landwirtschaftliche Zeitung*, 27, 729–730 and 739–740.

Fruwirth, C. (1907b) *Sorten, Saatfruchtbau und Pflanzenzüchtung in Württemberg*, Plieningen, Friedrich Find.

Fruwirth, C. (1908) Pflanzenzüchtung in Württemberg. *Deutsche Landwirtschaftliche Presse*, 35, 543–545.

Fruwirth, C. (1910) Formen des Pflanzenzüchtungsbetriebes. *Fühlings Landwirtschaftliche Zeitung*, 59, 641–651.

Fruwirth, C. (1911a) *Die Beziehungen der Pflanzenzüchtung zu den Wirtschaftsbetrieben*, Leipzig, Reichenbach'sche Verlagsbuchhandlung.

Fruwirth, C. (1911b) Formen des Pflanzenzüchtungsbetriebes. *Fühlings Landwirtschaftliche Zeitung*, 60, 174–178 and 455–456.

Fruwirth, C. (1912) Uber den Unterricht in der landwirtschaftlichen Pflanzenzucht. *Land- und forstwissenschaftlichen Unterrichtszeitung*, 26, 1–27.

Fruwirth, C. (1919a) Die gegenwärtige Organisation der Pflanzenzüchtung in Deutschland und Oesterreich-Ungarn. *Nachrichten der Deutschen Landwirtschafts-Gesell. für Oesterreich*, 3, 35–39.

Fruwirth, C. (1919b) Zur Steigerung der pflanzlichen Erzeugung auf bäuerlichen Betrieben. *Nachrichten der Deutschen Landwirtschafts-Gesell. für Oesterreich*, 3, 895–899.

Fujisaka, S. (1994) Will farmer participatory research survive in the International Agricultural Research Centres? In Scoones, I. and Thompson, J. (eds) *Beyond Farmer First: Rural People's Knowledge, Agricultural Research and Extension Practice.* London, Intermediate Technology Publications.

Fukuda-Parr, S. (ed.) (2007a) *The Gene Revolution: GM Crops and Unequal Development*, London, Earthscan.

Fukuda-Parr, S. (2007b) Institutional changes in Argentina, Brazil, China, India and South Africa. In Fukuda-Parr, S. (ed.) *The Gene Revolution: GM Crops and Unequal Development.* London, Earthscan.

Fukuda-Parr, S. (2007c) Introduction: genetically modified crops and development priorities. In Fukuda-Parr, S. (ed.) *The Gene Revolution: GM Crops and Unequal Development.* London, Earthscan.

Funke, C. (2008) Hafer, Avena sativa L. Bemühungen um das ehemals wichtigste Futtergetreide. In Röbbelen, G. (ed.) *Die Entwicklung der Pflanzenzüchtung in Deutschland (1908–2008). Hundert Jahre GFP, e.V. – eine Dokumentation.* Göttingen, Gesellschaft für Pflanzenzüchtung.

Gäde, H. (1993) *Beiträge zur Geschichte der Pflanzenzüchtung und Saatgutwirtschaft in den fünf neuen Bundesländern Deutschlands*, Berlin, Paul Parey.

Gassner, A. (1927) Ueber die Entwicklung und Tätigkeit der Bayerischen Landessaatzuchtanstalt in den verflossenen 25 Jahren. *Wochenblatt Bayern*, 117, 808–811.

Gaud, W. S. (1968) The Green Revolution: accomplishments and apprehensions. paper delivered at the meeting of the Society for International Development. Washington, DC, 8 March.

Gayon, J. and Zallen, D. (1998) The role of the Vilmorin Company in the promotion and diffusion of the experimental science of heredity in France, 1840–1920. *Journal of the History of Biology*, 31, 241–262.

Gerhard, G. (2008) Food and genocide: Nazi agrarian politics in the occupied territories of the Soviet Union. *Contemporary European History*, 18, 45–65.

Gerlach (1908) Das Kaiser Wilhelms-Institut für Landwirtschaft in Bromberg. *Landwirtschaftliche Jahrbücher*, 37, 181–200.

Gerschenkron, A. (1966) *Bread and Democracy in Germany*, New York, Howard Fertig.

Gessner, D. (1981) The dilemma of German agriculture during the Weimar Republic. In Bessel, R. and Feuchtwanger, E. J. (eds) *Social Change and Political Development in Weimar Germany*. London, Croom Helm.

GFP (ed.) (1933) *25 Jahre Gesellschaft zur Förderung deutscher Pflanzenzucht, e.V.*, [place not given], Gesellschaft zur Förderung deutscher Pflanzenzucht.

Gieryn, T. (1999) *Cultural Boundaries of Science: Credibility on the Line*, Chicago, University of Chicago Press.

Gies, H. (1979) Aufgaben und Probleme der nationalsozialistischen Ernährungswirtschaft 1933–1939. *Vierteljahrschrift für Sozial- und Wirtschaftsgeschichte*, 66, 466–499.

Gies, H. (2000) Reagrarisierung oder Industrialisierung? Programmatik und Realität in der nationalsozialistischen Agrar- und Wirtschaftspolitik. *Zeitschrift für Agrargeschichte und Agrarsoziologie*, 48, 145–160.

Gisevius, P. (1901) *Die Sortenfrage in den Nordostprovinzen, ein Führer für die Auswahl passender Getreide- und Kartoffel-Sorten...* Berlin, Parey.

Gisevius, P. (1907) Die Organisation der Saatzucht. *Hannoversche Land- und Forstwirtschaftliche Zeitschrift*, 476–479 and 499–502.

Gisevius, P. (1908) Die Züchtungseinrichtungen in Saatzüchtungsanstalten mit Berücksichtigung der Svalöfer Züchtungsbuchführung. In Anon. (ed.) *Pflanzenzüchtung (= Arbeiten der Landwirtschaftskammer für die Provinz Sachsen, vol. 15).* Halle, Verlag der Landwirtschaftskammer für die Provinz Sachsen.

Gisevius, P. (1912) Staatliche und sonstige Förderungsmittel zur Hebung der Pflanzenzüchtung. *Beiträge zur Pflanzenzucht*, 2, 6–14.

Glaser (1902) Welche Sorten landwirtschaftlicher Kulturpflanzenwerden in Süddeutschland besonders gebaut und was ist für ihre Vervbesserung geschehen? *Jahrbuch der DLG*, 17, 236–242.

Glover, D. (2010) The corporate shaping of GM crops as a technology for the poor. *Journal of Peasant Studies*, 37, 67–90.

Goeschl, T. and Swanson, T. (2002) Forecasting the impact of genetic use restriction technologies: a case study on the impact of hybrid crop varieties. In Swanson, T. (ed.) *Biotechnology, Agriculture and the Developing World: the Distributional Implications of Technological Change.* Cheltenham, Edward Elgar.

Goldman, M. (2005) *Imperial Nature: the World Bank and Struggles for Social Justice in the Age of Globalization*, New Haven, Yale University Press.

Goldsmith, A. (1997) The Rockefeller Foundation's Indian Agricultural Program: Why it worked. In Hewa, S. and Hove, P. (eds) *Philanthropy and Cultural Context: Western*

Philanthropy in South, East, and Southeast Asia in the 20th Century. Lanham, University Press of America.
von der Goltz, T. (1894) *Die agrarischen Aufgaben der Gegenwart*, Jena, Gustav Fischer.
von der Goltz, T. (1899) *Vorlesungen über Agrarwesen und Agrarpolitik*, Jena, Gustav Fischer.
von der Goltz, T. (1903) *Geschichte der Deutschen Landwirtschaft, vol. 2*, Stuttgart, J. G. Cotta.
von der Goltz, T. (1904) *Die Entwicklung der Landwirtschaft im 19. Jahrhundert*, Bonn, Verlag der Landwirtschaftskammer für die Rheinprovinz.
Goodell, G. (1984) Challenges to international pest management research and extension in the third world: Do we really want IPM to work? *Bulletin of the Entomological Society of America*, 30, 18–26.
Goodman, R. M. (1985) Bringing new technology to old world agriculture. *Bio/Technology*, 3, 708–709.
Gothein, G. (1910/11) *Agrarpolitisches Handbuch*, Berlin, Liebheit & Thiesen.
Gottwald, H. (1984) Deutscher Landwirtschaftsrat (DLR), 1872–1933. In Fricke, D. *et al.* (eds) *Lexikon zur Parteiengeschichte, vol. 2.* Leipzig, VEB Bibliographisches Institut.
Gouse, M. (2007) South Africa: Revealing the potential and obstacles, the private sector model and reaching the traditional sector. In Fukuda-Parr, S. (ed.) *The Gene Revolution: GM Crops and Unequal Development.* London, Earthscan.
Gow, D. D. and Morss, E. R. (1988) The notorious nine: critical problems in project implementation. *World Development*, 16, 1399–1418.
Graff, D. G. *et al.* (2003) The public-private structure of intellectual property ownership in agricultural biotechnology. *Nature Biotechnology*, 21, 989–995.
Grantham, G. (1984) The shifting locus of agricultural innovation in 19th century Europe: the case of the agricultural experiment stations. *Research in Economic History*, suppl. 3, 191–214.
Gregg, S. (2004) Uncovering the subsistence economy in the 20th century South. *Agricultural History*, 78, 417–437.
Griesbeck, A. (1924) Pflanzenzüchtung im bäuerlichen Betrieb. *Wochenblatt Bayern*, 114, 236–237.
Griffin, K. (1974) *The Political Economy of Agrarian Change: an Essay on the Green Revolution*, London, Macmillan.
Griffin, K. (1999) *Alternative Strategies for Economic Development*, London, Macmillan.
Griffin, K. and Khan, A. R. (1978) Poverty in the third world: Ugly facts and fancy theories. *World Development*, 6, 295–304.
Grigg, D. B. (1984) The agricultural revolution in Western Europe. In Bayliss-Smith, T. P. and Wanmali, S. (eds) *Understanding Green Revolutions: Agrarian Change and Development Planning in South Asia.* Cambridge, Cambridge University Press.
Grote, G. (1937?) Gegenwarts- und Zukunftsaufgaben der deutschen Pflanzenzucht. In [apparently the proceedings of a conference at Goslar]. copy in R68IV/4, BAB.
Guggenheim, S. (2006) Roots: reflections of a 'Rocky doc' on social science in the CGIAR. In Cernea, M. M. and Kassam, A. H. (eds) *Researching the Culture in Agri-Culture: Social Research for International Development.* Wallingford, CABI Publishing.
Guimarães, E. P., Kueneman, E. and Carena, M. J. (2006) Assessment of national plant breeding and biotechnology capacity in Africa and recommendations for future capacity building. *HortScience*, 41, 50–52.
Gupta, A. (1989) Scientists' views of farmers' practices in India: barriers to effective

interaction. In Chambers, R., Pacey, A. and Thrupp, L. A. (eds) *Farmer First: Farmer Innovation and Agricultural Research.* London, Intermediate Technology Publications.

Gupta, A. (2009) Nationale Armut, globale Armut u. Neoliberalismus: Eine anthropologische Kritik. In Büschel, H. and Speich, D. (eds) *Entwicklungswelten. Globalgeschichte der Entwicklungszusammenarbeit.* Frankfurt a.M., Campus Verlag.

Häcker (1902) Stand und Entwicklung der Landwirtschaft in Baden und Elsass-Lothringen seit dem Jahre 1890. *Jahrbuch der DLG*, 17, 134–147.

Hagen, W. (1980) *Germans, Poles and Jews: the Nationalist Conflict in the Prussian East, 1772–1914*, Chicago, University of Chicago Press.

Hall-Matthews, D. (2011) Can historians assist development policy-making or just highlight its faults? In Bayly, C. A., Rao, V., Szreter, S. and Woolcock, M. (eds) *History, Historians and Development Policy: a Necessary Dialogue.* Manchester, Manchester University Press.

Hanisch, E. (2002) Die Politik und die Landwirtschaft. In Bruckmüller, E. *et al.* (eds) *Geschichte der österreichischen Land- und Forstwirtschaft im 20. Jahrhundert: Politik, Gesellschaft, Wirtschaft.* Vienna, Ueberreuter.

Hansen, J. and Fischer, G. (1936) *Geschichte der Deutschen Landwirtschafts-Gesellschaft*, Berlin, Deutsche Landwirtschafts-Gesellschaft.

Hardin, C. M. (1955) *Freedom in Agricultural Education*, Chicago, University of Chicago Press.

Harrar, J. G. (1955) Food for the future. *Science*, 122, 313–316.

Harrar, J. G. (1963) *Strategy for the Conquest of Hunger*, New York, Rockefeller Foundation.

Harriss, J. (1987) Capitalism and peasant production: the Green Revolution in India. In Shanin, T. (ed.) *Peasants and Peasant Societies.* Oxford, Blackwell.

Harriss, J. (2005) Great promise, hubris and recovery: a participant's history of development studies. In Kothari, U. (ed.) *A Radical History of Development Studies: Individuals, Institutions and Ideologies.* London, Zed.

Harwood, J. (1997) The reception of genetic theory among academic plant-breeders in Germany, 1900–1930. *Sveriges Utsaedesfoerenings Tidskrift (Journal of the Swedish Seed Association)*, 107, 187–195.

Harwood, J. (2002) Politische Ökonomie der Pflanzenzucht in Deutschland, ca. 1870–1933. In Heim, S. (ed.) *Autarkie und Ostexpansion: Pflanzenzucht und Agrarforschung im Nationalsozialismus.* Gottingen, Wallstein.

Harwood, J. (2004) Linkage before Mendelism? Plant-Breeding Research in Central Europe, ca. 1880–1910. In Rheinberger, H.-J. and Gaudillière, J.-P. (eds) *Classical Genetic Research and its Legacy: The Mapping Cultures of Twentieth Century Genetics.* London, Routledge.

Harwood, J. (2005) *Technology's Dilemma: Agricultural Colleges between Science and Practice in Germany, 1860–1934*, Bern, Peter Lang.

Harwood, J. (2008) Research and extension in political context: Rural unrest and the origins of the Prussian chambers of agriculture. In Vivier, N. (ed.) *The State and Rural Societies: Policy and Education in Europe, 1750–2000.* Turnhout, Brepols.

Harwood, J. (2009) Peasant-friendly plant-breeding and the early years of the Green Revolution in Mexico. *Agricultural History*, 83, 384–410.

Harwood, J. (2010) The fate of peasant-friendly plant-breeding in Nazi Germany. *Historical Studies in the Natural Sciences*, 40, 570–604.

Harwood, J. (forthcoming) Schiemann's mentor: Erwin Baur and the institutionalization of genetics in Germany. To appear in a volume on the life and work of Elisabeth Schiemann, edited by E. Höxtermann.

Harwood, J. and Sturdy, S. (2010) What can development policy learn from the history of development? *Food Security*, 2, 285–290.

Harwood, R. (2000) The evolution of FSR-E in Asia through the mid 1970s: a view from IRRI. In Collinson, M. (ed.) *A History of Farming Systems Research.* Wallingford and Rome, CABI Publishing and FAO.

Haushofer, H. (1958) *Ideengeschichte der Agrarwirtschaft und Agrarpolitik im deutschen Sprachgebiet, vol. 2: vom ersten Weltkrieg bis zur Gegenwart*, Munich, Bayerischer Landwirtschaftsverlag.

Haushofer, H. (1960) *Die Furche der Deutschen Landwirtschafts-Gesellschaft, 1885–1960*, Frankfurt a.M., DLG-Verlag.

Haushofer, H. (1963) *Die deutsche Landwirtschaft im technischen Zeitalter*, Stuttgart, Ulmer.

Haushofer, H. (1969) *Ein halbes Jahrhundert im Dienste der bayerischen Landwirtschaft und der Volksernährung – Zur Geschichte des Bayerischen Staatsministeriums für Ernährung, Landwirtschaft und Forsten*, Munich, BLV Verlagsgesellschaft.

Haushofer, H. (1975) Zur Geschichte der Bayerischen Landesanstalt für Bodenkultur und Pflanzenbau. *Landwirtschaftliches Jahrbuch für Bayern*, 52, 17–77.

Haushofer, H. (1986) Bäuerliche Führungsschichten in Bayern im 19.-20. Jahrhundert. In Pankraz, F. and Zorn, W. (eds) *Aus der Bayerischen Agrargeschichte, 1525–1978. Gesammelte Beiträge zur Bayerischen Agrargeschichte von Heinz Haushofer.* Munich, BLV Verlagsgesellschaft.

Hayami, Y. (1971) Elements of induced innovation: a historical perspective for the Green Revolution. *Explorations in Economic History*, 8, 445–472.

Hayami, Y. (1981) Induced innovation, green revolution and income distribution: Comment. *Economic Development and Cultural Change*, 30, 169–176.

Hayami, Y. (1998) The peasant in economic modernization. In Eicher, C. K. and Staatz, J. M. (eds) *International Agricultural Development.* Baltimore, Johns Hopkins University Press.

Hayami, Y. *et al.* (1975) *A Century of Agricultural Growth in Japan*, Tokyo and Minneapolis, Tokyo University Press and University of Minnesota Press.

Hayami, Y. and Ruttan, V. (1985) *Agricultural Development: an International Perspective*, Baltimore, Johns Hopkins University Press.

Hayter, T. (2005) Secret diplomacy uncovered: research on the World Bank in the 1960s and 1980s. In Kothari, U. (ed.) *A Radical History of Development Studies: Individuals, Institutions and Ideologies.* London, Zed.

Hazell, P. (2003) Green revolution. In Mokyr, J. (ed.) *The Oxford Encyclopedia of Economic History.* Oxford, Oxford University Press.

Hazell, P. and Ramasamy, C. (eds) (1991) *The Green Revolution Reconsidered: The Impact of High-Yielding Rice Varieties in South India*, Baltimore, Johns Hopkins University Press.

Hazell, P. *et al.* (2007) The Future of Small Farms for Poverty-Reduction and Growth (2020 Discussion Paper No. 42). Washington, DC, IFPRI.

Hazell, P. *et al.* (2010) The future of small farms: Trajectories and policy priorities. *World Development*, 38, 1349–1361.

Hecht, M. (1898) Zur landwirtschaftlichen Betriebsstatistik in Baden. *Wochenblatt Württemberg*, 676–679.

Hecht, M. (1912) Die Landwirtschaft in Baden. In Rebmann, E., Gothein, E. and

Jagemann, E. V. (eds) *Das Grossherzogtum Baden in allgemeiner, wirtschaftlicher und staatlicher Hinsicht dargestellt.* Karlsruhe, G. Braunsche Hofbuchdruckerei.

Heim, S. (2003) *Kalorien, Kautschuk, Karrieren: Pflanzenzüchtung und landwirtschaftliche Forschung in Kaiser-Wilhelm-Instituten 1933–1945*, Göttingen, Wallstein.

Heisey, P., Srinivasan, C. S. and Thirtle, C. (2001) Public sector plant-breeding in a privatizing world. Washington, DC, USDA – Economic Research Service.

Henning, F.-W. (1978) *Landwirtschaft und ländliche Gesellschaft in Deutschland, vol. 2: 1750–1976*, Paderborn, Ferdinand Schöningh.

Henning, F.-W. (1996) *Handbuch der Wirtschafts- und Sozialgeschichte Deutschlands, vol. 2: Deutsche Wirtschafts- und Sozialgeschichte im 19. Jahrhundert*, Paderborn, Ferdinand Schöningh.

Henriksen, I. (1993) The transformation of Danish agriculture, 1870–1914. In Persson, K. G. (ed.) *The Economic Development of Denmark and Norway since 1870.* Aldershot, Edward Elgar.

Henry, R. J. (1995) Technology transfer and its constraints: Early warnings from agricultural development in colonial India. In Macleod, R. and Kumar, D. (eds) *Technology and the Raj: Western Technology and Technical Transfers to India, 1700–1947.* New Delhi, Sage.

Herring, R. J. (2007) The genomics revolution and development studies: Science, poverty and politics. *Journal of Development Studies*, 43, 1–30.

Herrmann, K. (1994) Die Deutsche Landwirtschafts-Gesellschaft (DLG) und die Modernisierung der ostelbischen Gutswirtschaft. In Reif, H. (ed.) *Ostelbische Agrargesellschaft im Kaiserreich und in der Weimarer Republik. Agrarkrise – junkerliche Interessenpolitik – Modernisierungsstrategien.* Berlin, Akademie Verlag.

Herrmann, S. (2009) Auf dem Weg zur Superknolle. *Süddeutsche Zeitung*, 21 April, *16.*

Hightower, J. (1973) *Hard Tomatoes, Hard Times*, Cambridge, MA, Schenkman.

Hillmann, P. (1907) Die Tätigkeit der Saatzuchtabteilung. *Mitteilungen der DLG*, 22, 423–426.

Hillmann, P. (1909) Entwicklung und Stand der deutschen landwirtschaftlichen Pflanzenzüchtung. *Jahrbuch der DLG*, 24, 1–9.

Hillmann, P. (ed.) (1910a) *Die deutsche landwirtschaftliche Pflanzenzucht (= Arbeiten der DLG, vol. 168)*, Berlin, DLG-Verlag.

Hillmann, P. (1910b) Einleitung. In Hillmann, P. (ed.) *Die deutsche landwirtschaftliche Pflanzenzucht (= Arbeiten der DLG, vol. 168).* Berlin, DLG-Verlag.

Hiltner, L. (1911) Ueber die Sortenanbauversuche der Kgl. Agrikulturbotanischen Anstalt. *Die Scholle*, 11.

Hochberger, A. (1991) *Der Bayerische Bauernbund, 1893–1914*, Munich, C.H. Beck.

Hodge, J. M. (2007) *Triumph of the Expert: Agrarian Doctrines of Development and the Legacies of British Colonialism*, Athens, Ohio, Ohio University Press.

Hodge, J. M. (2010) British colonial expertise, post-colonial careering and the early history of international development. *Journal of Modern European History*, 8, 24–45.

Hödl, G. (2011) Chronic underachievement. Austria's development aid in historical perspective, unpublished paper, Department of African Studies, University of Vienna.

Hogg, D. (2000) *Technological Change in Agriculture: Locking in to Genetic Uniformity*, Basingstoke, Macmillan.

Holdefleiss, P. (1908) *Landwirtschaftliche Pflanzenzüchtung*, Hannover, Verlagsbuchhandlung Dr. Max Jänecke.

Holmen, H. (2005) Spurts in production – Africa's limping Green Revolution. In

Djurfeldt, G. *et al.* (eds) *The African Food Crisis: Lessons from the Asian Green Revolution.* Wallingford, CABI Publishing.
Holt, J. B. (1936) *German Agricultural Policy, 1918–1934* (Ph.D. diss., University of Heidelberg, 1934), Chapel Hill, University of North Carolina Press.
Holt-Gimenez, E., Altieri, M. A. and Rosset, P. (2006) Ten Reasons Why the Rockefeller and the Bill and Melinda Gates Foundations' Alliance for another Green Revolution will not solve the problems of poverty and hunger in Sub-Saharan Africa (Food First Policy Brief no. 12). Oakland, CA, Food First/Institute for Food and Development Policy.
Holtmeier-Schomberg, [?]. (1908) Die Entwicklung und Organisation der Pflanzenzüchtung in Dänemark, Schweden und der Probstei. *Landwirtschaftliche Jahrbücher* 37, 311–380.
Howard, A. (1940) *An Agricultural Testament*, London, Oxford University Press.
Hron, R. (1981) Getreidebau und Getreidezüchtung in Oesterreich. In Bundesanstalt für Pflanzenbau und Samenprüfung in Wien. (ed.) *100 Jahre Bundesanstalt für Pflanzenbau und Samenprüfung in Wien, 1881–1981.* Vienna, Bundesanstalt für Pflanzenbau und Samenprüfung in Wien.
Huang, J. *et al.* (2002) Plant biotechnology in China. *Science*, 295, 674–677.
Huang, J. *et al.* (2007) China: Emerging public sector model for GM crop development. In Fukuda-Parr, S. (ed.) *The Gene Revolution: GM Crops and Unequal Development.* London, Earthscan.
Hubbell, B. J. and Welsh, R. (1998) Transgenic crops: Engineering a more sustainable agriculture? *Agriculture and Human Values*, 15, 43–56.
Hucho, F. *et al.* (2005) *Gentechnologiebericht. Analyse einer Hochtechnologie in Deutschland*, Munich, Elsevier.
Hudde, W. (1965) Geschichte des landwirtschaftlichen Bildungswesens. In Hudde, W. and Schmiel, M. (eds) *Handbuch des landwirtschaftlichen Bildungswesens.* Munich, Bayerischer Landwirtschaftsverlag.
Huffman, W. E. and Evenson, R. E. (1993) *Science for Agriculture: a Long-Term Perspective*, Ames, Iowa State University Press.
Huizer, G. (1996) Social movements in the underdevelopment of development dialectic: a view from below. In Chew, S. C. and Denemark, R. A. (eds) *The Underdevelopment of Development: Essays in Honour of A. G. Frank.* London, Sage.
Hundhammer, A. (1926) *Die landwirtschaftliche Berufsvertretung in Bayern* (doctoral dissertation), Munich, University of Munich.
Hunt, J. C. (1974) Peasants, grain tariffs, and meat quotas: imperial German protectionism reexamined. *Central European History*, 7, 311–331.
Hurt, D. (2000) *Agricultural Technology in the 20th Century*, Manhattan, Kansas, Sunflower University Press.
Hussain, A. and Tribe, K. (1981) *Marxism and the Agrarian Question, vol. 1: German Social Democracy and the Peasantry, 1890–1907*, London, Macmillan.
IFAD (2001) Rural Poverty Report. Rome, International Fund for Agricultural Development.
Iiunuma, J. (1973) The introduction of American and European agricultural science into Japan in the Meiji era. In Shand, R. T. (ed.) *Technical Change in Asian Agriculture.* Canberra, Australian National University Press.
Ingersent, K. A. and Rayner, A. J. (1999) *Agricultural Policy in Western Europe and the United States*, Cheltenham, Edward Elgar.
Ishikawa, S. (1967) *Economic Development in Asian Perspective*, Tokyo, Kinokuniya.

Ishikawa, S. (1981) *Essays on Technology, Employment and Institutions in Economic Development: Comparative Asian Experience*, Tokyo, Kinokuniya.

Ishikawa, S. and Ohkawa, K. (1972) Significance of Japan's experience – technological changes in agricultural production and changes in agrarian structure. In Japan Economic Research Centre. (ed.) *Agriculture and Economic Development – Structural Readjustment in Asian Perspective.* Tokyo, Japan Economic Research Centre.

Jahnel, H. and Ludwig, H. (1961) Friedrich Nobbe, der Begründer der Samenkontrolle. *Proceedings of the International Seed Testing Association*, 26, 127–139.

Jain, H. K. (1982) Plant breeders' rights and genetic resources. *Indian Journal of Genetics and Plant Breeding*, 42, 121–128.

Janssen, K. and Gupta, A. (2009) Anticipating the future: 'Biotechnology for the poor' as unrealised promise? *Futures*, 41, 436–445.

Jas, N. (2001) *Au Carrefour de la Chimie et de l'Agriculture: les sciences agronomiques en France et en Allemagne, 1840–1914*, Paris, Edition des archives contemporaines.

Jatzlauk, M. (1994) Diskussionen und Untersuchungen über die Agrarverhältnisse im Verein für Sozialpolitik in den letzten Jahrzehnten des 19. Jahrhunderts. In Reif, H. (ed.) *Ostelbische Agrargesellschaft im Kaiserreich und in der Weimarer Republik.* Berlin, Akademie Verlag.

Jefferson, R. A. (1993) Beyond model systems: New strategies, methods and mechanisms for agricultural research. *Annals of the New York Academy of Sciences*, 700, 53–73.

Jenkins, M. (1940) The segregation of genes affecting yield of grain in maize. *Journal of the American Society of Agronomy*, 32, 55–63.

Jennings, B. H. (1988) *Foundations of International Agricultural Research: Science and Politics in Mexican Agriculture*, Boulder, Westview Press.

Jiggins, J. (2008) Talking was always going to be hard to do. *New Scientist*, 2650, 16–17.

Johnson, N. and Ruttan, V. (1994) Why are farms so small? *World Development*, 22, 691–706.

Johnston, B. (1951) Agricultural productivity and economic development in Japan. *Journal of Political Economy*, 59, 498–513.

Johnston, B. (1970) The Japanese 'model' of agricultural development: its relevance to developing nations. In Ohkawa, K., Johnston, B. and Kaneda, H. (eds) *Agriculture and Economic Growth: Japan's Experience.* Princeton, NJ and Tokyo, Princeton and Tokyo University Presses.

Johnston, B. (1983) The design of rural development: a policy perspective. In Nobe, K. and Sampath, R. K. (eds) *Issues in Third World Development.* Boulder, Westview Press.

Johnston, B. and Kilby, P. (1975) *Agriculture and Structural Transformation: Economic Strategies in Late-Developing Countries*, New York, Oxford University Press.

Jones, J. W. (1936) Improvement in rice. *Yearbook of Agriculture 1936* (U.S. Dept. of Agriculture), 415–454.

Jowit, J. and Vidal, J. (2010) Genetically modified crops could bring benefits, says environment secretary. *Guardian*, 5 June, 7.

Juma, C. (1989) *The Gene Hunters: Biotechnology and the Scramble for Seeds*, London/Princeton, NJ, Zed Books/Princeton University Press.

Just, R. E. and Rausser, G. (1993) The governance structure of agricultural science and agricultural economics: a call to arms. *American Journal of Agricultural Economics*, 75, 69–83.

Kappen, H. and Henkelmann, W. (1932) Düngung und Pflanzenzucht als Mittel der Produktionskostensenkung. In Beckmann, F. *et al.* (eds) *Deutsche Agrarpolitik im Rahmen der inneren und äusseren Wirtschaftspolitik, Teil I.* Berlin, Reimar Hobbing.

Kassam, A. (2006) Agricultural institutions and receptivity to social research: the case of CGIAR. In Cernea, M. M. and Kassam, A. H. (eds) *Researching the Culture in Agri-Culture: Social Research for International Development.* Wallingford, CABI Publishing.

Kaul (1938) Die Bedeutung von Hochzuchtsaatgut für den Ernteertrag. *Mitteilungen für die Landwirtschaft*, 53, 811–812.

Keeley, J. and Scoones, I. (2000) Knowledge, power and politics: the environmental policy-making process in Ethiopia. *Journal of Modern African Studies*, 38, 89–120.

Keith, D. (2008) Why walking out was our only option. *New Scientist*, 2650, 17–18.

Kenney, M. (1986) *Biotechnology: the University-Industrial Complex*, New Haven, Yale University Press.

Kiers, E. T. *et al.* (2008) Agriculture at a crossroads. *Science*, 320, 320–321.

Kiessling, L. (1906) Die Organisation einer Landessaatgutzüchtung in Bayern. *Fühlings Landwirtschaftliche Zeitung*, 55, 329–338.

Kiessling, L. (1910a) Die Königliche Bayerische Landessaatzuchtanstalt. In Hillmann, P. (ed.) *Die deutsche landwirtschaftliche Pflanzenzucht.* Berlin, Deutsche Landwirtschafts-Gesellschaft.

Kiessling, L. (1910b) Saatgutzüchtung und Saatgutwirtschaften. *Wochenblatt Bayern*, issues 7, 8 and 9.

Kiessling, L. (1911) Ackerbauvereine (Ein Mittel zur Förderung des Acker- und Pflanzenbaues). *Haus- und Landwirtschafts-Kalender des Landwirtschaftlichen Vereins in Bayern.*

Kiessling, L. (1912) Die züchterische Bearbeitung der Landsorten in Bayern. *Beiträge zur Pflanzenzucht*, 2, 74–96.

Kiessling, L. (1913) *Ackerbauvereine und einschlägige Organisationsmassnahmen zur Förderung des Acker- und Saatfruchtbaues* Berlin, Paul Parey.

Kiessling, L. (1914) Erfolge und Aufgaben der Pflanzenzüchtung mit besonderer Berücksichtigung der bayerischen Verhältnisse. *Bericht über die 48. Wanderversammlung bayerischer Landwirte in Erlangen vom 23.-27. Mai 1914.*

Kiessling, L. (1916) Einiges über die gegenwärtige Förderung des Kleesamenbaues in Bayern. *Mitteilungen der DLG*, 31, 191–192.

Kiessling, L. (1917) Inwiefern hat die deutsche Pflanzenzucht zum wirtschaftlichen Durchhalten beigetragen? *Landwirtschaftliches Jahrbuch für Bayern*, 7, 24–32.

Kiessling, L. (1918a) Geheimer Hofrat Professor Dr. C. Kraus. *Landwirtschaftliches Jahrbuch für Bayern*, 8, 539–546.

Kiessling, L. (1918b) C. Kraus. *Zeitschrift für Pflanzenzüchtung*, 6, 222–225.

Kiessling, L. (1922a) Die Entwicklung der Landeskultur im Austellungsgau Bayern seit der letzten Wanderaustellung der DLG. *Jahrbuch der DLG*, 37, 109–122.

Kiessling, L. (1922b) Einige Gegenwartsfragen der Pflanzenzüchtung und des Saatbaues. *Mitteilungen der DLG*, 37, *454–460.*

Kiessling, L. (1936) Gemeinschaftsarbeit in der deutschen Pflanzenzüchtung. *Der Forschungsdienst,* Sonderheft 3, 41–48.

Kiessling, L. (1938) Dr.-Ing.h.c. Jakob Ackermann, Irlbach. *Der Züchter*, 10, 102–103.

Kiessling, L. and Gassner, A. (1927) Entwicklung und Tatigkeit der Bayerischen Landessaatzuchtanstalt im ersten Vierteljahrhundert ihres Bestehens. *Landwirtschaftliches Jahrbuch für Bayern*, 10–12, 1–94.

King, F. H. (1911) *Farmers of Forty Centuries: or Permanent Agriculture in China, Korea and Japan*, Madison, WI, Mrs. F. H. King.

Kittlitz, E. V. (2005) Entwicklungsgeschichte der Landessaatzuchtanstalt in Hohenheim,

1905–2005. *Hohenheimer Themen: Zeitschrift für kulturwissenschaftliche Themen*, 13 (Sonderband), 1–219.

Klein, E. (1973) *Geschichte der deutschen Landwirtschaft im Industriealter*, Wiesbaden, Franz Steiner.

Klemm, V. (1992) *Agrarwissenschaften in Deutschland: Geschichte – Tradition von den Anfängen bis 1945*, St. Katharinen, Scripta Mercaturae Verlag.

Klemm, V. (1994) *Agrarwissenschaften im 'Dritten Reich'. Aufstieg oder Sturz? (1933–1945)*, Berlin, Fördergesell. A.D. Thaer and Fachgebiet Sozialgeschichte der Agrarentwicklung der Humboldt-Universität zu Berlin.

Kloppenburg, J. R. (1st ed. 1988) *First the Seed: the Political Economy of Plant Biotechnology, 1492–2000*, Cambridge, Cambridge University Press.

Kloppenburg, J. R. (2nd ed. 2004) *First the Seed: the Political Economy of Plant Biotechnology*, Madison, WI, University of Wisconsin Press.

Kluge, U. (1989) *Vierzig Jahre Agrarpolitik in der Bundesrepublik Deutschland*, Hamburg, Parey.

Kluge, U. (2000) Deutsche Agrarpolitik im 20. Jahrhundert zwischen Protektionismus und wirtschaftlicher Modernisierung: Ausklang des Agrarischen? In Muenkel, D. (ed.) *Der lange Abschied vom Agrarland: Agrarpolitik, Landwirtschaft und ländliche Gesellschaft zwischen Weimar und Bonn.* Göttingen, Wallstein.

Kluge, U. (2005) *Agrarwirtschaft und ländliche Gesellschaft im 20. Jahrhundert*, Munich, R. Oldenbourg Verlag.

Knight, J. (2003) A dying breed. *Nature*, 421, 568–570.

Kochendörfer, M. (1996) Landeshoheitliche und genossenschaftliche Getreidezüchtungsbestrebungen zur Jahrhundertwende. *Zeitschrift fur Agrargeschichte und Agrarsoziologie*, 44, 174–193.

Koernicke, F. (1885) *Die Arten und Varietäten des Getreides*, Berlin, Parey.

Koning, N. (1994) *The Failure of Agrarian Capitalism: Agrarian Politics in the United Kingdom, Germany, the Netherlands and the USA, 1846–1919*, London, Routledge.

Kopsidis, M. (2006) *Agrarentwicklung: Historische Agrarrevolutionen und Entwicklungsökonomie*, Stuttgart, Franz Steiner.

Kothari, U. (2005) From colonial administration to development studies: a post-colonial critique of the history of development studies. In Kothari, U. (ed.) *A Radical History of Development Studies: Individuals, Institutions and Ideologies.* London, Zed.

Kothari, U. (2006) From colonialism to development: reflections of former colonial officers. *Journal of Commonwealth and Comparative Politics*, 44, 118–136.

Kothari, U. (2011) History, time and temporality in development discourse. In Bayly, C. A., Rao, V., Szreter, S. and Woolcock, M. (eds) *History, Historians and Development Policy: a Necessary Dialogue.* Manchester, Manchester University Press.

Kousari, K. (2004) Deep-rooted commodity trap lies behind Africa's poverty. *The Guardian*, 15 March.

Kratzsch, G. (2008) Pflanzenzüchtung und Saatgutwesen in der sowjetischen Besatzungszone und der Deutschen Demokratischen Republik (1945–1990). Entwicklungen im volkseigenen Wirtschaftssystem. In Röbbelen, G. (ed.) *Die Entwicklung der Pflanzenzüchtung in Deutschland (1908–2008). Hundert Jahre GFP, e.V. – eine Dokumentation.* Göttingen, Gesellschaft für Pflanzenzüchtung.

Kraus, C. (1897) Der gegenwärtige Zustand des niederbayerischen Gerstenbaues und die Massregeln zur Hebung desselben. *Zeitschrift für das gesamte Brauwesen*, 20, 11–18.

Kraus, C. (1899a) Förderung der Landwirtschaft in Bayern durch praktische Anbau- u.

Düngungsversuche. *Vierteljahresschrift des Bayerischen Landwirtschaftrathes*, 4, Beiheft, 332–353.
Kraus, C. (1899b) Wie erzielt man die beste Braugerste. *Jahrbuch der DLG*, 14, 220–231.
Kraus, C. (1900) [untitled talk]. In Anon. (ed.) *Bericht über die xxxvi. Wanderversammlung bayerischer Landwirte in Nördlingen vom 28. und 29. Mai 1900.* Augsburg, Ph. J. Pfeiffer.
Kraus, C. (1903) *Sortenwahl und Saatgutzüchtung bei niederbayerischen Getreiden*, Munich, Pössenbacher'sche Buchdruckerei.
Kraus, C. (1905a) Die Gliederung des Gersten- und Haferhalmes und deren Beziehungen zu den Fruchtständen... *Naturwissenschaftliche Zeitschrift für Land- und Forstwirtschaft*, Beiheft 1, 1–153.
Kraus, C. (1905b) Ueber die Saatgutzüchtung in Bayern. *Jahrbuch der DLG*, 20, 340–351.
Kraus, C. (1907) Inwieweit hat die Getreidezüchtung auf die Landrassen Rücksicht zu nehmen und welche Massnahmen sind geeignet, die Saatgutzüchtung in wirksamster Weise zu fördern? In Anon. (ed.) *VIII. Internationaler landwirtschaftlicher Kongress Wien.* Vienna, F.A. Wachtl.
Kraus, C. (1909a) Die Bestrebungen zur Hebung des Gerstenbaues in Bayern. *Zeitschrift für das gesamte Brauwesen*, 32, 601–605 and 617–622.
Kraus, C. (1909b) Züchtungen von Gerste und Hafer 1899–1908. *Fühlings Landwirtschaftliche Zeitung*, 58, 465–487.
Kraus, C. (1914) Lokalzüchtung und Sorteneinheitsgebiete in Bayern. *Landwirtschaftliches Jahrbuch für Bayern*, 4, 89–101.
Kraus, C. (n.d.) Einrichtungen zur Beschaffung und Verbreitung von Saatfrucht geeigneter Sorten der Gerste, *Mitteilung der Kgl. Saatzuchtanstalt Weihenstephan.*
Kraus, C. and Kiessling, L. (1918) Die Landsortenzüchtung in Bayern. *Deutsche Landwirtschaftliche Presse*, 45, 217–218.
Krausmann, F. (2006) Vom Kreislauf zum Durchfluss: Oesterreichs Agrarmodernisierung als sozialökologischer Transformationsprozess. In Dix, A. and Langthaler, E. (eds) *Grüne Revolutionen. Agrarsysteme und Umwelt im 19. und 20. Jahrhundert.* Innsbruck, StudienVerlag.
Krech (1904) Die Einrichtung von Saatbauvereinen im nordöstlichen Deutschland, insbes. des westpreussischen Saatbauvereins Althausen. *Jahrbuch der DLG*, 19, 240–243.
Kreutz, H. (1942) Ludwig Kiessling. *Der Züchter*, 14, 96–98.
Krishnamoorthy, C. (1982) Assumptions and approach of the All-India Coordinated Research Project for Dryland Agriculture. In Anderson, R. S. *et al.* (eds) *Science, Politics and the Agricultural Revolution in Asia.* Boulder, Westview Press for the AAAS.
Krohn, H. (1934) Aufgaben der Saatzucht in der Erzeugungsschlacht. *Mitteilungen für die Landwirtschaft*, 49, 1119–1120.
Krueger, R. (2001) Agricultural biotechnology: our past and our future. In Nelson, G. (ed.) *Genetically Modified Organisms in Agriculture: Economics and Politics.* San Diego, Academic Press.
Kryzymowski, R. (1913) Beziehungen zwischen der Betriebsintensität und der Sortenfrage. *Jahrbuch der DLG*, 28, 456–467.
Kryzymowski, R. (1951) *Geschichte der deutschen Landwirtschaft*, Stuttgart, Ulmer.
Kuckuck, H. (1951) *Entwicklung und Probleme neuzeitlicher Pflanzenzüchtung: Mendel oder Lyssenko*, Berlin, Parey.
Kuckuck, H. (1978) Pflanzenzüchtung im Wandel der vergangenen 50 Jahre (unpublished

lecture given 30 June at Max-Planck Institut für Züchtungsforschung, Cologne-Vogelsang).

Kühle, L. (1925) Eröffnung der Wanderversammlung im staatlichen botanischen Institut. *Beiträge zur Pflanzenzucht*, 8, 9–18.

Kühle, L. (1926) Der Stand und die Lage der deutschen Pflanzenzucht. *Mitteilungen der DLG*, 41, 856–865.

Kühle, L. (1933a) Eröffnungsrede auf der GFP-Wanderversammlung am 3. Juli 1932 in Müncheberg/Mark. In Anon (ed.) *25 Jahre Gesell. z. Förderung der deutschen Pflanzenzucht, e.V.* [no place cited], Gesell. z. Förderung der deutschen Pflanzenzucht.

Kühle, L. (1933b) 25 Jahre Gesell. z. Förderung der deutschen Pflanzenzucht. In Anon. (ed.) *25 Jahre Gesell. z. Förderung der deutschen Pflanzenzucht, e.V.* [no place cited], Gesell. z. Förderung der deutschen Pflanzenzucht.

Kuhn, [?] (1900) Wie erzielt man eine gute Braugerste? *Wochenblatt Baden*, 2–4.

Kuhn, [?] (1902) Neues über Züchtung des Spelzes. *Wochenblatt Baden*, 251–255.

Kulisch, P. (1913) Die staatliche Förderung der Saatzucht und des Saatgutbaues in Elsass-Lothringen. *Jahrbuch der DLG*, 28, 467–487.

Kutz, M. (1984) Kriegserfahrung und Kriegsvorbereitung. Die agrarwirtschaftliche Vorbereitung des 2. Weltkriegs in Deutschland vor dem Hintergrund der Weltkrieg I – Erfahrung (parts 1 and 2). *Zeitschrift für Agrargeschichte und Agrarsoziologie*, 32, 59–82 and 135–164.

Kydd, J. and Dorward, A. (2001) The Washington Consensus on poor country agriculture: Analysis, prescription and institutional gaps. *Development Policy Review*, 19, 467–478.

Ladejinsky, W. (1973) Agrarian reform in Asia: the Green Revolution and its reform effects. In Shand, R. T. (ed.) *Technical Change in Asian Agriculture.* Canberra, Australian National University Press.

Ladejinsky, W. (1977) Advancing human welfare. In Walinsky, L. J. (ed.) *The Selected Papers of Wolf Ladejinsky: Agrarian Reform as Unfinished Business.* New York, Oxford University Press for the World Bank.

Lal, D. (1975) The agrarian question. *South Asian Review*, 8, 389–400.

Lang, F. (1927) Die bayerische Landwirtschaft. In Anon. (ed.) *Das Land Bayern. Seine kulturelle und wirtschaftliche Bedeutung für das Reich.* Munich, München-Augsburger Abendzeitung.

Lang, H. (1907) Zu der Frage: Gefahren und Nachteile des modernen Pflanzenzuchtbetriebes. *Deutsche Landwirtschaftliche Presse*, 34, 759–760.

Lang, H. (1909a) Bemerkungen zu den Saatfruchtmärkten. *Wochenblatt Baden*, 178–179.

Lang, H. (1909b) Das erste Geschäftsjahr der Grossh. Saatzuchtanstalt. *Wochenblatt Baden*, 786–787.

Lang, H. (1909c) Die Anerkennung selbständiger Saatzuchtbetriebe durch die Grossh. Saatzuchtanstalt Hochburg. *Wochenblatt Baden*, 703.

Lang, H. (1909d) Die Landespflanzenzüchtung. *Wochenblatt Baden*, 77, 973–974.

Lang, H. (1909e) Einiges über vergleichende Sortenanbauversuche zugleich Aufruf zur Beteiligung. *Wochenblatt Baden*, 403–404.

Lang, H. (1910a) Die züchterische Bearbeitung des Mais. *Wochenblatt Baden*, 149–152.

Lang, H. (1910b) Mitteilung der Grosshzgl. Saatzuchtanstalt Hochburg. *Wochenblatt Baden*, 79–80.

Lang, H. (1910c) *Theorie und Praxis der Pflanzenzüchtung*, Stuttgart, Eugen Ulmer.

Lang, H. (1911a) Die Landsorten. *Wochenblatt Baden*, 1139–1141.

Lang, H. (1911b) *Erster Tätigkeitsbericht der Grossh. Badischen Saatzuchtanstalt*, Karlsruhe, Macklot'sche Buchdruckerei.

Lang, H. (1911c) Reiseeindrücke aus Dänemark und Schweden. *Wochenblatt Baden*, 79, 1066–67, 1221–1224 and 1316–1318.

Lang, H. (1912a) Die Förderung des Futterbaues in Gegenden mit Klein- und Mittelbetrieb. *Illustrierte Landwirtschaftliche Zeitung*, 32, 5–7.

Lang, H. (1912b) Reiseeindrücke aus Dänemark und Schweden. *Wochenblatt Baden*, 80, 197–199 and 230–231.

Lang, H. (1913) *Jahresbericht der Grossh. Badischen Saatzuchtanstalt ... für das Jahr 1912*, Karlsruhe, Macklot'sche Buchdruckerei.

Lang, H. (1914a) Bericht über die zweite ordentliche Generalversammlung der österreichischen Gesell. für Pflanzenzüchtung und über die daran sich anschliessenden Veranstaltungen. *Wochenblatt Baden*, 25, 638–640.

Lang, H. (1914b) *Die Pflanzenzüchtung als Mittel der Landwirtschaftspflege*, Emmendingen, Druck- u. Verlagsgesellschaft vorm. Dölter.

Lang, H. (1914c) Mitteilung der Grossh. Saatzuchtanstalt Hochburg. *Wochenblatt Baden*, 719.

Lang, K. (1971) *Die Landwirtschaftsverwaltung in Württemberg im 19. Jahrhundert bis zum Ende des ersten Weltkrieges*, Stuttgart, Ulmer.

Lange, F. (1917) *Landwirtschaftlich-Statistischer Atlas*, Berlin, Dietrich Reimer (Ernst Vohsen).

Lappé, F. M. and Collins, J. (1986) *World Hunger: Twelve Myths*, New York, Grove Press.

Lappé, M. and Bailey, B. (1999) *Against the Grain: the Genetic Transformation of Global Agriculture*, London, Earthscan.

Laube, W. (2004) Aus einer Chronik der F. von Lochow-Petkus GmbH. *Vorträge für Pflanzenzüchtung*, 65, 41–42.

Laur (1910) Die Betriebsberatung in der bäuerlichen Wirtschaft. In Anon. (ed.) *Neuere Erfahrungen auf dem Gebiete des landwirtschaftlichen Betriebswesens.* Berlin, Deutsche Landwirtschafts-Gesellschaft.

Lehmann, H. G. (1970) *Die Agrarfrage in der Theorie und Praxis der deutschen und internationalen Sozialdemokratie...* Tübingen, J.C.B. Mohr (Paul Siebeck).

Leonard, D. K. (1986) Putting the farmer in control: building agricultural institutions. In Berg, R. J. and Whitaker, J. S. (eds) *Strategies for African Development.* Berkeley and London, University of California Press.

Lepenies, P. H. (2009) Lernen vom Besserwisser: Wissenstransfer in der "Entwicklungshilfe" aus historischer Perspektive. In Büschel, H. and Speich, D. (eds) *Entwicklungswelten. Globalgeschichte der Entwicklungszusammenarbeit.* Frankfurt a.M., Campus Verlag.

Lerman, K. (2001) Wilhelmine Germany. In Breuilly, J. (ed.) *Nineteenth Century Germany: Politics, Culture and Society, 1780–1918.* London, Arnold.

Leverenz, C. (n.d. [but *c.*1924]) *Die meistgebauten landwirtschaftlichen Pflanzenzuchten Deutschlands*, [presumably Halle, Landwirtschaftskammer für die Provinz Sachsen].

Levins, R. (1986) Science and progress: seven developmentalist myths in agriculture. *Monthly Review*, 38, 13–20.

Lewis, D. (2005) Individuals, organizations and public action: trajectories of the 'non-governmental' in development studies. In Kothari, U. (ed.) *A Radical History of Development Studies: Individuals, Institutions and Ideologies.* London, Zed.

Lewis, J. (2000) Leveraging partnerships between the public and private sector – Experience of USAID's agricultural biotechnology program. In Persley, G. J. and Lantin, M. M. (eds) *Agricultural Biotechnology and the Poor.* Washington, DC, CGIAR.

Lewontin, S. (1983) *The Green Revolution and the Politics of Agricultural Development in Mexico since 1940*, Ph.D. diss., University of Chicago.

Lichter, J. (1994) *Landwirtschaft und Landwirtschaftskammer in der Rheinprovinz am Vorabend des Ersten Weltkriegs*, Cologne, Botermann und Botermann.

Lichter, J. (1996) Zwangslagen der nationalsozialistischen Agrarpolitik von 1933 bis 1939. In Schulz, G. (ed.) *Von der Landwirtschaft zur Industrie. Wirtschaftlicher und gesellschaftlicher Wandel im 19. u. 20. Jahrhundert.* Paderborn, Ferdinand Schöningh.

Lieber, D. (1924) 15 Jahre Pflanzenzüchtung in Baden. *Deutsche Landwirtschaftliche Presse*, 51, 49–51.

Lieber, D. (1932) Die Pflanzenzucht im Lichte badischer Züchterarbeit. *Mitteilungen der DLG*, 47, 386–387.

Liebscher, G. (1892) Getreide Anbauversuche: Haferernte 1891... *Jahrbuch der DLG*, 7, (241) - (313).

Liebscher, G. (1893) Fortschritte in der Pflanzenzüchtung. *Jahrbuch der DLG*, 8, 152–162.

Liebscher, G. (1896) Die Getreidezüchtung, ein Mittel von grosser Bedeutung für die Rentabilität des Getreidebaues. *Wochenblatt Württemberg*, 217–218.

Lines, G. and Uphoff, N. (2005) A remarkable civil society contribution to food and nutrition security in Madagascar and beyond. In Schulz, M. and Kracht, U. (eds) *Food and Nutrition Security in the Process of Globalization.* Münster, Lit-Verlag.

Lipton, M. (1978) Inter-farm, inter-regional and farm-nonfarm income distribution: The impact of the new cereal varieties. *World Development*, 6, 319–337.

Lipton, M. (2001) Reviving global poverty-reduction: What role for genetically-modified plants? *Journal of International Development*, 13, 823–846.

Lipton, M. (2007) Plant breeding and poverty: Can transgenic seeds replicate the 'Green Revolution' as a source of gains for the poor? *Journal of Development Studies*, 43, 31–62.

Lipton, M. (2010) From policy aims and small-farm characteristics to farm science needs. *World Development* 38, 1399–1412.

Lipton, M. with Longhurst, W. R. (1989) *New Seeds and Poor People*, London, Unwin Hyman.

von Lochow, F. (1904) Die Aufgaben der Saatzucht-Abteilung. *Jahrbuch der DLG*, 19, 25–31.

von Lochow, F. (1913) 25 Jahre deutsche Pflanzenzucht. *Illustrierte Landwirtschaftliche Zeitung*, 448–450.

Love, H. H. and Reisner, J. H. (1964) *The Cornell-Nanking Story (= Cornell International Agricultural Development Bulletin no. 4)*, Ithaca, New York State College of Agriculture.

Lovin, C. R. (1969) Agricultural reorganization in the Third Reich: the Reich Food Corporation (Reichsnährstand), 1933–1936. *Agricultural History*, 43, 447–461.

Lovin, C. R. (1974) Die Erzeugungsschlacht 1934–1936. *Zeitschrift für Agrargeschichte und Agrarsoziologie*, 22, 209–220.

Lucas, W. (1931) *Die Berufsausbildung des landwirtschaftlichen Akademikers und seine Betätigungsmöglichkeiten*, Berlin, Parey.

Lundgreen, P. *et al.* (1986) *Staatliche Forschung in Deutschland, 1870–1980*, Frankfurt a.M., Campus.

Lupton, F. G. H. (1992) *Changes in Varietal Distribution of Cereals in Central and Western Europe (=* vol. 4 of *Agro-ecological Atlas of Cereal-Growing in Europe)*, Wageningen, Pudoc.

Maat, H. (2001) *Science Cultivating Practice: a History of Agricultural Science in the Netherlands and its Colonies, 1863–1986*, Wageningen, Ponsen & Looijen.

Maat, H. (2007) Is participation rooted in colonialism? Agricultural innovation systems and participation in the Netherlands Indies. *IDS Bulletin*, 38, 50–60.

Maehl, W. H. (1980) German Social Democratic agrarian policy, 1890–1895, reconsidered. *Central European History*, 13, 121–157.

Magdoff, F., Foster, J. B. and Buttel, F. H. (eds) (2000) *Hungry for Profit: the Agribusiness Threat to Farmers, Food and the Environment*, New York, Monthly Review Press.

Mai, U. (2002) *'Rasse und Raum'. Agrarpolitik, Sozial- und Raumplanung im NS-Staat*, Paderborn, Schöningh.

Mann, C. (1997) Reseeding the Green Revolution. *Science*, 277, 1038–1043.

Mann, C. (1999) Crop scientists seek a new revolution. *Science*, 283, 310–314.

Manning, R. (2000) *Food's Frontier: the Next Green Revolution*, Berkeley, University of California Press.

Marcus, A. and Lowitt, R. (eds) (1990) *The U.S. Department of Agriculture in Historical Perspective (*special issue of *Agricultural History, vol. 64, no. 2)*, Berkeley, University of California Press.

Marglin, S. A. (1996) Farmers, seedsmen and scientists: Systems of agriculture and systems of knowledge. In Apffel-Marglin, F. and Marglin, S. A. (eds) *Decolonizing Knowledge: from Development to Dialogue.* Oxford, Clarendon Press.

Massing, P. (1967) *Rehearsal for Destruction: a Study of Political Anti-semitism in Imperial Germany*, New York, Howard Fertig.

Matchett, K. (2002) *Untold Innovation: Scientific Practice and Corn Improvement in Mexico, 1935–1965.* Ph.D. diss., University of Minnesota.

Matchett, K. (2006) At odds over inbreeding: an abandoned attempt at Mexico/United States collaboration to 'improve' Mexican corn, 1940–1950. *Journal of the History of Biology*, 39, 345–372.

Maurya, D. M. (1989) The innovative approach of Indian farmers. In Chambers, R., Pacey, A. and Thrupp, L. A. (eds) *Farmer First: Farmer Innovation and Agricultural Research.* London, Intermediate Technology Publications.

Mazoyer, M. and Roudart, L. (2006) *A History of World Agriculture from the Neolithic Age to the Current Crisis*, London, Earthscan.

McAfee, K. (2008) Exporting crop biotechnology: the myth of molecular miracles. In Otero, G. (ed.) *Food for the Few: Neo-liberal Globalism and Biotechnology in Latin America.* Austin, TX, University of Texas Press.

McCalla, A. F. and Brown, L. R. (2000) Feeding the developing world in the next millennium: a question of science? In Persley, G. J. and Lantin, M. M. (eds) *Agricultural Biotechnology and the Poor.* Washington, DC, CGIAR.

McCann, J. (2005) *Maize and Grace: Africa's Encounter with a New World Crop, 1500–2000*, Cambridge, Harvard University Press.

McCook, S. (2002) *States of Nature: Science, Agriculture and Environment in the Spanish Caribbean, 1760–1940*, Austin, University of Texas Press.

McIntyre, B. D., *et al.* (2009a) *Synthesis Report (International Assessment of Agricultural Knowledge, Science and Technology for Development)*, Washington, DC, Island Press for IAASTD.

McIntyre, B. D., *et al.* (2009b) *Global Report (International Assessment of Agricultural Knowledge, Science and Technology for Development)*, Washington, DC, Island Press for IAASTD.

McKie, R. (2011) GM crops essential to human survival, says UK science chief. *Guardian Weekly*, 28 November, 10.

McMichael, P. (2004) Biotechnology and food security: profiting on insecurity? In Beneria, L. and Bisnath, S. (eds) *Global Tensions: Challenges and Opportunities in the World Economy*. London, Routledge.

McNamara, R. S. (1987) Paupers of the world and how to develop them [orig. publ. 1973]. In Shanin, T. (ed.) *Peasants and Peasant Societies (2nd ed.)*. Oxford, Blackwell.

Meinberg, W. (1937) Steigerung der landwirtschaftlichen Erzeugung. *Der Vierjahresplan*, 4, 197–202.

Meinel, A. (2008) *Aufbruch in die wissenschaftliche Pflanzenzüchtung. Der Beitrag von Wilhelm Rimpau (1842–1903)*, Göttingen, Gesellschaft für Pflanzenzüchtung.

Mellor, J. W. (1998) Foreign aid and agriculture-led development. In Eicher, C. K. and Staatz, J. M. (eds) *International Agricultural Development*. Baltimore, Johns Hopkins University Press.

Mende, M. (1997) Revolutionierende Rüben: die Industrialisierung der Landwirtschaft uber den Anbau und die Verarbeitung von Zuckerrüben. *Blätter für Technikgeschichte*, 59, 162–183.

Merkel, F. (1910) Geschichtliches und Technisches über Saatenanerkennung. *Mitteilungen der DLG*, 25, 656–658.

Merkel, F. (1914) Saatbauorganisationen im Deutschen Reiche. *Mentzel und Lengerke*, 67, 310–318.

Merkel, F. (1919) Dr. h.c. F. von Lochow-Petkus zu seinem 70. Geburtstag. *Mitteilungen der DLG*, 34, 489–491.

Merl, S. (1994) Das Agrargenossenschaftswesen Ostdeutschlands 1878–1928. Die Organisation des landwirt. Fortschritts und ihre Grenzen. In Reif, H. (ed.) *Ostelbische Agrargesellschaft im Kaiserreich und in der weimarer Republik. Agrarkrise – junjkerliche Interessenpolitik – Modernisierungsstrategien*. Berlin, Akademie Verlag.

Meyer, K. (1934) Die Landmaschine, ein Volksproblem? *Mitteilungen für die Landwirtschaft*, 49, 552–555.

Meyer, K. (1935) Erzeugungsschlacht und Wissenschaft. In Anon. (ed.) *Erzeugungsschlacht 1934/35 (= Flugschriften des Reichsnährstandes, no. 35)*. Berlin, Reichsnährstand Verlags-Gesell.

Meyer, K. (1936) Rede anlässlich der 1. Sitzung des Forschungsrates. *Der Forschungsdienst*, 1, 8–14.

Meyer, K. (1937) Geleitwort. *Der Forschungsdienst*, 3, 1.

Meyer, K. (1938) Drei Jahre Forschungsdienst. In Meyer, K. (ed.) *Forschung für Volk und Nahrungsfreiheit* (= *Der Forschungsdienst*, supplem. vol. 8*)*. Neudamm/Berlin, J. Neumann.

Meyer, K. (1939a) Grundsätze und Ziele nationalsozialistischer Agrarpolitik. In Meyer, K. (ed.) *Gefüge und Ordnung der deutschen Landwirtschaft*. Berlin, Reichsnährstand Verlags-Gesellschaft.

Meyer, K. (1939b) Unsere Forschungsarbeit. *Der Forschungsdienst*, 7, 183–235.

Meyer, K. (1970) W. Rimpau (1842–1903). In Franz, G. and Haushofer, H. (eds) *Grosse Landwirte*. Frankfurt a.M., DLG-Verlag.

Middendorf, G. *et al.* (2000) New agricultural biotechnologies: the struggle for democratic choice. In Magdoff, F., Foster, J. B. and Buttel, F. H. (eds) *Hungry for Profit: the Agribusiness Threat to Farmers, Food and the Environment*. New York, Monthly Review Press.

Milatz, R. (1937) Weg und Arbeit des deutschen Sortenregisters. *Der Forschungsdienst*, 3, 345–357.

Millikan, M. and Hapgood, D. (eds) (1967) *No Easy Harvest: the Dilemma of Agriculture in Underdeveloped Countries*, Boston, Little-Brown.

Millikan, M. and Rostow, W. W. (1998) Notes on foreign economic policy. In Simpson, C. (ed.) *Universities and Empire*. New York, New Press.

Minogue, M. and Kothari, U. (2002) Conclusion: orthodoxy and its alternatives in contemporary development. In Kothari, U. and Minogue, M. (eds) *Development Theory and Practice: Critical Perspectives*. Basingstoke, Palgrave.

Mitchell, T. (2002) *Rule of Experts: Egypt, Techno-Politics, Modernity*, Berkeley, University of California Press.

Möckl, K. (1972) *Die Prinzregentenzeit. Gesellschaft und Politik während der Aera des Prinzregenten Luitpold in Bayern*, Munich, R. Oldenbourg.

Modrow, W. (1904) Saatzucht im nordöstlichen Deutschland. *Jahrbuch der DLG*, 19, 243–255.

Moeller, R. (1981) Peasants and tariffs in the *Kaiserreich*: how backward were the *Bauern*? *Agricultural History*, 55, 370–384.

Moeller, R. (1986a) *German Peasants and Agrarian Politics, 1914–1924: The Rhineland and Westphalia*, Chapel Hill, University of North Carolina Press.

Moeller, R. (1986b) Introduction: locating peasants and lords in modern German historiography. In Moeller, R. (ed.) *Peasants and Lords in Modern Germany: Recent Studies in Agricultural History*. Boston, Allen & Unwin.

Moeller, R. (ed.) (1986c) *Peasants and Lords in Modern German History: Recent Studies in Agricultural History*, London, Allen & Unwin.

Mommsen, W. (1995) *Imperial Germany, 1867–1918: Politics, Culture, and Society in an Authoritarian State*, London, Edward Arnold.

Moon, S. (2004) Empirical knowledge, scientific authority, and native development: the controversy over sugar/rice ecology in the Netherlands East Indies, 1905–1914. *Environment and History*, 10, 59–81.

Moon, S. (2007) *Technology and Ethical Idealism: a History of Development in the Netherlands East Indies*, Leiden, CNWS Publications.

Mooser, J. (1992) Preussische Agrarreformen, Bauern und Kapitalismus. Bemerkungen zu Hartmut Harnischs Buch ‘Kapitalistische Agrarreform und industrielle Revolution’. *Geschichte und Gesellschaft*, 18, 533–554.

Morris, M., Edmeades, G. and Pehu, E. (2006) The global need for plant breeding capacity: what roles for the public and private sectors? *HortScience*, 41, 30–39.

Morris, M. and Ekasingh, B. (2002) Plant breeding research in developing countries: What roles for the public and private sectors? In Byerlee, D. and Echeverria, R. (eds) *Agricultural Research Policy in an Era of Privatization*. Wallingford, CABI Publishing.

Moseman, A. H. (1966) International needs in plant breeding research. In Frey, K. J. (ed.) *Plant Breeding: a Symposium Held at Iowa State University*. Ames, Iowa State University Press.

Moser, P. (2000) Eine “Sache des ganzen Volkes”? Ueberlegungen zum Prozess der Vergesellschaftung der bäuerlichen Landwirtschaft in der Industriegesellschaft. *Traverse*, 64–78.

Moser, P. (2003) *züchten, säen, ernten: Agrarpolitik, Pflanzenzucht und Saatgutwesen in der Schweiz, 1860–2002*, Baden, Switzerland, Hier und Jetzt Verlag.

Moser, P. (2006) Wie aus Züchtern Vermehrer gemacht wurden: Saatzüchtung und

Saatgutwesen in der Schweiz im 20. Jahrhundert. In Dix, A. and Langthaler, E. (eds) *Grüne Revolutionen (= Jahrbuch für Geschichte des ländlichen Raumes 2006).* Innsbruck, StudienVerlag.

Müller, F. (1927) Einiges über die Aussentätigkeit der Landessaatzuchtanstalt in den letzten 25 Jahren ihres Bestehens (1902–1927). *Bayerische Ackerbauzeitung* (= Sonderbeilage z. *Wochenblatt Bayern*, Nr. 46), 70–72.

Müller, H.-H. (1989) Landwirtschaft u. industrielle Revolution – am Beispiel der Magdeburger Börde. In Pierenkemper, T. (ed.) *Landwirtschaft und industrielle Entwicklung: zur ökonomischen Bedeutung von Bauernbefreiung, Agrarreform und Agrarrevolution.* Stuttgart, Franz Steiner Verlag.

Müller, P. (1969) *Zum Problem der Mittelbindung in der Entwicklungshilfe*, Hamburg, Deutsches Uebersee Institut.

Münkel, D. (1996) *Nationalsozialistische Agrarpolitik und Bauernalltag*, Frankfurt a.M., Campus Verlag.

Murphy, D. J. (2007) *Plant Breeding and Biotechnology: Societal Context and the Future of Agriculture*, Cambridge, Cambridge University Press.

Muth, H. (1975) Die Führungsschichten der Bauernverbände. In Franz, G. (ed.) *Bauernschaft und Bauernstand, 1500–1970.* Limburg, C.A. Starke Verlag.

Mütter, B. (1988) Agrarmodernisierung in Deutschland zwischen Reichsgründung und erstem Weltkrieg: das Beispiel Oldenburg. *Geschichte in Wissenschaft und Unterricht*, 11, 651–673.

Myren, D. T. (1970) The Rockefeller Foundation program in corn and wheat in Mexico. In Wharton, C. R. (ed.) *Subsistence Agriculture and Economic Development.* London, Cass.

Nair, K. (1969) *The Lonely Furrow: Farming in the United States, Japan and India*, Ann Arbor, University of Michigan Press.

Nair, K. (1979) *In Defense of the Irrational Peasant: Indian Agriculture after the Green Revolution*, Chicago, University of Chicago Press.

Nathusius, L. (1955) *Theodor Roemer: Lebensabriss und bibliographischer Ueberblick*, Halle, Universitäts- und Landesbibliothek.

Ndiritu, C. G. (2000) Kenya: biotechnology in Africa: Why the controversy? In Persley, G. J. and Lantin, M. M. (eds) *Agricultural Biotechnology and the Poor.* Washington, DC, CGIAR.

Nicholls, W. (1964) Agricultural policy: the place of agriculture in economic development. In Berrill, K. (ed.) *Economic Development with Special Reference to East Asia.* London, Macmillan.

Nicolaisen, W. (1944) Die Pflanzenzucht in der Erzeugungsschlacht. *Der Forschungsdienst*, 17, 499–512.

Niggl, L. (1953) *Die Geschichte der deutschen Grünlandbewegung, 1914–1945*, [no place of publ. cited], privately published.

Nipperdey, T. (1990) *Deutsche Geschichte 1866–1918. Band 1: Arbgeitswelt und Bürgergeist*, Munich, C.H. Beck.

Norman, D. (1978) Farming systems research to improve the livelihood of small farmers. *American Journal of Agricultural Economics*, 65, 813–818.

Norman, D. (2000) FSR: a personal evolution. In Collinson, M. (ed.) *A History of Farming Systems Research.* Wallingford and Rome, CABI Publishing and FAO.

Normile, D. (1999) Crossing rice strains to keep Asia's rice bowls brimming. *Science*, 283, 313.

Nwalozie, M. *et al.* (2007) West and Central Africa: Strategizing biotechnology for food security and poverty reduction. In Fukuda-Parr, S. (ed.) *The Gene Revolution: GM Crops and Unequal Development.* London, Earthscan.

Oasa, E. K. (1987) The political economy of international agricultural research: a review of the CGIAR's response to criticisms of the green revolution. In Glaser, B. (ed.) *The Green Revolution Revisited: Critique and Alternatives.* London, Allen & Unwin.

Oasa, E. K. and Jennings, B. (1982) Science and authority in international agricultural research. *Bulletin of Concerned Asian Scholars*, 14, 30–44.

Oberkrome, W. (2009) *Ordnung und Autarkie. Die Geschichte der deutschen Landbauforschung, Agrarökonomie und ländlichen Sozialwissenschaft im Spiegel von Forschungsdienst und DFG (1920–1970)*, Stuttgart, Franz Steiner.

Ogura, T. (ed.) (1967) *Agricultural Development in Modern Japan*, Tokyo, Fuji Publishing.

Olby, R. (2000) Mendelism: from hybrids and trade to a science. *Comptes rendus de l'Académie des sciences, série III*, 323, 1043–1051.

Olea-Franco, A. (2001) One Century of Higher Agricultural Education and Research in Mexico (1850s–1960s)... Ph.D. diss., Harvard University.

Olmstead, A. L. and Rhode, P. W. (2008) *Creating Abundance: Biological Innovation and American Agricultural Development*, Cambridge, Cambridge University Press.

Opitz, K. (1913) Der Titel: Saatgutwirtschaft. *Illustrierte Landwirtschaftliche Zeitung*, 77, 24 September.

Otero, G. (2008) Neoliberal globalism and the biotechnology revolution: Economic and historical context. In Otero, G. (ed.) *Food for the Few: Neo-liberal Globalism and Biotechnology in Latin America.* Austin, University of Texas Press.

Overton, M. (1996) *Agricultural Revolution in England: the Transformation of the Agrarian Economy, 1500–1850*, Cambridge, Cambridge University Press.

Paarlberg, D. (1981) The land grant colleges and the structure issue. *American Journal of Agricultural Economics*, 63, 129–134.

Paarlberg, R. (2000) The global food fight. *Foreign Affairs*, 79, 24–38.

Paarlberg, R. (2008) *Starved for Science: How Biotechnology is being kept out of Africa*, Cambridge, MA, Harvard University Press.

Palladino, P. (1990) The political economy of applied research: plant breeding in Great Britain, 1910–1940. *Minerva*, 28, 446–468.

Palladino, P. (2002) *Plants, Patients and the Historian: (Re)membering in the Age of Genetic Engineering*, New Brunswick, NJ, Rutgers University Press.

Pammer, F. (1952) Oesterreichische Sortenliste landwirtschaftlicher Kulturpflanzen – Getreide. *Die Bodenkultur*, special issue 3, 266–277.

Pammer, G. (1902) *Ueber Sortenauswahl und Getreidezüchtung mit Rücksicht auf die Hebung des Getreidebaues in Niederösterreich* (lecture on 25 February 1902 to k.k. Landwirtschafts-Gesell. in Vienna), Vienna, Verlag der k. k. Samenkontroll-Station in Wien.

Pammer, G. (1903) Getreidesorten-Anbauversuche in Niederösterreich. *Landwirtschaftliche Zeitschrift (der k.k. Landwirtschafts-Gesell. in Wien)*, 29, 27–28.

Pammer, G. (1904) Die Aktion der Getreidezüchtung in Oesterreich. *Wiener Landwirtschaftliche Zeitung*, 54, 745.

Pammer, G. (1905) Ueber Veredelungszüchtungen mit einigen Landsorten des Roggens in Niederösterreich. *Zeitschrift des landwirtschaftlichen Versuchswesens in Oesterreich*, 8, 1015–1053.

Pammer, G. (1907) Inwieweit hat die Getreidezüchtung auf die Landrassen Rücksicht zu nehmen und welche Massnahmen sind geeignet, die Saatgutzüchtung in wirksamster

Weise zu fördern? In Anon. (ed.) *VIII. Internationaler landwirtschaftlicher Kongress Wien.* Vienna, F.A. Wachtl.

Parayil, G. (1992) The Green Revolution in India: a case study of technological change. *Technology and Culture*, 33, 737–756.

Pardey, P. *et al.* (2006) *Agricultural Research: a Growing Global Divide?* Washington, DC, Int. Food Policy Research Institute.

Paris, T. *et al.* (2006) Rice for the poor: the call and opportunity for social research. In Cernea, M. M. and Kassam, A. H. (eds) *Researching the Culture in Agri-Culture: Social Research for International Development.* Wallingford, CABI Publishing.

Paul, D. and Kimmelman, B. (1988) Mendel in America: Theory and practice, 1900–1919. In Rainger, R., Benson, K. R. and Maienschein, J. (eds) *The American Development of Biology.* Philadelphia, University of Pennsylvania Press.

Paul, H. and Steinbrecher, R. (2003) *Hungry Corporations: Transnational Biotechnology Companies Colonize the Food Chain*, London, Zed.

Peal, D. (1988) Self-help and the state: rural coops in Imperial Germany. *Central European History*, 21, 244–266.

Pearse, A. (1980) *Seeds of Plenty, Seeds of Want: Social and Economic Implications of the Green Revolution*, Oxford, Clarendon Press.

Perelman, M. (1977) *Farming for Profit in a Hungry World: Capital and the Crisis in Agriculture*, Montclair, NJ, Allenheld – Osmun.

Perkins, J. A. (1981) The agricultural revolution in Germany, 1850–1914. *Journal of European Economic History*, 10, 71–118.

Perkins, J. A. (1984) The political economy of sugar beet in Imperial Germany. In Albert, B. and Graves, A. (ed.) *Crisis and Change in the International Sugar Economy, 1860–1914.* Norwich, ISC Press.

Perkins, J. A. (1986) Dualism in German agrarian historiography. *Comparative Studies in Society and History*, 28, 287–306.

Perkins, J. A. (1990) The organization of German industry, 1850–1930: the case of beet-sugar production. *Journal of European Economic History*, 19, 549–574.

Perkins, J. H. (1990) The Rockefeller Foundation and the Green Revolution, 1941–1956. *Agriculture and Human Values*, 7, 6–18.

Perkins, J. H. (1997) *Geopolitics and the Green Revolution: Wheat, Genes and the Cold War*, New York, Oxford University Press.

Perrin, R. and Winkelmann, D. (1976) Impediments to technical progress on small vs. large farms. *American Journal of Agricultural Economics*, 58, 888–894.

Peters, C. (1986) *Ferdinand von Lochow: ein bedeutender Pflanzenzüchter und Landwirt: Leben, Lebenswerk, Auswirkungen* (2 vols.), doctoral diss., Humboldt-Universität zu Berlin.

Petzina, D. (1968) *Autarkiepolitik im Dritten Reich. Der nationalsozialistische Vierjahresplan*, Stuttgart, Deutsche Verlagsanstalt.

von Pfetten, S. (n.d.) *Kurzer Bericht über Ackerbau, Saatbau und Saatzucht der Freiherrl. v. Pfetten'schen Oekonomien Nieder- und Oberarnbach in den Jahren 1880–1914*, [no place or publisher cited].

Pflug, Rittergutsbesitzer (1911) Zu Fruwirth: 'Formen des Pflanzenzüchtungsbetriebs'. *Fühlings Landwirtschaftliche Zeitung*, 60, 28–30 and 452–455.

Pflug, Rittergutsbesitzer (1912) Züchtung von Gründüngungspflanzen. Staatliche Züchtungsaufträge. .... *Mitteilungen der DLG*, 27, 648–650.

Pflug, R. and Kiessling, L. (1917) Inwiefern hat die deutsche Pflanzenzucht zum wirtschaftlichen Durchhalten beigetragen? *Landwirtschaftliches Jahrbuch für Bayern*, 7, 4–32.

Pimentel, D. *et al.* (1989) Benefits and risks of genetic engineering in agriculture. *BioScience*, 39, 606–609.
Pincus, J. (2009) The post-Washington consensus and lending operations in agriculture: new rhetoric and old operational realities. In Fine, B., Lapavitsas, C. and Pincus, J. (eds) *Development Policy in the 21st Century: Beyond the Post-Washington Consensus.* London, Routledge.
Pingali, P. L. and Traxler, G. (2002) Changing locus of agricultural research: will the poor benefit from biotechnology and privatization trends? *Food Policy* 27, 223–238.
Pinstrup-Andersen, P. and Cohen, M. J. (2001) Modern agricultural biotechnology and developing country food security. In Nelson, G. (ed.) *Genetically Modified Organisms in Agriculture: Economics and Politics.* San Diego, Academic Press.
Pinstrup-Andersen, P. and Hazell, P. (1985) The impact of the green revolution and prospects for the future. *Food Reviews International*, 1, 1–25.
Pinstrup-Andersen, P. and Schiöler, E. (2001) *Seeds of Contention: World Hunger and the Global Controversy over GM Crops*, Baltimore, Johns Hopkins University Press.
Poitras, M. (2008) Unnatural growth: the political economy of biotechnology in Mexico. In Otero, G. (ed.) *Food for the Few: Neo-liberal Globalism and Biotechnology in Latin America.* Austin, University of Texas Press.
Polanyi, K. (2001; orig. publ. 1944) *The Great Transformation: the Political and Economic Origins of our Time*, Boston, Beacon Press.
Pollack, A. (2008) Less water, more biotech. *New York Times*, 23 October, B1 and B6.
Pommer, E. (ed.) (1900) *Jahresbericht über die Erfahrungen und Fortschritten auf dem Gesammtgebiete der Landwirtschaft* (vol. 13, 1898), Braunschweig, Vieweg.
Popplow, M. (2008) Gelehrte zwischen Obst und Gemüse. Historische Traditionslinien des ökonomischen Gartens Johann Metzgers auf dem Heidelberger Schloss. In Rink, C. (ed.) *Mit Spaten und Feder. Johann Metzger (1789–1852). Landschaftsarchitekt, Botaniker und Gestalter des Heidelberger Schlossgartens (=Archiv und Museum der Universität Heidelberg, Schriften 13).* Ubstadt-Weiher, Verlag Regionalkultur.
Popplow, M. (2009) Economizing agricultural resources in the German economic enlightenment. In Klein, U. and Spary, E. (eds) *Materials and Expertise in Early Modern Europe: Between Market and Laboratory.* Chicago, University of Chicago Press.
Popplow, M. (2010) Die ökonomische Aufklärung als Innovationskulltur des 18. Jahrhunderts zur optimierten Nutzung natürlicher Ressourcen. In Popplow, M. (ed.) *Landschaften agrarisch-ökonomischen Wissens: Strategien innovativer Ressourcennutzung in Zeitschriften und Sozietäten des 18. Jahrhunderts.* Münster, Waxmann.
Porter, D., Allen, B. and Thompson, G. (1991) *Development in Practice: Paved with Good Intentions*, London, Routledge.
Pray, C. (1996) The impact of privatizing agricultural research in Great Britain: an interim report on PBI and ADAS. *Food Policy*, 21, 305–318.
Pray, C. (2002) The growing role of the private sector in agricultural research. In Byerlee, D. and Echeverria, R. (eds) *Agricultural Research Policy in an Era of Privatization.* Wallingford, CABI Publishing.
Pray, C. *et al.* (2002) Five years of Bt cotton in China – the benefits continue. *The Plant Journal*, 31, 423–430.
Pray, C. and Naseem, A. (2007) Supplying crop biotechnology to the poor: Opportunities and constraints. *Journal of Development Studies*, 43, 192–217.
Press, E. and Washburn, J. (2000) The kept university. *Atlantic Monthly*, 285, 39–54.
Pretty, J. N. (1997) The sustainable intensification of agriculture. *Natural Resources Forum*, 21, 247–256.

Pretty, J. N. and Chambers, R. (1994) Towards a learning paradigm: new professionalism and institutions for a sustainable agriculture. In Scoones, I. and Thompson, J. (eds) *Beyond Farmer First: Rural People's Knowledge, Agricultural Research and Extension Practice.* London, Intermediate Technology Publications.

Pretty, J. N. *et al.* (2006) Resource-conserving agriculture increases yields in developing countries. *Environmental Science and Technology*, 40, 1114–1119.

Pribyl, L. *et al.* (eds) (1890) *Das Goldene Buch der Land- und Forstwirtschaft in Oesterreich-Ungarn*, Vienna, Selbstverlag der Herausgeber.

Pringle, P. (2003) *Food, Inc.: Mendel to Monsanto – the Promises and Perils of the Biotech Harvest*, New York, Simon and Schuster.

Pritchett, L. (2011) Remembering the forgetting in schooling. In Bayly, C. A., Rao, V., Szreter, S. and Woolcock, M. (eds) *History, Historians and Development Policy: a Necessary Dialogue.* Manchester, Manchester University Press.

von Proskowetz, E. (1924) Vorwort. In Deutsche Landwirtschaftsgesellschaft in Brünn. (ed.) *Beiträge zum landwirtschaftlichen Pflanzenbau, insbesondere Getreidebau. Festschrift zum 70. Geburtstage Prof. Franz Schindler.* Berlin, Paul Parey.

von Proskowetz, E. (1937) *Memorabilien*, Prague, [no publisher cited].

Pruns, H. (1979) *Staat und Agrarwirtschaft, 1800–1865. Subjekte und Mittel der Agrarverfassung und Agrarverwaltung im Frühindustrialismus (= Berichte über Landwirtschaft*, 194. Sonderheft), Hamburg, Parey.

Puente-Rodríguez, D. (2010) *Redesigning Genomics – Reconstructing Societies: Local Sustainable Biotechnological Developments* (Ph.D. diss., Vrije Universiteit, Amsterdam), Zutphen, Wöhrmann Print Service.

Puhle, H.-J. (1966) *Agrarische Interessenpolitik und preussischer Konservatismus im Wilhelminischen Reich (1893–1914)*, Hannover, Verlag für Literatur und Zeitgeschehen.

Puhle, H.-J. (1975) *Politische Agrarbewegungen in kapitalistischen Industriegesellschaften. Deutschland, USA und Frankreich im 20. Jahrhundert*, Göttingen, Vandenhoeck & Ruprecht.

Puhle, H.-J. (1986) Lords and peasants in the Kaiserreich. In Moeller, R. G. (ed.) *Peasants and Lords in Modern Germany: Recent Studies in Agricultural History.* Boston, Allen & Unwin.

Pyta, W. (1991) *Landwirtschaftliche Interessenpolitik im deutschen Kaiserreich. Der Einfluss agrarischer Interessen auf die Neuordnung der Finanz- und Wirtschaftspolitik am Ende der 1870er Jahre am Beispiel von Rheinland und Westfalen*, Stuttgart, Franz Steiner Verlag.

Qaim, M. and Zilberman, D. (2003) Yield effects of genetically modified crops in developing countries. *Science*, 299, 900–902.

Raina, R. S. (2006) Researching the drylands. *Seminar*, 546, 25–29.

Raina, R. S. *et al.* (2006) The soil sciences in India: Policy lessons for agricultural innovation. *Research Policy*, 35, 691–714.

Ramaswami, C. and Pray, C. (2007) India: Confronting the challenge – the potential of genetically modified crops for the poor. In Fukuda-Parr, S. (ed.) *The Gene Revolution: GM Crops and Unequal Development.* London, Earthscan.

Ramm, E. (ed.) (1925) *Deutsche Hochzuchten, Bd.. 5: Sommergerste, Wintergerste und Hafer*, Berlin, Paul Parey.

Ramm, E. (ed.) (1926) *Deutsche Hochzuchten, Bd. 6: Sommerroggen, Winterroggen, Sommerweizen u. Winterweizen*, Berlin, Paul Parey.

Randerson, J. (2008) Science minister attempts to reopen the debate on GM crops. *The Guardian*, 22 September.

von Rathlef, H. (1925) Wie lässt sich die Leistung der deutschen Pflanzenzüchterischen Arbeit erhöhen? *Deutsche Landwirtschaftliche Presse*, 52, 244–245.

Raum, H. (1912) Züchtung und Saatbau des Fichtelgebirgshafers. *Landwirtschaftliches Jahrbuch für Bayern*, 2, 841–940.

Raum, H. (1959) Die akademischen Lehrgestalten der bayerischen Landwirtschaft. *Bayerisches landwirtschaftliches Jahrbuch*, 36, 741–761.

Rausser, G., Simon, L. and Ameden, H. (2000) Public-private alliances in biotechnology: Can they narrow the knowledge gaps between rich and poor? *Food Policy*, 25, 499–513.

Raven, P. (2003) Public-good plant-breeding: what are the international priorities? lecture at Natural History Museum, London, 22 May.

RdP (ed.) (1942) *Ratgeber für Saatgutbeschaffung u. Sortenwahl 1942*, Berlin, Reichsverband der Pflanzenzucht.

Redclift, M. (1983) Production programs for small farmers: Plan Puebla as myth and reality. *Economic Development and Cultural Change*, 31, 551–570.

Reeves, T. G. and Cassaday, K. (2002) History and past achievements of plant breeding. *Australian Journal of Agricultural Research*, 53, 851–863.

Reif, H. (1994) Einleitung. In Reif, H. (ed.) *Ostelbische Agrargesellschaft im Kaiserreich und in der Weimarer Republik. Agrarkrise – junkerliche Interessenpolitik – Modernisierungsstrategien.* Berlin, Akademie Verlag.

Reitemeier, A. (1904) *Geschichte der Züchtung landwirtschaftlicher Kulturpflanzen*, doctoral diss., Universität Breslau.

Remy, T. (1907a) Das Saatgut und seine Gewinnung unter besonderer Berücksichtigung der Wirksamkeit der Saatbauvereine. *Deutsche Landwirtschaftliche Presse*, 34, 296–297.

Remy, T. (1907b) Einige Gedanken über die Gefahren und Nachteile des modernen Pflanzenzuchtbetriebes. *Deutsche Landwirtschaftliche Presse*, 34, 687–688.

Remy, T. (1910) Bonn-Poppelsdorf: Institut für Boden- u. Pflanzenbaulehre… In Hillmann, P. (ed.) *Die deutsche landwirtschaftliche Pflanzenzucht (= Arbeiten der DLG*, vol. 168). Berlin, DLG-Verlag.

Reynolds, M. P. and Borlaug, N. E. (2006) Applying innovations and new technologies for international collaborative wheat improvement. *Journal of Agricultural Science*, 144, 95–110.

Rhoades, R. E. (2006) Seeking half our brains: constraints and incentives in the social context of interdisciplinary research. In Cernea, M. M. and Kassam, A. H. (eds) *Researching the Culture in Agri-Culture: Social Research for International Development.* Wallingford, CABI Publishing.

Richards, P. (1985) *Indigenous Agricultural Revolution: Ecology and Food Production in West Africa*, London, Hutchinson.

Richards, P. (1994) The shaping of biotechnology: Institutional culture and ideotypes. *Biotechnology and Development Monitor*, 18, 24.

Richey, W. and Strich, M. (eds) (1984) *Der Honig ist nicht weit vom Stachel: Sprichwörter, Redensarten, Wetterregeln und Rätsel aus dem Bauernleben*, Leipzig, Koehler & Amelang.

Richter, W. (1908) Bemerkungen zu dem Artikel 'Gefahren und Nachteile des modernen Pflanzenzuchtbetriebes', *Deutsche Landwirtschaftliche Presse*, 35, 75.

Rist, G. (2008) *The History of Development: from Western Origins to Global Faith*, London, Zed Books.

Rivera, W. M. (1996) Agricultural extension in transition worldwide: structural, financial and managerial strategies for improving agricultural extension. *Public Administration and Development*, 16, 151–161.

Röbbelen, G. (ed.) (2002) *Biographisches Lexikon zur Geschichte der Pflanzenzüchtung, 2. Folge*. Göttingen, Gesellschaft für Pflanzenzüchtung.

Roemer, T. (1914) *Mendelismus und Bastardzüchtung der landwirtschaftlichen Kulturpflanzen* (= *Arbeiten der DLG*, no. 266), Berlin, DLG Verlag.

Roemer, T. (1926) *Beobachtungen auf dem Gebiete des Ackerbaues in den Vereinigten Staaten von Nordamerika*, Berlin, Parey.

Roemer, T. (1932) Steigerung der Ernten durch Düngung und Züchtung. In Sering, M. (ed.) *Die deutsche Landwirtschaft unter volks- und weltwirtschaftlichen Gesichtspunkten.* Berlin, Parey.

Roemer, T. and Rudorf, W. (eds) (1941ff.) *Handbuch der Pflanzenzüchtung*, 5 vols., Berlin, Paul Parey.

Rolfes, M. (1976) Landwirtschaft, 1850–1914. In Aubin, H. and Zorn, W. (eds) *Handbuch der deutschen Wirtschafts- und Sozialgeschichte*, vol. 2. Stuttgart, Ernst Klett.

Rosenberg, H. (1969) Zur sozialen Funktion der Agrarpolitik im Zweiten Reich. In Rosenberg, H. (ed.) *Probleme der deutschen Sozialgeschichte.* Frankfurt a.M., Suhrkamp.

Rösener, W. (1997) *Einführung in die Agrargeschichte*, Darmstadt, Primus Verlag.

Rosset, P. (2002) Taking seriously the claim that genetic engineering could end hunger: a critical analysis. In Bailey, B. and Lappé, M. (eds) *Engineering the Farm: Ethical and Social Aspects of Agricultural Biotechnology.* Washington, Island Press.

Rossiter, M. (1979) The organization of the agricultural sciences. In Oleson, A. and Voss, J. (eds) *The Organization of Knowledge in Modern America, 1860–1920.* Baltimore, Johns Hopkins University Press.

Royal Society (2009) *Reaping the Benefits: Science and Sustainable Intensification of Global Agriculture*, London, Royal Society.

Rudorf, W. (1955) Bahnbrechende Fortschritte in der Pflanzenzüchtung. In Gerl, F. (ed.) *Mehr Forschen, Besser Leben. Bahnbrechende Leistungen der Landbauwissenschaften.* Bad Godesberg, Land- und Forstwirtschaftlicher Forschungsrat e.V.

von Rümker, K. (1889) Ueber neue Methoden und Hilfsmittel für Vervollkommnung unserer Kulturgewächse durch Züchtung. *Jahrbuch der DLG*, 4, 62–79.

von Rümker, K. (1894) Einiges über Zuckerrübenzüchtung. *Blätter für Zuckerrübenbau*, 1, 194–197.

von Rümker, K. (1895) Die Rassenzüchtung landwirtschaftlicher Kulturpflanzen als Forschungsgebiet und Lehrgegenstand. In Anon. (ed.) *Festschrift zum 70. Geburtstags J. Kühn.* Berlin, Parey.

von Rümker, K. (1898) Der wirtschaftliche Mehrwert guter Kulturvarietäten und auserlesenen Saatgutes. *Arbeiten der DLG*, 36, 127–153.

von Rümker, K. (1899) Kann Deutschland seinen Getreidebedarf noch selbst decken? *Mitteilungen der landwirt. Institute der königlichen Universität Breslau*, 1, 152–195.

von Rümker, K. (1900) Ueber die Bedeutung der landwirt. Versuchsstationen und speciell über den Wert und die Organisation von Versuchswirtschaften in historischer Beleuchtung. *Journal für Landwirtschaft*, 48, 179–203.

von Rümker, K. (1904) Erster Bericht über das landwirtschaftliche Versuchsfeld der Königlichen Universität Breslau in Rosenthal. *Mitteilungen der landwirtschaftlichen Institute der Königlichen Universität Breslau*, 2, 817–911.

von Rümker, K. (1905) *Der Saatbau und die Saatbauvereine*, Berlin, Parey.

von Rümker, K. (1907a) Die systematische Einteilung und Benennung der Getreidesorten für praktische Zwecke. *Fühlings Landwirtschaftliche Zeitung*, 56, 545–566.
von Rümker, K. (1907b) *Ueber Sortenauswahl bei Getreide (= Tagesfragen aus dem modernen Ackerbau*, no. 5*)*, Berlin, Parey.
von Rümker, K. (1909) *Ueber Organisation der Pflanzenzüchtung*, Berlin, Parey.
von Rümker, K. (1910) Was können wir von der Saatzucht des Auslandes lernen? *Jahrbuch der DLG*, 25, 47–59.
von Rümker, K. (1911a) *Das landwirtschaftliche Versuchs- und Unterrichtswesen in den USA*, Berlin, Parey.
von Rümker, K. (1911b) Ueber Bedeutung und Methoden der Saatgutzucht. *Mentzel und Lengerke*, 64–90.
von Rümker, K. (1912) Die Abteilung für Pflanzenzucht am Institut für landwirt. Pflanzen-produktionslehre der Königlichen Universität Breslau. *Deutsche Landwirtschaftliche Presse*, 27 January, 75–78.
von Rümker, K. (1912–13) Die Entwicklung der landwirtschaftlichen Pflanzenzucht in Deutschland und ihre betriebswirtschaftliche Aufgaben. *Zeitschrift für Pflanzenzüchtung*, 1, 329–346.
von Rümker, K. (1913) *Die Entwicklung der Landwirtschaft in den letzten 25 Jahren*, Berlin, Parey.
von Rümker, K. (1914) Das landwirtschaftliche Versuchs- und Bildungswesen. *Journal für Landwirtschaft*, 62, 97–114.
von Rümker, K. (1925a) Die wirtschaftliche Lage der Pflanzenzüchter und deren Auswirkung auf die allgemeine Produktion der deutschen Landwirtschaft. *Deutsche Landwirtschaftliche Presse*, 52, 664–666.
von Rümker, K. (1925b) Dr. h.c. Ferdinand von Lochow-Petkus. Seine Bedeutung als Züchter und sein Einfluss auf die Entwicklung der deutschen Pflanzen- und Tierzucht. *Mitteilungen der DLG*, 40, 270–273.
von Rümker, K. (1928a) 20 Jahre Gesellschaft zur Förderung deutscher Pflanzenzucht. *Mitteilungen der DLG*, 43, 248–250.
von Rümker, K. (1928b) Pflanzenzüchtung. In Keiser, F. (ed.) *Die Landwirtschaft (= Die Deutsche Wirtschaft und ihre Führer*, vol. 7*)*. Gotha, Flamberg Verlag.
von Rümker, K. (1929) Gedanken über Aufgaben von Wissenschaft und Praxis auf dem Gebiete der Pflanzenzucht. *Mitteilungen der DLG*, 44, 921–923.
von Rümker, K. (1937) *Mein Leben*, Berlin, typescript.
von Rümker, K. and von Tschermak, E. (1910) *Landwirtschaftliche Studien in Nordamerika mit besonderer Berücksichtigung der Pflanzenzüchtung*, Berlin, Parey.
Ruttan, V. (1977) The Green Revolution: Seven generalizations. *International Development Review*, 4, 16–23.
Ruttan, V. (1980) Bureaucratic productivity: the case of agricultural research. *Public Choice*, 35, 529–547.
Ruttan, V. (1983) Reforming the global agricultural research support system. In Nobe, K. and Sampath, R. K. (eds) *Issues in Third World Development*. Boulder, Westview Press.
Ruttan, V. (1989) The international agricultural research system. In Compton, J. L. (ed.) *The Transformation of International Agricultural Research and Development*. Boulder, Lynne Rienner.
Ruttan, V. W. (1999) Biotechnology and agriculture: a skeptical perspective. *AgBioForum*, 2, 54–60.
Saatstelle (1932) *Verzeichnis der in das DLG-Hochzuchtregister für landwirtschaftliche Pflanzenzuchten eingetragenen Sommersaaten u. Hackfrüchte*, Berlin, DLG

Saatstelle (1933) *Verzeichnis der in das DLG-Hochzuchtregister für landwirtschaftliche Pflanzenzuchten im Herbst 1933 eingegangenen Winter-Getreidesorten*, Berlin, DLG

Saatzuchtstelle (1905a) Das DLG-Hochzuchtregister. *Mitteilungen der DLG*, 20, 408–410.

Saatzuchtstelle (1905b) Einrichtungen der DLG zur Förderung der Pflanzenzüchtung und des Pflanzenbaues. *Mitteilungen der DLG*, 20, 104–108.

Saatzuchtstelle (1924) DLG-Hochzuchtregister für landwirtschaftliche Pflanzenzuchten. *Mitteilungen der DLG*, 39, 821–822.

Saatzuchtstelle (1925) DLG-Hochzuchtregister. *Mitteilungen der DLG*, 40, 168.

Sample, I. (2008) Hunger in Africa blamed on western rejection of GM food. *Guardian*, 8 September, 9.

Sanghi, N. K. (1989) Changes in the organization of research on dryland agriculture. In Chambers, R., Pacey, A. and Thrupp, L. A. (eds) *Farmer First: Farmer Innovation and Agricultural Research.* London, Intermediate Technology Publications.

Santaniello, V. (2002) Biotechnology and traditional breeding in sub-Saharan Africa. In Swanson, T. (ed.) *Biotechnology, Agriculture and the Developing World: the Distributional Implications of Technological Change.* Cheltenham, Edward Elgar.

Santayana, G. (1906) *The Life of Reason: or the Phases of Human Progress, vol. 5: Reason in Science*, London, Constable.

Saraiva, T. (2010) Fascist Labscapes: Geneticists and the new landscapes of fascism in Italy and Portugal. *Historical Studies in the Natural Sciences*, 40, 457–498.

Sawada, S. (1965) Innovation in Japanese agriculture, 1880–1935. In Lockwood, W. (ed.) *The State and Economic Enterprise in Japan.* Princeton, Princeton University Press.

Scharnagel, T. (1914) Die bayerische Pflanzenzüchtung mit besond. Berücksichtigung der Qualitätsfrage bei Getreide. *Mitteilungen des Oberbayer. Kreisackerbau-Verbandes*, 3.

Scharnagel, T. (1928) Bayerische Pflanzenzüchtungen. *Wochenblatt Bayern*, 118, 492–494.

Scharnagel, T. (1929) Die bayerische Saatzüchtung. *Mitteilungen der DLG*, 44, 495–496.

Scharnagel, T. (1930) Alte und neue Wege der Pflanzenzüchtung. *Bayerische Ackerbauzeitung* (Beilage z. *Wochenblatt Bayern* 120), 57–59.

Scharnagel, T. (1936/37) Bayerische Landessaatzuchtanstalt Weihenstephan. *Praktische Blätter für Pflanzenbau und Pflanzenschutz*, 14, 35–90.

Scharnagel, T. (1953) Die Bedeutung der Lokalsortenzüchtung für Bayern. *Landwirtschaftliches Jahrbuch für Bayern*, 30, 174–181.

Scheibe, A. (1942) Ludwig Kiessling. *Zeitschrift für Pflanzenzüchtung*, 24, 592–598.

Scheibe, A. (1961) Die Entwicklung der Pflanzenzucht in Deutschland: ein historischer Ueberblick. In Anon. (ed.) *Saatgut in der Bundesrepublik Deutschland.* Bonn, Land- und Hauswirtschaftliche Auswertungs- und Informationsdienst, e.V.

Scheibe, A. (1987) *Bedeutung der wissenschaftlichen Institute für die private Pflanzenzüchtung (= Berichte über Landwirtschaft*, Sonderband no. 200), Hamburg, Paul Parey.

Schick, R. (1958) 100 Jahre Landwirtschaftliche Pflanzenzüchtung in Deutschland. *Sitzungsberichte der Deutschen Akademie der Landwirtschaftswissenschaften zu Berlin*, 7, 3–21.

Schiffer, W. (2001) *Bauern-Bilder. Fotographien aus 50 Jahren Landwirtschaft* (ed. G. Strotdrees and H.-G. Topüth), Münster-Hiltrup, Landwirtschaftsverlag.

Schikorra, M. (1914a) Grundlagen der Futterpflanzenzüchtung. *Deutsche Landwirtschaftliche Presse*, 41, 327–328 and 339–340.

Schikorra, M. (1914b) Erfolge der Pflanzenzüchtung. *Illustrierte Landwirtschaftliche Zeitung*, 34, 307–308 and 317–318.

Schimmelpfennig, D. E., Pray, C. and Brennan, M. F. (2004) The impact of seed industry concentration on innovation: a study of U.S. biotech market leaders. *Agricultural Economics*, 30, 157–167.

Schindler, F. (1907a) Inwieweit hat die Getreidezüchtung auf die Landrassen Rücksicht zu nehmen und welche Massnahmen sind geeignet, die Saatgutzüchtung in wirksamster Weise zu fördern? In Anon. (ed.) *VIII. Internationaler landwirtschaftlicher Kongress Wien.* Vienna, F.A. Wachtl.

Schindler, F. (1907b) Ueber die Bedeutung des Getreidebaues für die mitteleuropäische Landwirtschaft. In Anon. (ed.) *VIII. Internationaler landwirtschaftlicher Kongress Wien.* Vienna, F.A.Wachtl.

Schindler, F. (1909) *Der Getreidebau auf wissenschaftlicher und praktischer Grundlage*, Berlin, Parey.

Schindler, F. (1928) Ueber die Notwendigkeit der Erforschung und Erhaltung der Getreidelandrassen im Hinblick auf ihre züchterische und wirtschaftliche Bedeutung. In Doerell, E. G. (ed.) *Festschrift anlässlich des 70. Geburtstages Prof. Dr. Julius Stoklasa.* Berlin, Parey.

Schindler, F. (1937) *Rückblick auf mein Leben und mein Wirken im Dienste der landwirtschaftlichen Lehre u. Forschung (Quellen und Grundlagen, Reihe B, Nr. 6)*, Prague, Czechoslovakian Academy of Agriculture.

Schissler, H. (1978) *Preussische Agrargesellschaft im Wandel. Wirtschaftliche, gesellschaftliche, und politische Transformationsprozesse von 1763–1847*, Göttingen, Vandenhoeck & Ruprecht.

Schlögl, A. (1954) *Bayerische Agrargeschichte: die Entwicklung der Land- und Forstwirtschaft seit Beginn des 19. Jahrhunderts*, Munich, Bayerischer Landwirtschafts-Verlag.

Schmiel, M. (1987) Landwirtschaftliches Bildungswesen. In Jeismann, K.-E. and Lundgreen, P. (eds) *Handbuch der deutschen Bildungsgeschichte, Bd III: 1800–1870.* Munich, C.H. Beck.

Schmiel, M. (1991) Landwirtschaftliche Berufsbildung. In Berg, C. (ed.) *Handbuch der deutschen Bildungsgeschichte, Bd. 4.* Munich, C.H. Beck.

Schmitt, G. (1991) Why is the agriculture of advanced western economies still organized by family farms? Will this continue to be so in future? *European Review of Agricultural Economics*, 18, 443–458.

Schneider, J. (2002) Selecting with farmers: the formative years of cereal breeding and public seed in Switzerland (1889–1936). In Cleveland, D. and Soleri, D. (eds) *Farmers, Scientists and Plant Breeding.* Wallingford, CAB International.

Schönberg, M. (1936) Pflanzenbautagung Weihenstephan 1936. *Der Forschungsdienst*, 2, 117–120.

Schreiner, G. (1974) Ziele und Mittel der Agrarstrukturpolitik seit 1871 im Deutschen Reich und in der Bundesrepublik Deutschland, doctoral diss., Universität Bonn.

von der Schulenburg, H. (2002) Hans Lang. In Röbbelen, G. (ed.) *Biographisches Lexikon zur Geschichte der Pflanzenzüchtung, 2. Folge*, Göttingen, Gesell. für Pflanzenzüchtung.

von der Schulenburg, H. (2004) Ende des Sortenwirrwarrs bei Futterrüben. *Vorträge für Pflanzenzüchtung*, 65, 43–48.

Schultz, T. (1983, 1st ed. 1964) *Transforming Traditional Agriculture*, Chicago, University of Chicago Press.

Schwartz, M. (1990) Zwischen "Reich" und "Nährstand". Zur soziostrukturellen und politischen Heterogenität der Landesbauernschaft Westfalen im "Dritten Reich". *Westfälische Forschungen*, 40, 303–336.

Schweiger, G. (1929) Vom Gerstenbauverein zur Ackerbauorganisation. In Dörfler, H. (ed.) *Die Landwirtschaft in Bayern: eine Sammlung von Darstellungen über Entwicklung und Stand der bayerischen Landwirtschaft.* Freising, Datterer.

Schweiger, G. and Balz, P. (1927) Zum 25-jährigen Bestehen der Landessaatzuchtanstalt. *Bayerische Ackerbauzeitung* (= Sonderbeilage z. *Wochenblatt Bayern*, Nr. 46).

Scoones, I. (2002) Can agricultural biotechnology be pro-poor? A sceptical look at the emerging 'consensus'. *IDS Bulletin*, 33, 114–119.

Scoones, I. (2006) *Science, Agriculture and the Politics of Policy: the Case of Biotechnology in India*, New Delhi, Orient Longman.

Scoones, I. (2008) Mobilizing against GM crops in India, South Africa and Brazil. *Journal of Agrarian Change*, 8, 315–344.

Scoones, I. (2009) The politics of global assessment: the case of the IAASTD. *Journal of Peasant Studies*, 36, 547–571.

Scoones, I. and Thompson, J. (eds) (1994a) *Beyond Farmer First: Rural People's Knowledge, Agricultural Research and Extension Practice*, London, Intermediate Technology Publications.

Scoones, I. and Thompson, J. (1994b) Introduction. In Scoones, I. and Thompson, J. (eds) *Beyond Farmer First: Rural People's Knowledge, Agricultural Research and Extension Practice.* London, Intermediate Technology Publications.

Scott, J. C. (1998) *Seeing like a State: How Certain Schemes to Improve the Human Condition Have Failed*, New Haven, Yale University Press.

Seitzer, J. F. (2008) Globalisierung in der Pflanzenzüchtung. In Röbbelen, G. (ed.) *Die Entwicklung der Pflanzenzüchtung in Deutschland (1908–2008). Hundert Jahre GFP, e.V. – eine Dokumentation.* Göttingen, Gesellschaft für Pflanzenzüchtung.

Sempolowsky, A. (1903) Einiges über die Getreidezüchtung im Königreich Polen. *Deutsche Landwirtschaftliche Presse*, 30, 219–220 and 227.

Serageldin, I. and Persley, G. J. (2000) *Promethean Science: Agricultural Biotechnology, the Environment, and the Poor*, Washington, DC, CGIAR.

Sessous, G. (1936) Aus der Reichsarbeitsgemeinschaft "Pflanzenbau" des Forschungsdienstes. *Der Forschungsdienst*, 2, 2–9.

Shanin, T. (1987a) Introduction: peasantry as a concept. In Shanin, T. (ed.) *Peasants and Peasant Societies.* Oxford, Blackwell.

Shanin, T. (1987b) Short historical outline of peasant studies. In Shanin, T. (ed.) *Peasants and Peasant Societies.* Oxford, Blackwell.

Shen, X. (2010) Understanding the evolution of rice technology in China – from traditional agriculture to GM rice today. *Journal of Development Studies*, 46, 1026–1046.

Shepherd, C. (2005) Imperial science: the Rockefeller Foundation and agricultural science in Peru, 1940–1960. *Science as Culture*, 14, 113–137.

Shepherd, C. (2006) From in vitro to in situ: on the precarious extension of agricultural science in the indigenous 'third world'. *Social Studies of Science*, 36, 399–426.

Shrum, W. (2000) Science and story in development: the emergence of non-governmental organizations in agricultural research. *Social Studies of Science*, 30, 95–124.

Siegl, G. (n.d.) Die Geschichte der Landesanstalt für Pflanzenzucht und Samenprüfung in Rinn (Tirol) von ihrer Gründung 1939 bis zu ihrer Auflösung 1999. unpublished paper, Institut für Geschichtswissenschaften, Universität Innsbruck.

Sierig, E. (1908) Zu der Frage: 'Gefahren u. Nachteile des modernen Pflanzenzuchtbetriebes'. *Deutsche Landwirtschaftliche Presse*, 35, 15.

Silveira, J., da Borges, J. and de Carvalho, I. (2007) Brazil: Confronting the challenges of

global competition and protecting biodiversity. In Fukuda-Parr, S. (ed.) *The Gene Revolution: GM Crops and Unequal Development.* London, Earthscan.
Simmonds, N. (1979) *Principles of Crop Improvement*, New York, Longman.
Simmonds, N. (1990) The social context of plant-breeding. *Plant Breeding Abstracts*, 60, 337–341.
Simmonds, N. (1991) Selection for local adaptation in a plant breeding programme. *Theoretical and Applied Genetics*, 82, 363–367.
Sirks, M. J. (ed.) (1936) *Congrès International de Sélectionneurs de Plantes*, Netherlands 22–27 June, Leiden, E. J. Brill.
Sisler, D. G. and Colman, D. (1979) *Poor Rural Households, Technical Change, and Income Distribution in Developing Countries: Insights from Asia*, Ithaca, Department of Agricultural Economics, Cornell University.
Skidelsky, R. (2009) *Keynes: Return of the Master*, New York, Public Affairs.
Smith, E. (2009) Imaginaries of development: the Rockefeller Foundation and rice research. *Science as Culture*, 18, 461–482.
Smith, J. (2009) *Science and Technology for Development*, London, Zed.
Snell, K. (1939) Sortenschutz durch Registrierung. *Der Züchter*, 11, 22–24.
Soleri, D. *et al.* (2008) Testing assumptions underlying economic research on transgenic food crops for Third World farmers: evidence from Cuba, Guatemala and Mexico. *Ecological Economics*, 67, 667–682.
Speich, D. (2009a) The Kenyan style of 'African socialism': developmental knowledge claims and the explanatory limits of the cold war. *Diplomatic History*, 33, 449–466.
Speich, D. (2009b) Der Blick vom Lake Success: Das Entwicklungsideal der frühen UNO als "lokales Wissen". In Büschel, H. and Speich, D. (eds) *Entwicklungswelten. Globalgeschichte der Entwicklungszusammenarbeit.* Frankfurt a.M., Campus Verlag.
Spencer, D. S. C. (1986) Agricultural research: lessons of the past, strategies for the future. In Berg, R. J. and Whitaker, J. S. (eds) *Strategies for African Development.* Berkeley, University of California Press.
Spennemann, F. (ed.) (n.d., *c.*1938) *Die Neuordnung des deutschen Saatgutwesens (= Arbeiten des Reichsnährstandes*, no. 50), Berlin, Reichsnährstand-Verlag.
Sperling (1908) Bemerkungen zu dem Artikel 'Gefahren und Nachteile des modernen Pflanzenzuchtbetriebes'. *Deutsche Landwirtschaftliche Presse*, 35, 73–75.
Sperling, I. *et al.* (2001) Participatory plant breeding: a framework for analyzing diverse approaches. In Anon. (ed.) *An Exchange of Experiences from South and Southeast Asia: Proceedings of an International Symposium on Participatory Plant Breeding and Participatory Plant Genetic Resource Enhancement.* Cali/Colombia, CGIAR.
Spielman, D. J., Cohen, J. and Zambrano, P. (2006) Will agbiotech applications reach marginalized farmers? Evidence from developing countries. *AgBioForum*, 9, 23–30.
Spielman, D. J. and Grebmer, K. V. (2004) *Public-Private Partnerships in Agricultural Research: an Analysis of Challenges Facing Industry and the CGIAR (= Environment and Production Technology Division* working paper #113*)*, Washington, DC, IFPRI.
Spillane, C. (2002) Agricultural biotechnology and developing countries: proprietary knowledge and diffusion of benefits. In Swanson, T. (ed.) *Biotechnology, Agriculture and the Developing World: the Distributional Implications of Technological Change.* Cheltenham, Edward Elgar.
Staatz, J. M. and Eicher, C. K. (1990) Agricultural development ideas in historical perspective. In Eicher, C. K. & Staatz, J. M. (eds) *Agricultural Development in the Third World.* Baltimore, Johns Hopkins University Press.
Staatz, J. M. and Eicher, C. K. (1998) Agricultural development ideas in historical

perspective. In Eicher, C. K. and Staatz, J. M. (eds) *International Agricultural Development*. Baltimore, Johns Hopkins University Press.

Stadelmann (1924) Saatzüchterische Tagesfragen. *Wochenblatt Bayern*, 114, 233–234.

Stakman, E. C., Bradfield, R. and Mangelsdorf, P. (1967) *Campaigns Against Hunger*, Cambridge, MA, Harvard University Press.

Staples, A. (2006) *The Birth of Development: How the World Bank, FAO and World Health Organization Changed the World, 1945–1965*, Kent, OH, Kent State University Press.

Staub, W. and Blase, M. (1971) Genetic technology and agricultural development. *Science* 173, 119–123.

Staus, L. (2011) Die "innere Kolonisation" als Spiegel der Debatte um die Bedeutung der Landwirtschaft in Wirtschaft u. Gesellschaft Deutschlands an der Wende vom 19. zum 20. Jahrhundert. *Arbeitskreis für Agrargeschichte Newsletter*, 28, 3–9.

Stavis, B. (1975) *Making Green Revolution: The Politics of Agricultural Development in China*, Ithaca, Rural Development Committee, Cornell University.

Steglich, B. (1891) Einrichtung, Ziel und Aufgaben der Versuchsstation für Pflanzenkultur in Dresden. In Anon. (ed.) *Mitteilungen der Oekonomischen Gesellschaft im Kgr. Sachsen, 1890–1891*. Dresden, G. Schönfeld.

Steglich, B. (1893) *Ueber Verbesserung und Veredelung landwirtschaftlicher Kulturgewächse durch Züchtung* (lecture, 2 December 1892 at Oekonomische Gesell. im Kgr. Sachsen), Leipzig, Reichenbach.

Steglich, B. (1896) *Sortenauswahl und Züchtung des Getreides* (lecture 8 November 1895 to Oekonomische Gesell. im Kgr. Sachsen), Leipzig, Reichenbachsche Verlagsbuchhandlung.

Steglich, B. (1905) *Einrichtung und Tätigkeit der Saatzuchtanstalt Svalöf in Schweden und Fortschritte auf dem Gebiet der Pflanzenzüchtung* (lecture 2 December 1904 to Oekonomische Gesell. im Kgr. Sachsen zu Dresden), Leipzig, Reichenbach.

Steglich, B. (1907) *Förderung und Hebung der Rentabilität des Pflanzenbaus durch Organisation der Pflanzenzüchtung* (lecture 8 November 1907 to the Oekonomische Gesell. im Kgr. Sachsen zu Dresden), Leipzig, Reichenbach.

Steglich, B. (1910) Dresden: Landwirtschaftliche Abteilung d. Königlich Sächsischen Pflanzenphysiologischen Versuchsstation. In Hillmann, P. (ed.) *Die deutsche landwirtschaftliche Pflanzenzucht*. Berlin, DLG.

Steinbeck, N. (1997) *Bürokratisierung und betriebliche Rationalisierung: die preussischen Landwirtschaftskammern und die Entwicklung der Agrarverfassung im wilhelminischen Kaiserreich*, Bern, Peter Lang.

Stephani, W. (1911) Die Pflanzenzuchtstation des landwirtschaftlichen Instituts an der Universität Halle. *Landwirtschaftliche Umschau*, 3, 1–4.

Steuert, L. (1905) *Die Königliche Bayerische Akademie Weihenstephan und ihre Vorgeschichte. Festschrift zur Jahrhundertfeier 1905*, Berlin, Parey.

Stiefel, K. (1977) *Baden 1648–1952*, Karlsruhe, Verein für oberrheinische Rechts- und Verwaltungsgeschichte.

Stokstad, E. (2008) Dueling visions for a hungry world. *Science*, 319, 1474–1476.

Stoll, P. H. (1905) Entwicklung und Stand der Getreidezüchtung. *Wochenblatt Baden*, 70–71, 84–85 and 98–100.

Stone, G. D. (2002) Fallacies in the genetic-modification wars, implications for developing countries, and anthropological perspectives. *Current Anthropology*, 43, 611–630.

Stoop, W., Uphoff, N. and Kassam, A. (2002) A review of agricultural research issues raised by the system of rice intensification (SRI) from Madagascar: opportunities for improving farming systems for resource-poor farmers. *Agricultural Systems*, 71, 249–274.

Strampelli, N. (1933) *Early Ripening Wheats and the Advance of Italian Wheat Cultivation*, Rome, [no publ. cited].

Strosnider, R. (1984) A private seed company's views of the roles of public and private breeders – cooperation and support. *HortScience*, 19, 800–802.

Stross, R. E. (1986) *The Stubborn Earth: American Agriculturalists on Chinese Soil, 1898–1937*, Berkeley, University of California Press.

Stubbe, H. (1982) *Geschichte des Instituts für Kulturpflanzenforschung Gatersleben d. Deutschen Akad. der Wissenschaften zu Berlin, 1943–1968*, Berlin, Akademie-Verlag.

Stutzer, A. and Gisevius, P. (1904) *Der Wettbewerb der dänischen und der schwedischen Landwirte mit Deutschland. Reiseerinnerungen*, Stuttgart, Eugen Ulmer.

Sucker, U. (1998) *Anfänge der modernen Phytomedizin. Die Gründungsgeschichte der Biologischen Bundesanstalt für Land- und Forstwirtschaft (1898 bis 1919)*... Berlin, Parey.

Swanson, T. (2002) Biotechnologies and developing countries: how will the anticipated industrial changes in agriculture affect developing countries? In Swanson, T. (ed.) *Biotechnology, Agriculture and the Developing World: the Distributional Implications of Technological Change.* Cheltenham, Edward Elgar.

Swanson, T. and Goeschl, T. (2002) Policy options for the biotechnology revolution: what can be done to address the distributional implications of biotechnologies? In Swanson, T. (ed.) *Biotechnology, Agriculture and the Developing World: the Distributional Implications of Technological Change.* Cheltenham, Edward Elgar.

Teich, M. (2000) *Bier, Wissenschaft und Wirtschaft in Deutschland 1800–1914*, Vienna, Boehlau.

Tendler, J. (1993) Tales of dissemination in small-farm agriculture: Lessons for institution builders. *World Development*, 21, 1567–1582.

Teubal, M. (2008) Genetically modified soybeans and the crisis of Argentina's agriculture model. In Otero, G. (ed) *Food for the Few: Neo-liberal Globalism and Biotechnology in Latin America.* Austin, TX, University of Texas Press

Teuteberg, H.-J. (1977) *Die deutsche Landwirtschaft beim Eintritt in die Phase der Hochindustrialisierung*, Cologne, Forschungsinstitut für Sozial- u. Wirtschaftsgeschichte an der Universität zu Köln.

Thiel, H. (1904) Zum Gedächtnis des Amtsrats Dr. Wilhelm Rimpau-Schlanstedt. *Jahrbuch der DLG*, 19, 6–15.

Thomas, A. (2000) Development as practice in a liberal capitalist world. *Journal of International Development*, 12, 773–787.

Thomson, J. C. (1969) *While China Faced West: American Reformers in Nationalist China, 1928–1937*, Cambridge, MA, Harvard University Press.

Tilley, H. (2003) African environments and environmental sciences: the African Research Survey, ecological paradigms and British colonial development, 1920–1940. In Beinart, W. and Mcgregor, J. (eds) *Social History and African Environments.* Oxford, James Currey.

Tilley, H. (2010) Global histories, vernacular science, and African genealogies. *Isis*, 101, 110–119.

Tiltman, H. H. (1934) *Peasant Europe*, London, Jarrolds.

Tirrell, S. R. (1951) *German Agrarian Politics after Bismarck's Fall: the Formation of the Farmers' League*, New York, Columbia University Press.

Toenniessen, G. and Delmer, D. (2005) The role of intermediaries in maintaining the public sector's essential role in crop varietal improvement, report for the Science Council of the CGIAR, Washington, DC, CGIAR.

Tomlinson, I. (2011) A rock and a hard place. *Living Earth* (newsletter of the Soil Association), 242, 18–21.

Tooze, A. (2006) *The Wages of Destruction: the Making and Breaking of the Nazi Economy*, London, Allen Lane.

Tornow, W. (1972) *Chronik der Agrarpolitik u. Agrarwirtschaft des Deutschen Reiches von 1933–1945*, Hamburg, Parey.

Tripp, R. (1997) Unproductive debates: the Green Revolution, biotechnology and agricultural diversity. *Biotechnology and Development Monitor*, 32, 24.

Tripp, R. (2001) Agricultural technology policies for rural development. *Development Policy Review*, 19, 479–489.

Tripp, R. (2002) Can the public sector meet the challenge of private research? Commentary on 'Falcon and Fowler' and 'Pingali and Traxler'. *Food Policy*, 27, 239–246.

Tripp, R. *et al.* (1990) Farming systems research revisited. In Eicher, C. K. and Staatz, J. M. (eds) *Agricultural Development in the Third World*, Baltimore, Johns Hopkins University Press.

Tripp, R. and Byerlee, D. (2000) Public plant breeding in an era of privatization. *Natural Resources Perspectives* (publ. by Overseas Development Institute, London), 57.

Tripp, R. *et al.* (2006) *Self-Sufficient Agriculture: Labour and Knowledge in Small-Scale Farming*, London, Earthscan.

Tschajanow, A. (1923) *Die Lehre von der bäuerlichen Wirtschaft. Versuch einer Theorie der Familienwirtschaft im Landbau*, Berlin, Parey.

von Tschermak, E. (1936–37) 25-jähriges Jubiläum der Gesell. für Pflanzenzüchtung "Z" in Wien. *Zeitschrift für Züchtung, Reihe A*, 21, 522–524.

von Tschermak, E. (1958) *Leben und Wirken eines österreichischen Pflanzenzüchters*, Berlin, Parey.

Tuchman, B. (1976) The green revolution and the distribution of agricultural income in Mexico. *World Development*, 4, 17–24.

Uekötter, F. (2010) *Die Wahrheit ist auf dem Feld. Eine Wissensgeschichte der deutschen Landwirtschaft*, Göttingen, Vandenhoeck & Ruprecht.

van Ufford, P. Q. (1993) Knowledge and ignorance in the practices of development policy. In Hobart, M. (ed.) *An Anthropological Critique of Development: the Growth of Ignorance*. London, Routledge.

Ulander, A. (1906) Die schwedische Pflanzenzucht zu Svalöf. Eine kurze Darstellung. *Journal für Landwirtschaft*, 54, 105–124.

Ullmann, H.-P. (1988) *Interessenverbände in Deutschland*, Frankfurt a.M., Suhrkamp.

Unger, C. R. (2010a) Industrialization vs. agrarian reform: West German modernization policies in India in the 1950s and 1960s. *Journal of Modern European History*, 8, 47–65.

Unger, C. R. (2010b) Histories of development and modernization: findings, reflections, future research. *H-Soz-u-Kult*, online, 9 December. Available at: http://hsozkult.geschichte.hu-berlin.de/forum/2010–12–001.

Uphoff, N. (2007) Agroecological alternatives: capitalizing on existing genetic potentials. *Journal of Development Studies*, 43, 218–236.

Ushiyama, K. (2008) The establishment of Japanese capitalism: 1888 to World War I. In Teruoka, S. (ed.) *Agriculture in the Modernization of Japan, 1850–2000*. New Delhi, Manohar.

Valdes, A. and Foster, W. (2010) Reflections on the role of agriculture in pro-poor growth. *World Development*, 38, 1362–1374.

Vanloqueren, G. and Baret, P. (2009) How agricultural research systems shape a technological regime that develops genetic engineering but locks out agroecological innovations. *Research Policy*, 38, 971–983.

Van Molle, L. (2008) A state for the peasants or the peasants for the state? The two faces of Belgian agricultural policy, ca 1830–1914. In Vivier, N. (ed.) *The State and Rural Societies: Policy and Education in Europe, 1750–2000*, Turnhout, Brepols.

Vanschoenbeek, G., Van Molle, L. and Vanhaute, E. (2002) The workers' party against the Farmers' League: Social Democracy and the peasantry in Belgium, 1893–1914. In Blok, A. *et al.* (eds) *Urban Radicals, Rural Allies: Social Democracy and the Agrarian Issue, 1870–1914.* Bern, Peter Lang.

Vascik, G. (1991) The German Peasant League and the limits of rural liberalism in Wilhelmian Germany. *Central European History*, 24, 147–175.

Victor, D. G. and Runge, C. F. (2002) Farming the genetic frontier. *Foreign Affairs*, 81, 107–121.

Vidal, J. (2008) Millions could starve as fertilizer prices soar, says UN. *Guardian*, 13 August, 21.

Vivier, N. (ed.) (2008) *The State and Rural Societies: Policy and Education in Europe, 1750–2000*, Turnhout, Brepols.

Voss, J. and Breuninger, W. (1944) *Weizensorten. Ihre Erkennung und Bewertung*, Berlin, Reichsnährstandsverlag.

Wacker, J. (1911) Förderung des Acker- und Pflanzenbaus auf gemeinschaftlichem Wege. *Wochenblatt Württemberg*, 34, 435–440.

Wacker, J. (1912) Rückblicke und Ausblicke in Bezug auf die Tätigkeit der Kgl. Saatzuchtanstalt Hohenheim. *Wochenblatt Württemberg*, 35, 262–266 and 284–287.

Wacker, J. (1918) Die Förderung des Dinkelbaus in Württemberg durch die Kgl. Saatzuchtanstalt. In Anon. (ed.) *Festschrift zur Feier des 100-jährigen Bestehens der Kgl. Württ. Landwirtschaftlichen Hochschule Hohenheim.* Stuttgart, Eugen Ulmer.

Wacker, J. (1925) Die Weizenzüchtung (Winter-, Sommer- und Spelzweizen) in Württemberg. *Illustrierte Landwirtschaftliche Zeitung*, 485–486.

Wacker, J. (1933) Die württembergische Landespflanzenzüchtung. *Zeitschrift für Züchtung, Reihe A*, 18, 153–162.

Wade, R. (1996) Japan, the World Bank, and the art of paradigm maintenance: the 'East Asian Miracle' in political perspective. *New Left Review*, 217, 3–36.

Wade, R. (2001) Showdown at the World Bank. *New Left Review*, 7, 124–137.

Warren, D. M. (1989) Linking scientific and indigenous agricultural systems. In Compton, J. L. (ed.) *The Transformation of International Agricultural Research and Development.* Boulder, Lynne Rienner.

Watts, J. (2011) Beijing warned over wheat production. *Guardian*, 29 June.

Watts, M. (1995) 'A new deal in emotions': Theory and practice and the crisis of development. In Crush, J. (ed.) *Power of Development.* London, Routledge.

Waugh, R., Hildebrand, P. E. and Andrew, C. (1989) Farming systems research and extension. In Compton, J. L. (ed.) *The Transformation of International Agricultural Research and Development.* Boulder, Lynne Rienner.

Weaver, W. (1970) *Scene of Change: a Lifetime in American Science*, New York, Charles Scribner's Sons.

Webb, S. (1982) Agricultural protection in Wilhelminian Germany: forging an empire with pork and rye. *Journal of Economic History*, 42, 309–326.

Webster, A. J. (1989) Privatisation of public sector research: the case of a plant breeding institute. *Science and Public Policy*, 16, 224–232.

Weck, D. (1927) Landwirtschaftliche Pflanzenzüchtung u. Vererbungswissenschaft. *Illustrierte Landwirtschaftliche Zeitung*, 670–671.

Weebadde, C. and Mensah, C. (2006) Report of breakout group 2: How will we provide improved varieties of specialty minor and subsistence crops in the future? *HortScience*, 4, 55.

Wehler, H.-U. (1983) *Das Deutsche Kaiserreich 1871–1918*, Göttingen, Vandenhoeck & Ruprecht.

Weis, T. (2007) *The Global Food Economy: the Battle for the Future of Farming*, London, Zed.

Weiss, F. (1918) Arbeiten der Saatzuchtanstalt Hohenheim auf dem Gebiet des Futterbaus. In Anon. (ed.) *Festschrift zur Feier des 100-jährigen Bestehens der Kgl. Württ. Landwirtschaftlichen Hochschule Hohenheim.* Stuttgart, Eugen Ulmer.

Weller, K. (1928) Der Stand und die Bedeutung der Futterpflanzenzüchtung in Süddeutschland. *Mitteilungen der DLG*, 43, 1028–1032.

Weller, K. (1951) Professor Thomas Scharnagel 70 Jahre. *Zeitschrift für Pflanzenzüchtung*, 29, 114–116.

Weller, K. (1953) Die Pflanzenzüchtung und Saatguterzeugung in Schweden im Vergleich zu Bayern. *Landwirtschaftliches Jahrbuch für Bayern*, 30, 242–260.

Weller, K. and Lechner, L. (1953) 50 Jahre Bayerische Pflanzenzüchtung. *Landwirtschaftliches Jahrbuch für Bayern* (Sonderheft), 30, 1–250.

Wellhausen, E. J. (1976) The agriculture of Mexico. *Scientific American*, 235, 129–150.

Wharton, C. R. (1969) The green revolution: Cornucopia or Pandora's box? *Foreign Affairs*, 47, 464–476.

White, F. C. (1986) The interrelationships of agricultural research and farm structure. In Busch, L. and Lacy, W. (eds) *The Agricultural Scientific Enterprise: a System in Transition.* Boulder, Westview Press.

White, S. C. (1996) Depoliticizing development: the uses and abuses of participation. *Development in Practice*, 6, 6–15.

Wick (1934) Die Neuordnung des Saatenanerkennungswesens. *Mitteilungen für die Landwirtschaft*, 49, 318–319.

Wieland, T. (2004) *'Wir beherrschen den pflanzlichen Organismus besser. . . .' Wissenschaftliche Pflanzenzüchtung in Deutschland 1889–1945*, Munich, Deutsches Museum.

Wield, D., Chataway, J. and Bolo, M. (2010) Issues in the political economy of agricultural biotechnology. *Journal of Agrarian Change*, 10, 342–366.

Wienhues, F. and Giessen, J.-E. (1957) Die Abstammung europäischer Weizensorten. *Zeitschrift für Pflanzenzüchtung*, 37, 217–230.

Wiggins, S. (2009) Can the smallholder model deliver poverty reduction and food security for a rapidly growing population in Africa? *How to Feed the World in 2050.* Economic and Social Development Dept. of the FAO, Rome, FAO.

Wiggins, S., Kirsten, J. and Llambi, L. (2010) The future of small farms. *World Development*, 38, 1341–1348.

Wilkinson, J. (1997) Regional integration and the family farm in the MERCOSUL countries: new theoretical approaches as supports for alternative strategies. In Goodman, D. and Watts, M. (eds) *Globalizing Food: Agrarian Questions and Global Restructuring.* London, Routledge.

Wilks, W. (ed.) (1906) *Report of the Third International Conference 1906 on Genetics: Hybridisation, the Cross-Breeding of Varieties, and General Plant-Breeding*, London, Royal Horticultural Society.

Williams, G. (1987) Rural development: a change in policy, why and how? In Shanin, T. (ed.) *Peasants and Peasant Societies.* Oxford, Blackwell.

Winkler, H.-A. (1979) Pluralismus oder Protektionismus? Verfassungspolitische Probleme des Verbandswesens im Deutschen Kaiserreich. In Winkler, H.-A. *Liberalismus und Antiliberalismus.* Göttingen, Vandenhoeck & Ruprecht.

Witcombe, J. R. (2001) Participatory varietal selection in high-potential production systems. In Anon. (ed.) *An Exchange of Experiences from South and Southeast Asia: Proceedings of an International Symposium on Participatory Plant Breeding and Participatory Plant Genetic Resource Enhancement.* Cali/Colombia, CGIAR.

Wittwer, S. (1983) The new agriculture: a view from the 21st century. In Rosenblum, J. W. (ed.) *Agriculture in the 21st Century.* New York, John Wiley.

Wohltmann, F. (1895) Die Lage unserer Pflanzenzüchtung. *Deutsche Landwirtschaftliche Presse*, 22, 79–80.

Wohltmann, F. (1911) Das landwirtschaftliche Institut der Universität zu Halle a.S., seine Entwicklung und Neugestaltung. *Kühn-Archiv*, 1, 1–59.

Wolf, E. C. (1986) *Beyond the Green Revolution: New Approaches for Third World Agriculture*, Washington, DC, Worldwatch Institute.

Wood, G. D. (1998) Projects as communities: consultants, knowledge and power. *Impact Assessment and Project Appraisal*, 16, 54–64.

Woodhouse, P. (2009) Technology, environment and the productivity problem in African agriculture: Comment on the World Development Report 2008. *Journal of Agrarian Change*, 9, 263–276.

Woolcock, M., Szreter, S. and Rao, V. (2011) How and why does history matter for development policy? *Journal of Development Studies*, 47, 70–96.

World Bank (2007) *World Development Report 2008: Agriculture for Development.* Washington, DC, World Bank.

Württembergisches Landessaatzuchtanstalt. (ed.) (n.d.) *Die Vereinigung württembergischen Pflanzenzüchter, ihre Betriebe u. ihre Züchtungen*, Stuttgart, Eugen Ulmer.

Wygodzinski, W. (1916) Die Landwirtschaftskammern. *Schmollers Jahrbuch*, 40, 275–334.

van Zanden, J. L. (1991) The first green revolution: the growth of production and productivity in European agriculture, 1870–1914. *Economic History Review*, 44, 215–239.

Zerbe, N. (2001) Seeds of hope, seeds of despair: towards a political economy of the seed industry in southern Africa. *Third World Quarterly*, 22, 657–673.

Ziai, A. (2007) Development discourse and its critics: an introduction to post-development. In Ziai, A. (ed.) *Exploring Post-development: Theory and Practice, Problems and Perspectives.* London, Routledge.

Zimmermann, C. (1998) Ländliche Gesellschaft und Agrarwirtschaft im 19. und 20. Jahrhundert. Transformationsprozesse als Thema der Agrargeschichte. In Trossbach, W. and Zimmermann, C. (eds) *Agrargeschichte. Positionen und Perspektiven.* Stuttgart, Lucius und Lucius.

Index